Climate Change Delusion and the Great Electricity Rip-off

CLIMATE CHANGE DELUSION AND THE GREAT ELECTRICITY RIP-OFF

Ian Plimer

Connor Court Publishing

Connor Court Publishing Pty Ltd

Copyright © Ian Plimer 2017

ALL RIGHTS RESERVED. This book contains material protected under International and Federal Copyright Laws and Treaties. Any unauthorised reprint or use of this material is prohibited. No part of this book may be reproduced or transmitted in any form or by any means, electronic or mechanical, including photocopying, recording, or by any information storage and retrieval system without express written permission from the publisher.

PO Box 7257
Redlands Bay
QLD 4165
sales@connorcourt.com
www.connorcourt.com

ISBN: 9781925501629 (pbk.)

Cover design by Ian James

Printed in Australia

About the Author

PROFESSOR IAN PLIMER is Australia's best-known geologist. He is Emeritus Professor of Earth Sciences at the University of Melbourne, where he was Professor and Head of Earth Sciences (1991-2005) after serving at the University of Newcastle (1985-1991) as Professor and Head of Geology. He was Professor of Mining Geology at The University of Adelaide (2006-2012) and in 1991 was also German Research Foundation research professor of ore deposits at the Ludwig Maximilians Universität, München (Germany). He was on the staff of the University of New England, the University of New South Wales and Macquarie University. He has published more than 120 scientific papers on geology and was one of the trinity of editors for the five-volume *Encyclopedia of Geology*. This is his eleventh book written for the general public, the best known of which are *Telling Lies for God* (Random House, 1994), *Milos-Geologic History* (Koan, 1999), *A Short History of Planet Earth* (ABC Books, 2000), *Heaven and Earth* (Connor Court, 2009), *How to Get Expelled from School* (Connor Court, 2011), *Not for Greens* (Connor Court, 2014) and *Heaven and Hell* (Connor Court, 2016).

He has won the Leopold von Buch Plakette (German Geological Society), the Clarke Medal (Royal Society of NSW), the Sir Willis Connolly Medal (Australasian Institute of Mining and Metallurgy). He is a Fellow of the Australian Academy of Technological Sciences and Engineering and an Honorary Fellow of the Geological Society of London. In 1995, he was Australian Humanist of the Year and later was awarded the Centenary Medal. He was Managing Editor of *Mineralium Deposita*, president of the Society for Geology Applied to Ore Deposits, president of International Association for the Genesis of Ore Deposits, president of the Australian Geoscience Council and sat on the Earth Sciences Committee of the Australian Research Council for many years. He won the Eureka Prize for the promotion of science, the Eureka Prize for *A Short History of Planet Earth* and the Michael Daley Prize (now a Eureka Prize) for science broadcasting. He was an advisor to governments and corporations and was a regular broadcaster.

Professor Plimer spent much of his life in the rough and tumble of the zinc-lead-silver mining town of Broken Hill where an interdisciplinary scientific knowledge intertwined with a healthy dose of scepticism and pragmatism are necessary. His time in the outback has introduced him to those who can immediately see the weaknesses of an argument. He is Patron of Lifeline Broken Hill and the Broken Hill Geocentre. He worked for North Broken Hill Ltd and was a director of CBH Resources Ltd, Ivanhoe Australia Ltd and Sun Resources NL. In his post-university career he is proudly a director of a number of ASX-listed public companies (Silver City Minerals Ltd, Niuminco Group Ltd, Lakes Oil NL) and various unlisted private Hancock Prospecting companies (Roy Hill Holdings Pty Ltd, Hope Downs Iron Ore Pty Ltd, Queensland Coal Pty Ltd, Hancock Beef [Shanghai], Hanrine Exploration SA).

A new Broken Hill mineral, plimerite $ZnFe_4(PO_4)_3(OH)_5$, was named in recognition of his contribution to Broken Hill geology. Ironically, plimerite is green and soft. It fractures unevenly, is brittle and insoluble in alcohol. A ground-hunting rainforest spider *Austrotengella plimeri* from the Tweed Range (NSW) has been named in his honour because of his "provocative contributions to issues of climate change". The author would like to think that *Austrotengella plimeri* is poisonous.

CONTENTS

1. WE'VE BEEN CONNED — 1

2. ELECTRICITY — 22
OLD TECHNOLOGY — 23
WHERE DOES ENERGY COME FROM? — 24
OLD FRIENDS — 27
Coal — 27
- Poverty — 27
- Industrial revolutions — 31
- The world is a better place — 33
- China — 36
- India — 39
- The United States — 40
- The United Kingdom — 42
- Australia — 42
- HELE power stations — 47
- What is this filthy black evil demonised substance? — 49
- Dirty coal — 53
- Coal seam gas — 54
- Oil shales — 57
- Carbon collection and storage — 58

Gas — 60
Nuclear — 64
Hydro — 71
Energy security — 73
Save the whales — 75

NEW FRIENDS	76
Wind	76
USA	90
Europe	92
United Kingdom	96
Australia	97
Solar	101
Fuel poverty	111
Bio power	118
Batteries	125
Energy history	128
In the long ago	128
The first flush of renewables	129
Energy in Australia	130
Australia's high carbon dioxide emissions	131
Emissions targets	136
National energy market	140
Carpetbaggers	150
South Australia, the land of sea breezes and sunbeams	153
Follow the money	156
How to kill off efficient industry	157
The myth of green dreamtime	161
3. TRANSPORT	**163**
DIESEL VEHICLES	163
Diesel or die	163
The UK disaster	164
The home of diesel	165

ELECTRIC CARS	166
A light bulb moment	167
Electric car emissions, costs and batteries	169
Solar cars, trams, trains and planes	174
Solar cars	174
Solar trams	175
Solar train	176
Solar plane	176
4. CLIMATE CHANGE	**178**
THE KEY QUESTION	178
SCIENTIFIC METHOD	181
PEER REVIEW	195
FRAUD, MISTAKES AND HOAXES	204
Incorrect publications	204
Fraud	205
Weak stomach publication	206
The Sokal hoax	207
The penis and climate change	208
Seinfeld public urination	211
MODELS	211
Failure of models	212
The pause	213
Adaption	214
Rainfall	214
Better models	216
The hot spot	216

Violent climate activists	218
Shots in the dark	218
Assassination	219
PREDICTIONS	221
SCIENTIFIC CONSENSUS	229
Incorrect consensus	229
The 97% consensus fraud	233
WHAT IS CLIMATE "SCIENCE"?	248
Geology and climate change	248
Et moi	250

5. TEMPERATURE 255

HOW IS GLOBAL TEMPERATURE MEASURED?	255
Who measures what and how?	255
Urban effects	269
ADJUSTED MEASUREMENTS	270
Tampered Australian temperatures	276
Kiwi "homogenisation"	284
Temperature tampering in the US	286
WHAT WARMING?	288
HOTTEST YEAR ON RECORD	289

6. SCARY SCENARIOS 299

WHAT IS NORMAL CLIMATE?	300
The dim distant past	300
The day before yesterday	309
Today	317

POPULAR SCARES	319
Carbon pollution	319
Sea level	321
Ocean acidification	326
Extinction	341
Coral reefs	349
Polar ice	356
Extreme weather	379
Drought	383
7. HOW TO STOP THE RORTS	**391**
CLIMATE GREENS	391
Environmentalism	391
Changing the world's politics	396
Nirvana	398
The bleak green world	400
Jobs	402
PARIS	404
Lip flapping	404
Twiddling the dials	405
Costs	406
Paris questions	407
Paris Agreement escapees	408
RENEWABLE ENERGY TARGETS	412
A little pain, 5% emissions reduction	413
Link between wind and solar power and your electricity bill	416

The Finkel Report	416
Subsidies and must-take	420
Fossil fuel subsidies	427
Nuclear power subsidies	427
Cheating (again)	428
Accountability	431
WHAT CAN I DO?	433

Acknowledgements

No book can be written without the inspiration, guidance and help of others. Over the years, I have had fruitful discussions on matters climate with Terry Barclay, the late Bob Carter, Tracey Lake, Ammun Luca, Jennifer Marohasy, John Nethery, Jo Nova, Phil Sawyer, Doug Sprigg and numerous scientists and engineers in the exploration and mining industry. Some of my ideas I have given the pub test in the front bar of the Junction Hotel (Broken Hill). This has shown that politicians have not heard the beat of the jungle drums and bureaucrats have no idea of how the world works.

Any document needs readers, referees, editors and harsh critics. Derek Wyness (Broken Hill) spent far too much of his valuable time reading and correcting the manuscript, as did Andrew Drummond (Perth) and Stephen Cable. These reviewers were very pedantic and picked up errors that an author's eye misses. My ever-patient wife Maja again had to give time for feeding and watering me while the intense stage of book writing was in progress. She also read various versions of the manuscript and provided critical comments. My publisher, Dr Anthony Cappello of Connor Court, was again left with impossible deadlines which he managed to achieve and was a font of information for presentation, marketing and promotion.

Much of my scientific information comes from subscribing to scientific journals and reading widely in the various fields of science. Many of us now no longer take our daily news from newspapers, radio or the television, especially as there some excellent blog sites in the ether such as blogs of Jo Nova, Andrew Bolt, Larry Pickering, *Friends of Science*, the *Global Warming Policy Foundation*, *Watts Up with That*, *Not a Lot of People*

Know That, *Tallbloke*, *Catallaxy Files*, *Science Matters*, *Breitbart*, *The Rational Optimist*, *No Tricks Zone*, *The Daily Caller*, etc. For an analysis of wind power, especially in South Australia, it's hard to go past *Stop These Things*.

My final thanks are to Michael Gilchrist for the layout, editing and his great patience as changes, corrections and additions were made by me while travelling. Even when there were so many versions, he managed to avoid becoming discombobulated.

1
WE'VE BEEN CONNED

You couldn't make this up.

Australia is the world's biggest exporter of energy as coal, liquefied natural gas, uranium and oil yet it has unreliable and increasingly expensive electricity on par with many Third World countries. Some households now have no electricity and employment-generating businesses are closing because of expensive energy. All this was predicted by independent scientists and engineers but, because they did not foolishly follow the false fad, they were silenced. Electricity is not a luxury. It is a necessity.

How did this happen? There was a genuine concern for the environment by most of the community. People acted in good intent trying to make the planet a better place. It all became pear-shaped when the environmental movement was taken over by activists with a political agenda, when scientists could see a way of staying alive by frightening people witless and when businesses could see a new way of making money. Climate political policy became unrelated to science and has now become the new fundamentalist green religion. We are wasting resources on invented issues. A belief in human-induced global warming is mired in totalitarianism, self-loathing, anti-capitalism and hypocrisy. Believers buy indulgences via subsidised wind and solar electricity and force non-believers and the poor to also pay indulgences.

Al Gore, the first to become a billionaire by scaring the

credulous about global warming, tells us that we must cut back on carbon dioxide emissions and reduce our electricity consumption or we'll all fry and die. Gore claims sea level will rise some 20 metres. He does not tell us that in 2016 his waterfront house devoured 34 times as much electricity as the average American household.[1]

I argue that the "science" of global warming is underpinned by scientific fraud, environmental followers are deluded and stick to religious beliefs about demonstrably wrong human-induced global warming and that renewable energy[2] is a futile attempt to address the mythical human-induced global warming. This has attracted armies of carpetbaggers. As a result we will be paying through the nose for electricity for a long time. We've been conned.

Politicians have given in to the cacophonous green mantra of a future climate catastrophe. Advisors with their snouts in the trough and with self-interest have reinforced the scare campaign. Serial exaggerators, climate "scientists" unemployable outside their ivory castles, the sensationalist media and businesses that will profit from changes in energy policy are each constantly giving politicians Chinese burns.

Members of Parliament are pushed from pillar to post by factions in their own party, by unions and pressure groups, most of whom are blessed with self-interest and are devoid of national interest. Whom is a politician to believe? It's simple. Believe the average householder who is in a world of pain.

Politicians and ministers are there for the short term whereas their non-independent green advisors in their departments

---

[1] Drew Johnson, "Al Gore's home devours 34 times more electricity than the average U.S. household", *Daily Caller*, 2 August 2017.
[2] Often known as ruinable energy because it ruins households, ruins employment, ruins businesses, ruins the environment and ruins political careers. Renewable energy is a feel-good change of language implying that energy is free and renewable. It is not.

are playing the long game and outlive their ministers. Outside politics, many politicians are eminently unskilled and unemployable, have neither science nor engineering training and are only interested in re-election to guarantee more power and bread and wine on their tables. Some politicians even grin when they are photographed with Al Gore, hoping this will assist in re-election. Australia's energy problems can be solved by going back to the drawing board, using common sense and listening to consumers of electricity.

Australia is one of the wealthiest and most energy-rich countries in the world yet the average household is struggling to pay electricity and gas bills, many people have had their electricity cut off because of non-payment of bills, others have a scheme of arrangement to pay their energy accounts over time, some pensioners are spending most of the day in buses and public buildings because they cannot afford to warm their houses and others have gone back to the "good old days" by wearing a few extra layers and collecting firewood because they can't afford heating.

In summer heat waves, people have not been able to afford to keep cool and, even if they could, the grids shut down because of overload. Shopping centres are filled with people trying to keep cool.

People open their gas and electricity bills now with great trepidation. Consumers are changing food shopping choices in order to pay electricity bills because electricity costs have risen quickly and now are a major component of household costs. This should not happen in an affluent country.

It is well known that the cheaper the energy, the more employment. The converse also applies. Because electricity costs have risen by up to 300% over the last few years, many small businesses have had to close (e.g. refrigeration, food preparation, recycling, foundries and plastics). Small business is the biggest employer in Australia, but people have lost their jobs, businesses have moved elsewhere and there is no end in sight. The pain will continue while desperate and nonsensical band-aid political solutions are applied.

The large energy-intensive industries such as manufacturing, food refrigeration, crop irrigation, mining, metal smelting and refining and transport are at the cross roads. In metals mining, the biggest electricity costs are the crushing and grinding. No metals can be extracted from solid rock without crushing and grinding. If the electricity cost doubles or triples, the mine becomes marginal or uneconomic and operations cease. People lose their jobs.

Mines are in rural Australia, unseen by city-based environmental activists who try to close them down concurrent with consuming the resources from these mines. Mine closure does not only affect the employees. For every worker at the mine, there are another five off the mine site involved in machinery maintenance, transport, fuel, food, accommodation and infrastructure. If a mine in a small rural town closes, the town dies.

Dead rural towns have no employment, sporting teams, banks, petrol filling stations, cafes, churches, supermarkets, hospital, dental clinic, general practitioner, police station and post office. They very quickly become welfare towns with drug, burglary and violence problems.

Smelting uses huge amounts of energy to melt rocks or concentrate to make metals. If a smelter closes, not only are the smelter workers out of work but there is a whole industry servicing the smelter and these workers also lose their jobs. Some towns (e.g. Roxby Downs, Mount Isa) only service mining, smelting and refining and are very exposed to energy costs and reliability. Australia is littered with mining ghost towns and if energy prices keep rising, there will be more.

Recent power failures in South Australia closed the Olympic

Dam mine, copper smelter and copper refinery, the Whyalla steelworks and the world's biggest lead smelter at Port Pirie. The cost to the employment-generating companies was hundreds of millions of dollars, jobs were lost and businesses are considering whether they should keep operating in South Australia. Ironically, the South Australian government is also trying to attract exploration and mining.

Australia is on the road to ruin because of the unproven assumption that human emissions of carbon dioxide drive global warming. Once the words emissions, sustainable, green, renewable, sequestration, carbon-neutral, carbon footprint, environment, subsidy and decarbonise are used in energy policy, then we know we are being conned because:

(a) renewable energy is unreliable,

(b) energy costs continue to skyrocket as a result of renewable energy, and

(c) once a government tries to solve a crisis, the solution becomes the next crisis. The solution to carbon dioxide emissions was sold as renewable energy which is now the crisis.

Solar and wind power are being spruiked to the community as renewable and sustainable. They are not. The higher the proportion of wind and solar energy, the higher the costs.

Yesterday when there was neither solar nor wind power, we had reliable, plentiful and cheap electricity for households and businesses. Cheap electricity produced jobs. By living in Australia, we won the lottery of life and to live in the late 20th and early 21st centuries is historically the very best time ever to be a human on planet Earth.

Countries like Australia that once had a surplus have now racked up massive debt and desperately need additional taxes to pay back debt. The thought of reducing expenditure does not

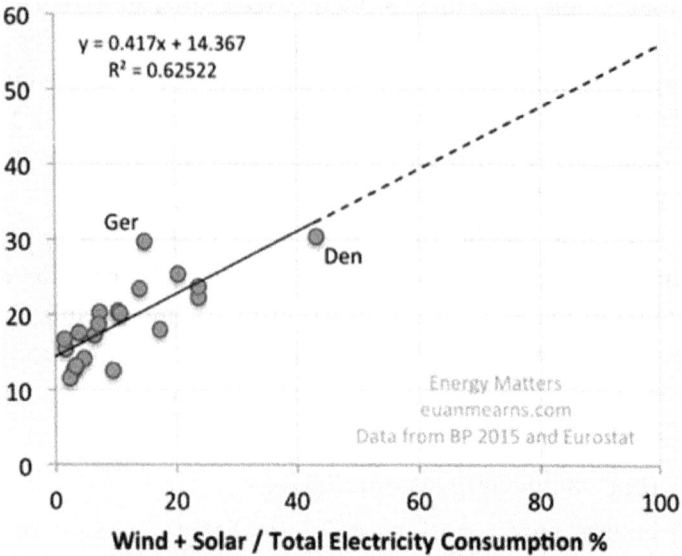

Figure 1: Diagram from Europe showing the higher the proportion of wind and solar energy, the higher the cost (from euanmeans.com). Denmark has the highest cost electricity because of its dependence on wind energy. South Australia now leads the world with costs higher than Denmark.

appear to be important when dealing with other people's money. Your ballooning electricity bill does not balance the books.

The rapacious banks are salivating because of untold fortunes to be made out of renewable energy and compliant governments. They have never been presented with such an opportunity given on a plate. And, as we all know, the banks are really concerned about the environment and do not hesitate to churn all their profits back into the environment. Major economic changes take only years to enact and decades to reverse following massive economic and human disruption.

State and Federal governments in Australia have deliberately embarked on irrational policies of making electricity expensive and unreliable in a futile attempt to reduce emissions of carbon dioxide. Poor political policy shows that there is a short time between being a wealthy country and a basket case.

Political policies in the former USSR and China led to hundreds of millions of deaths by starvation. Australia may be on the steep slippery economic slope to Venezuela[3]. Over the last decade, poor political policy in Australia resulted in spending billions for pink batts, NBN, Gonski 1, Gonski 2, school halls, loss and then control of borders and on useless submarines ($50 billion alone). Commonwealth debt ($600 billion) has skyrocketed, we are paying $1.25 billion a month on interest for this debt[4], not a single budget has been in surplus and we have ridiculously high electricity prices.

Here are a few fundamentals to contemplate:

1. It has never been shown that human emissions of carbon dioxide drive global warming.
2. A fundamentalist religious belief that human emissions of carbon dioxide drive global warming and must be reduced at all costs have led to:
    (a) lack of basic long-term planning for new base load power for when base-load coal-fired power stations reach the end of their lives,
    (b) overloaded electricity grids in eastern Australia

---

[3] Venezuela has the largest proven oil reserves in the world (297 billion barrels). It was a wealthy country. After election of Hugo Chavez in 1998, all major industries were nationalised and we saw democracy collapse in front of our eyes. After the death of Chavez death in 2013, Nicolás Maduro continued the same populist policies and imprisoned opposition leaders. The country is now desperately poor, there is a shortage of food, medicines, fuel and even toilet paper. Inflation is at 800%. There is such a shortage of food that 75% of Venezuelans have lost 8.6 kg in weight. Street gangs and violence are now out of control (Ben Kentish, 2017: Venezuelans lose average of 19lb in weight due to nationwide food shortages, study suggests. *The Independent*, 23rd February 2017).

[4] The figure touted in the press is $1 billion per month. John Stone, the former Secretary of the Treasury, told me on 3 September 2017 that the figure was more than likely $1.25 billion per month.

poised to fail on hot, cold, cloudy, windless or windy days or, in fact, any day that finishes with the letter "y",

(c) demonising of coal resulting in the closing of coal-fired power stations that produce cheap reliable electricity,

(d) wasting a king's ransom on "carbon" capture schemes,

(e) banning of exploring for and exploiting abundant onshore gas reserves that can produce cheap reliable electricity,

(f) refusal to build more dams for hydroelectricity, aquaculture, horticulture and food production,

(g) refusal to build nuclear power stations or even purchase off-the-shelf modular nuclear power stations for rural towns and isolated mines,

(h) littering scenic Australia with inefficient, unreliable, highly-subsidised bird- and bat-chomping polluting wind turbines that have driven electricity prices through the roof and pushed cheap reliable coal- and gas-fired generators out of the market,

(i) covering massive areas of food-producing land with inefficient, unreliable, subsidised, toxic solar panels that have driven electricity prices skywards and pushed cheap reliable coal- and gas-fired generators out of the market,

(j) creating more carbon dioxide emissions because renewable wind and solar energy needs 24/7 coal-fired station backup and the emissions from construction and maintenance of the renewable infrastructure far exceed those allegedly saved,

(k) adding to household electricity costs because

renewable energy plants are in remote areas that require construction of new transmission lines,

(l) panic measures to keep on the lights with carbon-dioxide emitting diesel generators that are far more expensive than any other power generation system,

(m) panic measures to install exorbitantly expensive battery systems that keep the lights on in an industrialised society for a few minutes,

(n) State and Federal governments fiddling around the edges with the broken model of renewable energy rather than scrapping the nightmare, stopping subsidies and letting power engineers sort out the mess, and

(o) changing of measured temperature ("homogenisation") to fraudulently create an apparent warming trend.

3. Climate has been changing since Earth formed 4,567 million years ago and there's nothing we can do about it.

4. Planet Earth is a warm wet volcanic planet. There has been ice on planet Earth for less than 20% of its existence. We are living in one of the cooler times on it.

5. Previous massive, rapid and cyclical climate changes took place before humans were emitting traces of carbon dioxide. Previous interglacials were warmer and longer than the current interglacial.

Rates of global warming and cooling today are slower than previous events of natural global warming and cooling. Humans currently live in boring climatic times on planet Earth and changes being experienced today are nothing unusual.

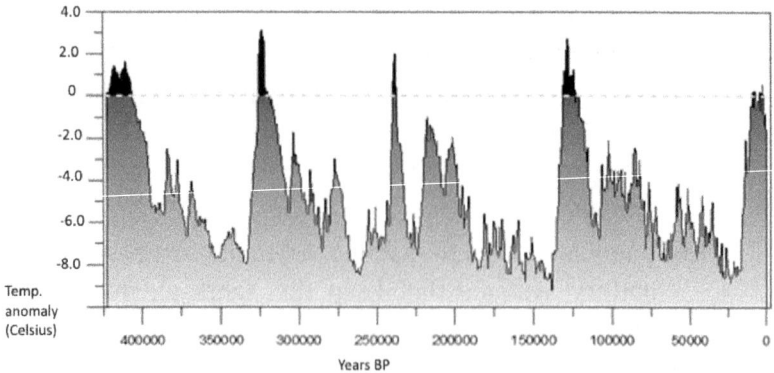

Figure 2: Reconstruction of temperature from Vostok (Antarctica) ice cores showing that the current interglacial is not as warm as past interglacials. Climate changed on 100,000-year cycles with 90,000 years of glaciation and 10,000 years of interglacials. Variations in temperature and rates of warming during glaciation were far greater than any modern variation in temperature. Humans and other mammals (e.g. polar bear) survived previous interglacials and glaciations.

7. Measured atmospheric carbon dioxide contents may have been far higher and more variable in the historical past. Historical carbon dioxide measurements show great variation whereas measurements by a different method show an increase over the last 50 years.

8. In the past, there has been no relationship between atmospheric carbon dioxide and temperature. Why should there be a relationship now? To claim that human emissions of carbon dioxide drive global warming is fraudulent. The data shows the exact inverse. When temperature increases after a glaciation, atmospheric carbon dioxide increases at least 800 years later.

9. Every great ice age was initiated when atmospheric carbon dioxide was higher than at present, at times hundreds of times higher. Atmospheric carbon dioxide is now historically very low. We are told by non-

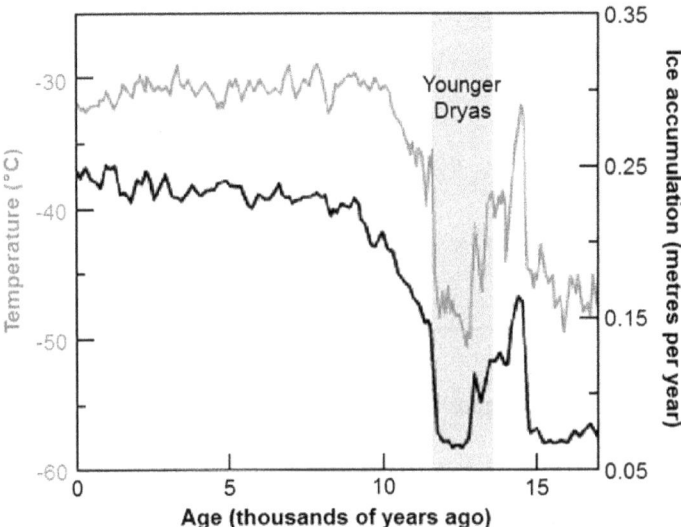

Figure 3: Rapid post-glacial climate change as represented by the Younger Dryas. Both the magnitude and rate of climate change were far faster than any modern changes.

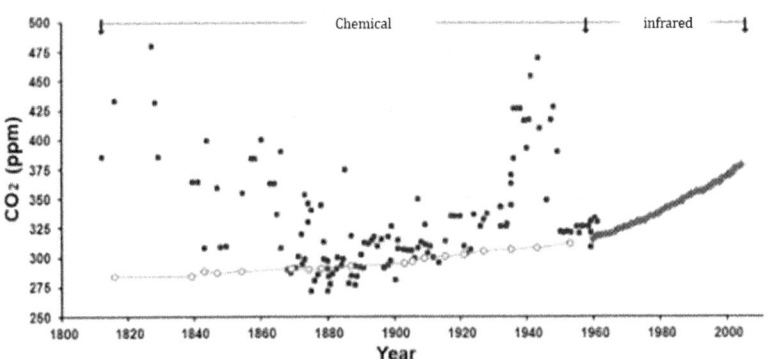

Figure 4: Atmospheric carbon dioxide measurements by chemical (1812-1960 Europe, accuracy 1-3%, black dots) and infrared methods (1959-2009 Mauna Loa, accuracy <0.1%, dotted line) compared with ice core measurements (circles). Ice core measurements have been "homogenised" and, despite slightly lower accuracy, the historical chemical measurements show a huge variation in atmospheric carbon dioxide.

Figure 5: Ice core data plot from Antarctica (Epica Dome C) of temperature versus carbon dioxide showing that rises in temperature stopped hundreds to thousands of years before carbon dioxide rises started. This plot shows that for the last 18,000 years, carbon dioxide has not driven warming and the inverse occurs: Warming drives an increase in carbon dioxide.

independent scientists, environmental activists, politicians and most of the media that a high atmospheric carbon dioxide drives global warming. This is fraud.

10. Annual human emissions (3% of the total) of carbon dioxide are meant to drive global warming. This has never been shown. If it could be shown, then it would also have to be shown that natural emissions (97%) don't drive global warming.

11. Historical ground temperature records have been changed ("homogenised") by adjusting older measurements downwards and more recent measurements upwards to show a warming trend. In science, raw data is sacrosanct. This is deliberate fraud and is undertaken by taxpayer-funded government agencies entrusted with collecting and preserving data.

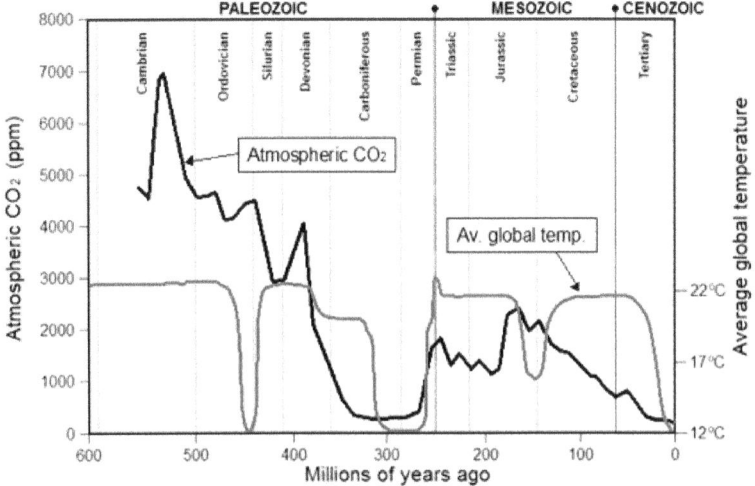

Figure 6: Average global temperature plotted against atmospheric carbon dioxide over the last 600 million years. The four periods of low temperature were ice ages, global temperature in the current ice age is lower than the long-term past global temperature average and there is no long-term relationship between carbon dioxide and temperature.

Imagine if a public company "homogenised" financial accounts from showing a loss to showing a profit. This fraud would send company officers to gaol.

12. Satellite measurements of temperature show cycles of warm and cool within a longer-term warming, large El Niño events, a 20-year pause in warming and current rapid cooling.

13. Within ice ages there are cold periods (glaciation) and warm periods (interglacials). We are currently in an interglacial and, unless a Paris or some other international talk fest trying to control our lives can change the orbit of the planet and the energy released by the Sun, we are heading for the next inevitable solar minimum, little ice age or glaciation.

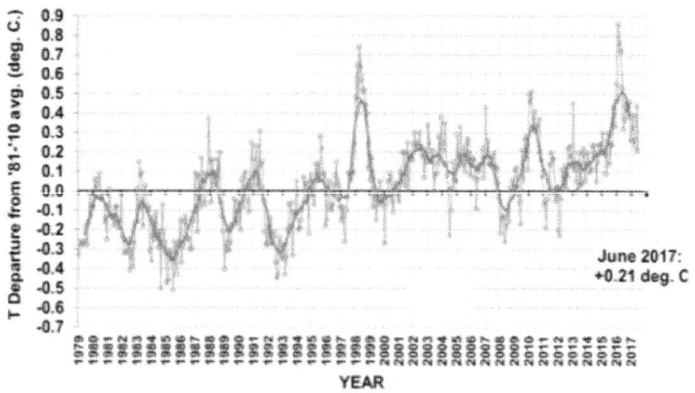

Figure 7: Cycles of temperature expressed as departure from the norm showing cyclical coolings and warmings (driven by ocean temperature oscillations), spikes (due to El Niño events) and a two decade hiatus where temperature did not rise.

14. Cold times create economic collapse, starvation, war and extinctions whereas warm times create economic booms, population increase, longevity and a rapid diversification of species.

15. The slight increase in temperature and carbon dioxide over the last century has been beneficial to the environment, crops and economies.

16. Within glaciations and interglacials, there are slightly warmer and colder cycles. For example, during the last 2,000 years of the current interglaciation, it was warm in Greek and Roman times (500 BC to 500 AD), cold in the Dark Ages (500 AD to 900 AD), warm in the Medieval Warming, (900 AD to 1300 AD), cold in the Little Ice age (1300 AD to 1850 AD) and warm in the Modern Warming (1850 AD to the present). There is a strong correlation between the activity of that great ball of fire in the sky and surface temperature on Earth.

15. Within warmer periods in an interglacial are cooling and warming cycles. For example, since 1850, there have been three cooling and three warming cycles. All warming events have been at the same rate showing that increased human emissions of carbon dioxide due to the industrial revolution in China, India and east Asia have not accelerated global warming. Carbon dioxide emissions therefore have had an immeasurable effect on global warming.

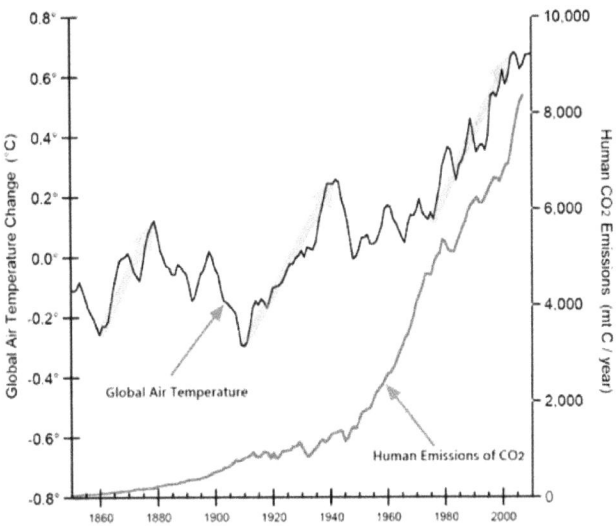

Figure 8: Warming and cooling since 1850 showing that the three warming events are at the same rate hence the massive increase in carbon dioxide emissions from Asia have no measurable effect on warming.

18. There is no relationship between human emissions of carbon dioxide and temperature. Without correlation then can be no causation.
19. Within periods of cooling there are short-lived warm times and in periods of warming there are short-lived cool times hence the daily, monthly and yearly weather are a poor guide to long-term climate.

20. Over the last two decades, there has been increased emission of carbon dioxide from developing countries and Europe yet global temperature has not increased. Environmental activists, most sectors of the media and politicians still refer to human-induced global warming, rapid warming or catastrophic warming yet no warming has taken place. This is fraud.

21. Sea level has risen 130 metres in the last 14,700 years in the current interglacial and is still slightly rising. There is a weak correlation between sea level change and temperature and sea level has risen in the past for a diversity of reasons unrelated to temperature. In many places, it is the land that is rising or sinking.

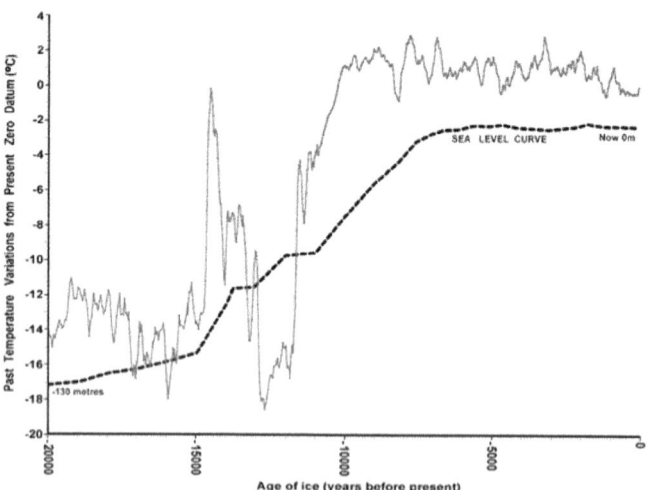

Figure 9: Post-glacial sea level rise of 130 metres (black dashes) and temperature changes (sold grey line; derived from ice core measurements) showing (a) lack of good correlation between temperature and sea level change (b) slight temperature decrease since the Holocene Maximum 6,000 years ago (c) very rapid natural global warming after the cold Younger Dryas 11,300 years ago (d) a lack of sea level rise during the Holocene maximum, and (e) the slight variations in temperature and sea level in modern times compared to the past.

22. There is a correlation between the rates of sea level change and solar activity.
23. Sea level changes in the past have been numerous, rapid and cyclical. For most of the last 20% of Earth's

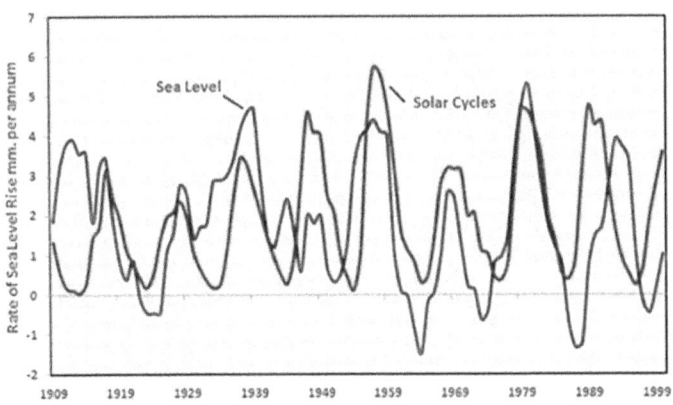

Figure 10: Correlation between rate of sea level change and solar cycles.

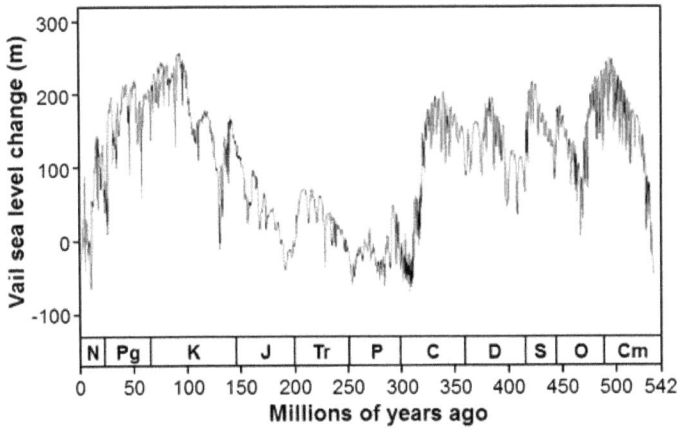

Figure 11: Sea level graph showing massive fluctuations in sea level over the last 500 million years. Current sea level on this graph is 0. For most of the last 500 million years, sea level has been considerably higher than now, past sea level changes have been rapid and greater than the worst-case computer predictions.

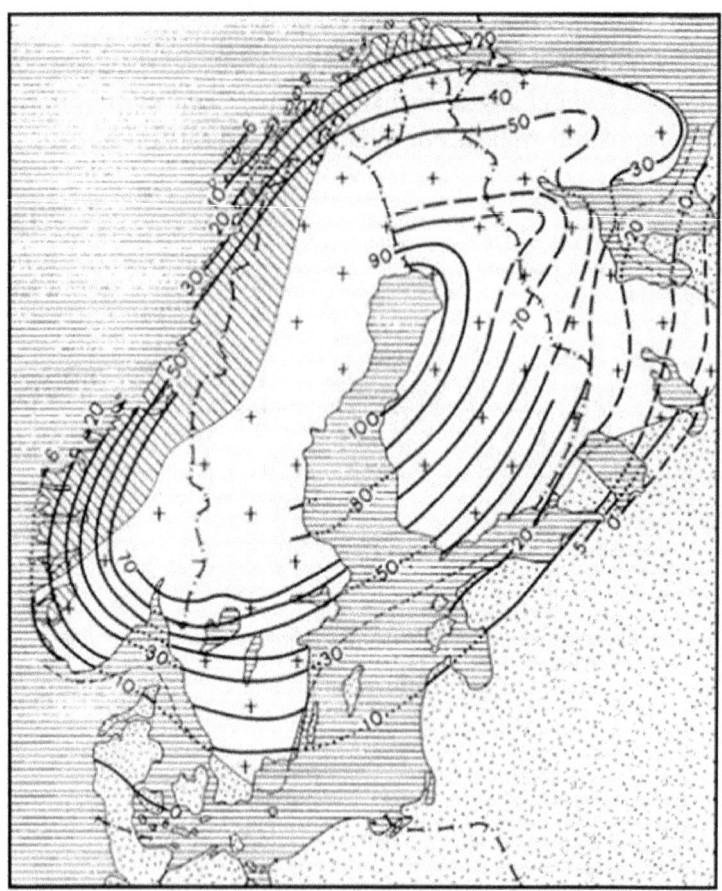

Figure 12: Contours showing the amount of uplift (in metres) over the last 6,000 years in Scandinavia (diagonal lines, folded rocks of the Caledonides; crosses crystalline rocks; dots shallow marine sediments; horizontal lines are water). Scandinavia was pushed down by a 5-kilometre thick ice sheet; this has melted and the area is now rebounding. The maximum uplift rate is 60 centimetres per year.

existence, sea level has been higher than at present. The land level rises and falls and there can be no conclusions about sea level change unless the land level changes are known first.

24. Very cold minimum temperatures measured in 2017 by government agencies are expunged from the record or changed to a warmer temperature to show a warming trend.
25. The fraudulent "homogenisation" of raw data, the inaccessibility of stored raw data, the deletion of large volumes of raw data, the opacity of modelling methods, the omission of validated past climate data, the uncertainty of measurement, claims of a consensus and the demonising of dissent all show that climate "science" has nothing to do with science but, rather, is post-modernist political activist thuggery. The historical temperature record may be nigh on useless.
26. If a public company "homogenised" their annual reports to show that a financial loss was "homogenised" to a profit, directors and officers of the company would go to gaol. "Homogenisation" of temperatures has resulted in massive electricity costs and billions of dollars of subsidies. What's good for the goose is good for the gander.
27. Taxpayers fund climate research and institutions like the Bureau of Meteorology, CSIRO and universities. We expect accountability and independent dispassionate information and should not have to pay crippling electricity bills because of fraudulent environmental activism.
28. Attempts to change global temperature or stop global climate change are deluded and are essentially attempts to control to every aspect of our lives by unelected opaque organisations such as the UN.
29. We are hairless apes and like it warm.
30. Climate change activism is the main weapon of the culture wars to deconstruct Western civilisation.

31. We mere mortals unfortunately can't turn water into wine. We also can't change global climate. In the 10th century, King Canute showed that even kings with great power cannot change nature. Who do we think we are?
32. When humans are extinct, the Sun will continue to drive and change climate.

The end result of these political policy disasters administered by the inept is that your electricity bill is far too high. You are being skinned alive for a false climate change ideology underpinned by fraud.

Politicians believed green activists' doom-and-gloom predictions and carpetbaggers swarmed in to make a killing out of renewable energy, batteries, desalination plants and whatever they can sell to the credulous. The electricity disaster was 20 years in the making and will take a generation to recreate a base-load reliable cheap electricity system. It was done before by power engineers, was destroyed by green ideology and needs to be done again by power engineers.

The global warming scare campaign was born in the early 1980s after a 30-year cooling trend. During this cooler period, the same people who are now frightening us with fry-and-die global warming scare stories were telling us that we were entering a new freeze-and-die ice age. Maybe the scare campaign is more related to power over people, money and research grants?

The monster could have been killed at birth but it was protected from the normal rigourous processes such as due diligence, scepticism, validation and the scientific method because it was sold as environmental concern for the welfare of the planet. As with all good science fiction stories, the global warming beast grew up to be a voracious uncontrollable monster.

On 31 October 1517 on the eve of All Saints' Day, Martin

Luther sent his *Disputatio pro declaratione virtutis*, commonly known as the *Ninety-five Theses*, to the Archbishop of Mainz. He may also have nailed copies of his *Theses* on the oak door of All Saints' Church and other churches in Wittenberg on 31 October or in mid-November.[5] The intent was to create academic debate, not a revolution, and Luther's religious beliefs did not differ from those of Rome.

Luther's writings divided the Catholic Church and changed the course of religious and cultural history in Western Civilisation. The theses mainly deal with indulgences offered by priests to Christians and questioned why the Pope, who was very rich, required money from the poor. Luther argued that indulgences led Christians to avoid true repentance and sorrow for sin by purchasing an indulgence certificate[6]. Other theses dealt with repentance and common beliefs about purgatory[7].

And so too with the modern green religion. Adherents purchase indulgences such as rooftop solar panels, electric cars, carbon credits, renewable energy certificates and try to reduce their carbon dioxide emissions as repentance for their modern comfortable Western lifestyle (i.e. inward punishment for sin). "Useful idiots" are everywhere evangelising that renewable energy is the way of the future. These indulgences are funded by the poor in their own community via high electricity costs. Adherents lecture sinners about our carbon dioxide emissions, do not lead by example and moralise about the lifestyle of those too poor to purchase indulgences.

You have the power to lower your electricity bill. You vote.

The only good politician is a frightened politician. Pester politicians, especially when they want you to keep them in a job.

---

[5] Hendrix, Scott H. 2015: *Martin Luther: Visionary Reformer*. Yale University Press.
[6] Jared Wicks, "Martin Luther's treatise on indulgences", *Theological Studies* 28 (3), 1967, 481-518.
[7] Scott C. Dixon, *The Reformation in Germany*. Blackwell, 2002.

2
Electricity

If you don't want to read all of this chapter, the messages are simple:

> (a) If you want cheap electricity, a strong economy and jobs, then mine coal for power generation. Globally, the greatest increase in use of any power generation source is coal.
>
> (b) Old technology is often the best because it is tried-and-proven.
>
> (c) No nation runs on wind and solar power.
>
> (d) Wind and solar power damage the environment, are unreliable, expensive, subsidised with your money and produce more emissions than coal-fired electricity.
>
> (e) The subsidies for wind and solar energy force cheap coal-fired electricity generators out of the market and your electricity costs rise.
>
> (f) A clever country with 30% of the world's uranium in South Australia alone, with large under-populated desert areas and an electricity shortage would have a totally vertically-integrated nuclear industry. This includes mining, beneficiation, fuel rod construction and leasing, fuel rod clean up and its own nuclear power industry to provide base-load electricity. This would provide generational employment and cheap electricity.

OLD TECHNOLOGY

We have long been told that coal-fired technology is old technology and the future rests with new technology. This is just lip flapping. What is not said is that some old technology has never been replaced by new technology because it is efficient, reliable, tried-and-proven and there is no better replacement.

A wheel is old technology some 5,500 years old[8] yet it has not been replaced. What new technology will replace the wheel? Electricity was probably discovered by the ancient Greeks around 600 BC. Is electricity an old technology that should be replaced with a new technology? If so, what is the new technology that will be used to replace electricity? The modern world could not operate without electricity.

Steam engines were invented in the late 17th century[9] and have become more efficient, especially at high temperature and pressure. High-pressure steam generated from heating water with the energy from burning coal drives turbines that spin magnets in a coil to produce electricity. A very large kettle is the basis of coal-fired electricity. Benjamin Franklin's rediscovery of electricity in 1752 led to a boom in experiments on electricity. Spinning magnets in a coil to produce electricity was published in 1832 by Michael Faraday[10] but it was known for a couple of decades earlier. This discovery changed the world.

Electricity in a coil used to spin magnets was a reversal of Faraday's discovery and the electric motor was invented in 1821.[11] Battery-powered electric cars were first driven in 1837. Electric motors were invented well before a continuously-

---
[8] Perhaps invented in 3,500 BC.
[9] Thomas Savery, 1698.
[10] Al-Khalili, 2015: The birth of electric machines: a comment on Faraday (132) "Experimental researches in electricity", *Phil. Trans A Math Phys Eng Sci* 373, doi: 10.1098/rsta.2014.0208
[11] Michael Faraday, 1821.

operating spark-ignition internal combustion engine in 1859.[12] Rudolf Diesel (1858-1913) patented his invention of the diesel engine in 1893. Is this new or old technology? I guess it depends upon whether one is a green environmental activist or not.

If we are reticent to use old technology, then we should totally forget electric cars because their technology is far older than that of diesel and petrol cars. We should also totally forget wheels because they are a 5,500-year old technology and electric cars should have no windows or windscreens because glass was invented about 1,500 BC by the Egyptians. Solar power uses photovoltaics discovered in 1839 whereas wind power has been used for at least the last 5,500 years so we can't use wind or solar energy because these are old technologies.

The two most modern technologies that should be promoted by those who want to ignore old technology are the diesel engine (1893) and nuclear power driven by radioactivity, discovered in 1895.[13]

As per usual, nonsense promoted by green environmental activists shows crass ignorance.

WHERE DOES ENERGY COME FROM?

The wind turbine turns work into electrical energy. Electricity turns electrical energy into work. Gravity can convert potential energy into electricity and heat (e.g. hydroelectricity). Chemicals, such as petrol, when ignited produce work, heat and sound. Sunlight induces particles in solar cell silicon to jump to a higher energy state and when they fall back to their original position, they release electrical energy. Large atoms, such as uranium 235, fall apart naturally into heat, particles and new isotopes. In all cases, energy is neither created nor destroyed.

---
[12] J. J. Étienne Lenoir.
[13] Henri Becquerel.

Time to bore the pants off you. The First Law of Thermodynamics states that the total energy of an isolated system is constant and energy can be transformed from one form to another but cannot be created or destroyed. The First Law is about the quantity of energy. The Second Law of Thermodynamics essentially states that as energy is transformed, more and more is wasted and there is a natural tendency of any isolated system to degenerate into a much more disordered state. We have known the essence of these laws for almost 200 years. We do not create energy from wind or solar, we convert one form of energy into another.

The Earth has three principal sources of energy. What we do as humans is convert one form of energy into another. The first source of energy is from very deep within the Earth and derives from the core and radioactivity. The core of some extraterrestrial bodies has frozen from liquid to solid (e.g. Mars, Moon) whereas the Earth's core is still molten because it has retained its heat and generates new heat from radioactivity. This heat is transferred to the mantle and crust mainly by convection through almost solid rock. Plumes of convected material move continents and plumes partially melt upper mantle rocks. These melts are buoyant and heat the Earth's crust.

Heat flow maps show that most of the present deep heat flow from the mantle occurs in the circum-Pacific 40,000 km-long Ring of Fire, the Mediterranean-transAsiatic Belt and the mid ocean ridges. Some of this heat is used in geothermal power stations such as in New Zealand, Philippines, Italy and Iceland. About 0.1% of the world's electricity is generated from geothermal systems.

The decay of potassium 40, uranium 238, uranium 235 and thorium 232 produces heat over an extraordinarily long time. It is this heat that has kept the Earth's core and mantle hot since the Earth formed. If radioactive minerals are concentrated by geological processes near the surface of the Earth, they can be mined. Natural decay of uranium has been used to try to generate hot dry rock electricity. Accelerating the decay of uranium 235 in a reactor produces breakdown products comprising daughter isotopes of lighter elements, particles and heat. The reactor heat is used to drive steam turbines to spin magnets to generate electricity.

The second principal source of energy is stored as carbon-rich materials in the top few thousand metres of the Earth. This energy is in the form of stored solar energy that is now present as combustible carbon-rich materials such as coal, oil, gas, tar sands and oil shales as fossil fuels. This energy derived from sunlight, was stored via life and was later compressed into gas, oil or rock. The chemical process of oxidation by burning converts carbon compounds into carbon dioxide and water vapour. This gives out heat that can be used to spin a turbine with steam or to drive an engine crankshaft. This process of hydrocarbon combustion drives the world.

The third principal source of energy on Earth is extraterrestrial. Solar radiation can be used directly to generate electricity. Radiation from the Sun is captured by the atmosphere and oceans as heat. Water evaporates from seas and the oceans and falls as rain. Lunar tides and the gravitational flow of rainwater from highlands can both be used to generate electricity. The transfer of atmospheric heat results in wind and storms. If wind flow is retarded with turbine blade, the captured energy can be used to generate electricity. Reduction of wind velocity creates energy and rain shadows behind turbines.

The Sun and rain stimulate plant growth and consequent biomass which can be used to produce steam to generate electricity. Without the Sun, there is no life on Earth (with the exception of a few weird bacteria). This third source of energy is most commonly referred to as green energy. It is not green. It reduces the supply of food, water and available energy to life on Earth, kills wildlife and consumes large amounts of fossil fuels for manufacture, construction, maintenance and backup.

If the greens were concerned about the planet they would support the use of fossil fuels because the amount of land used per unit energy is small (compared to solar, wind, hydro or biomass), the combustion processes for fossil fuels and biomass are the same, combustion adds the essential ingredients of life to the atmosphere (carbon dioxide and water vapour) and controlled fossil fuel burning does not kill wildlife. If the greens were really concerned about the land area used, emissions and damage to wildlife, they also would be great advocates of the nuclear industry.

Uranium 235 has the highest energy density of all forms of energy used to create electricity. If all the energy you needed to live your three score and ten years derived from uranium 235, the end result would be half a cup of spent fuel. To make 1,000 kWh of electricity, you need 0.03 grams of pure uranium 235, 265 litres of oil, 379 kg black coal, 1,000 one metre square solar panels or one 600-kW wind turbine operating flat out for 1.5 hours.

OLD FRIENDS

Make new friends, but keep the old. One is silver, the other is gold.

Coal

Poverty

Coal saved the Western world from poverty. As a result of the use of coal in the UK, Europe and USA in the industrial revolution of the 18th and 19th centuries, the benefits of economic development were spread from a small wealthy elite to the broader community. A middle class appeared. Coal raised most low-income earners out of poverty. Western coal-producing nations are now so wealthy that taxes can support activists to put workers in the coal industry out of work. The greatest dangers facing us on planet Earth today are probably not the rise of China, Islam or carbon dioxide emissions but global cooling and our own loss of faith in our inherited and superior Western civilisation.

It is claimed that times were good before the burning of coal. We sat around arm-in-arm happily singing, picking berries in the forest and living off locally produced "organic" natural tucker. We lived sustainably and had a very small carbon footprint. Since then, sinful modern humans have degenerated into selfish greedy capitalists deliberately emitting dangerous carbon dioxide into the atmosphere. This is rather like the fundamentalist Christian view of the world with original sin, the fall from grace, absolution and redemption.

Nothing could be further from the truth. In previous centuries, we clear felled forests, killed everything that might give us protein or was a perceived threat and polluted everything we touched. Ancient Egyptians could not drink Nile River water because it was so polluted. They had to make huge personal sacrifices and drink beer.[14]

Tens of thousands of years ago, if a human from a different tribe was competing to spear an animal for protein, we killed this person. We did not sing *Kumbaya*, cuddle the person, say peace and share. We killed. We died at an early age, normally from a big bacterial blast. Peace, longevity and security are not the natural state of affairs.

The greens want us to go back to this alleged "sustainable"

---

[14] Women and children drank "small beer" (i.e. low alcohol beer), a term we still use today.

life of poverty and have us using renewable energy. I don't. I am old enough and have lived in the "good old days" in a semi-rural environment and it was not that long ago. In every way, the modern world is far better than the "good old days". Solar power grew our crops, domesticated animals were vegans[15] that ate grass and crop stubble. We ate vegans and we still do. We were rescued from poverty by large amounts of cheap energy generated from coal.

Some people had reticulated town gas made from heating coking coal that produced gas and left a coke residue. Gasworks could be smelt kilometres away. The coke was also useful, being used for steel making and could also be used for home heating. We are now deliberately increasing the cost of electricity and are walking back down the path towards poverty. Ask the average person whether they can pay their latest electricity bill.

In the 1950s in Australia, there was an energy revolution. Electricity generated from coal became cheap, plentiful and reliable. It pulled people out of poverty. Diesel and petrol became cheaper and we could have a car, run a generator in the shed, pension off horses for a tractor and could transport food to the market by truck. All this saved back-breaking work. Electricity could be used for heating, cooking and refrigeration, lawns could be mowed by a machine that did not leave dung everywhere, there was less need for muscle power, there was more free time and foods that had to travel were available because we had refrigeration.

Cheap energy certainly allowed progress from a "sustainable" life to a comfortable life. We now have far more leisure time, better food, more food, more diverse food, fresh water on tap, cheap travel, self-regulating home heating and cooling, swimming pools, robotic vacuum cleaners and mops, automated

---

[15] In a speech on 3 September 2017, Warren Mundine jokingly stated that vegan is an Aboriginal word that means "Can't hunt". He got a good laugh from the audience.

sprinkler systems, satellite navigation systems, communication, digital books, portable communication devices such as 'phones and iPads, home theatres with surround sound and hard drives full of terabytes of music, movies and documentaries.

Modern society and economics allow us to spend our free time doing things we enjoy as opposed to being factory fodder or hunting and gathering just to survive. We now have more assets and are far wealthier. We can now afford to go to sporting events, concerts and restaurants. We can now afford to be greens, as long as it only hurts someone else's life. The greens don't care one *iota* if you can't pay your electricity bill.

Don't give me the "good old days", they weren't. No green is going to tell me that I should go back and live that frugal "sustainable" life again. I've done my bit in my childhood and it did not save the planet. Those greens who want this allegedly environmentally friendly romantic life are quite welcome to become retrogressive and go native. Somehow, I doubt if they could actually live a "sustainable" life, especially as most greens are based in cities. With no knowledge or experience of such a life, I doubt whether they would be capable of living the "sustainable" life that I had. And if they want "sustainability" and "renewables", they should have to pay for it themselves and leave me to enjoy my hard-earned comforts of life.

Until I see the major movers and shakers in the green movement living as I did when I was a child, then they are hypocrites with no credibility. They live in cities and huge amounts of vegetation was cleared and animals were killed for the building of the city.

I wait for the time when I can go to consult the green oracle, living in a cave in the bush, living off the land and using no trappings of the modern world.

Until then, get out of my life.

Industrial revolutions

Since the Industrial Revolution, life expectancy for people living in the Western World has doubled. This is directly due to coal and cheap energy. Ready access to cheap energy allows for better living conditions and easier lifestyles. Coal was the fuel of the Industrial Revolution and steam was its power. There was also a transport revolution to haul coal initially along canals and then later along railway systems from pits to steel and cotton mills. These transport systems allowed workers to travel short distances, enjoy a holiday at the seaside and to broaden horizons. Towns were lit by coal gas, train timetables meant that workers needed watches and inventors flourished to capitalise on the technological advances that came with the Industrial Revolution.

At that time, the UK parliament was concerned about protecting inventors' patents which were the backbone of the Industrial Revolution and didn't think about homosexual marriage or the hurt feelings of snowflakes.

Coal allowed a life of desperation to be replaced by a life of aspiration. In northern England, there were community gatherings, competitions, brass bands and outdoor activities away from the darkness and grime of an underground coal pit. Workers sought a better education. Women and children working in underground coal mines were replaced by pit ponies that later were replaced by machines.

Over a few generations, coal brought hundreds of millions of people from poverty to prosperity, despite population- and economy-destroying wars. The green activist Westerners are totally immoral telling billions of people in Asia, Africa and South America that they can't escape from poverty and develop to our standard of living using coal, gas and carbon-based energy as we have.

In today's world, people are better fed, better sheltered and better protected than in any time in history. Overall prosperity has increased in the last century. The average person's standard of living has improved ten-fold over this period of time, as has mine. Cheap inexpensive transport, a consequence of the Industrial Revolution, has revolutionised the spread of trade, services and ideas.

Without international trade, we could starve. For example, massive local rains in France in 1694 AD resulted in the third consecutive crop failure. Some 15% of people in France starved although plenty of food existed elsewhere in Europe. At that time, France was "sustainable", lived off the land and had very little trade with other countries. A convoy of 120 ships left Norway with grain for France. The convoy was captured by the Dutch and was recaptured by the French and escorted to Dunkirk. It was too little too late as there was still not enough grain to feed everyone in France. Starvation continued.

The World Bank reported that in 1981, 42% of people in the developing world had to live on less than a dollar a day. Some thirty years later, this percentage has been reduced to 14%. The access of low income countries to potable water has improved from 47% to 66%. This is a huge change in a short period of time. Such a change is unparalleled in history. The greens' actions attempt to reverse this trend. That is a moral conflict for the greens. Their solution: Ignore it.

The World Energy Outlook of the International Energy Agency shows that there is a near perfect correlation between global electricity from coal and gross domestic product. Consumption of coal grew by 5.4% in 2011 and is still growing at that rate. Asia leads the consumption race. The world is becoming a better place. The *per capita* food intake, longevity and wealth have increased whereas child mortality, disease and land area for food production has decreased.

There is still much to do but for a long time the trend has shown that the planet is becoming a better place. The same authority also shows that 3.6 billion people still have no or only partial access to electricity and the quickest and cheapest way to produce electricity is from coal. If we were really serious about the environment and the welfare of the poorest people in the world, we would flood India, China, Africa and South America with coal- and nuclear-fired power stations.

The annual BP Energy Review has shown that, in the last 100 years, energy from fossil fuels has remained at about 85% of the total energy consumed. Despite green noise, the fundamentals have not changed.

Cheap coal is still the backbone of electricity, health, longevity and food.

The world is a better place

The Copenhagen Consensus Centre and the World Bank show that the proportion of extremely poor people has more than halved over the last 30 years, from 42% of the global population in 1981 to 16% in 2016. In 1820, more than 80% of people were miserably poor. There are still 20% of the world's population who are illiterate. Although this sounds dreadful, some 70% of the world's population was illiterate in 1900. China has recorded the biggest improvement.

In 1950, South Korea and Pakistan had the same level of income and education. Today, the average South Korean has had 12 years of education, in Pakistan it has not yet reached six years. The South Korean *per capita* income has grown 23-fold whereas in Pakistan it is only 3-fold. Illiteracy costs. In 1900, global illiteracy cost 12% of global GDP. It is now 7% and the estimates are that in 2050 it will be 3.8%. These are world problems, we cannot ignore them. However, the world is a far better place than it was 100 years ago and global problems are being solved. But not by the greens.

There seems to be a view amongst the greens that we face crises for which there is no solution. If we look at the long history of experimentation, initiative, intuition, creativity, inventiveness, science and engineering, we can see the end product. We humans are creative and solve problems. Just fly in an Airbus A380 and admire the science and engineering that allow some 575 tonnes to fly at 12 km (39,000 feet) above sea level for 17 hours.

We humans are very good at overcoming hurdles to our existence and this we have shown since we began to live in villages more than 10,000 years ago. When we humans are dealt a wild card from the pack, we are pretty good at adapting and improving. We humans live on ice sheets, mountains, deserts, in the tropics, at high latitudes and at the seaside. If the need arose, no doubt civilisation would find a way to live under water. During interglacials, humans increase wealth which allows populations to grow whereas glaciation is associated with famine, starvation, disease and depopulation. Radiation is an impediment to living for a long time in space or on other planets.

We have already adapted to live in a great diversity of climates and if there were a change in climate, it would not create any problems for the human race as we already live in areas that vary from -40°C to +50°C. Who's to say what is an ideal climate for humans? The climate that the Eskimo prefers might not be the same climate that a jungle hunter in Borneo prefers. There is no ideal climate for humans and with energy, technology and innovation we have already shown that we can survive extremes. The world is a better place, increased affluence makes it an even better place because affluence enables us to financially solve any potential environmental problems.

One only has to travel to some African countries to see environmental degradation resulting from poverty, not from wealth. As these countries creep out of crippling poverty (often despite political despotism), the environmental degradation has decreased. It is only wealthy countries that have enough funding to address environmental problems, perceived or real. The best way to make the world a better place is to have more wealthy people and to have fewer greens stealing other people's money via tried-and-failed policies.

The world is now able to feed more mouths using less land (hence less forest has needed to be cleared), consuming more calories and spending less income proportionally on food than previously. Since 1970, the rate of increase in the global population has slowed. As the GDP *per capita* in a country increases, the fertility rate of women decreases and the demands on the environment also decrease. Over the decades, we have got better at everything.

We are living at a time in history when there are two nations with more than a billion consumers. They have endured thousands of years of gripping poverty and now there is a rapidly emerging large middle class. The world is in the middle of the biggest industrial revolution ever seen. Hundreds of millions of people are moving from rural China to the cities. This is the greatest diaspora in world history. Since 1979, coal has enabled China to pull 700 million people out of poverty in a new industrial revolution. This is nearly the population of Europe. These immigrants to Chinese cities want our Western standard of living. They have had a taste of it and there is no turning back. These aspirants use commodities and the consumption statistics from China and India are mind-boggling.

It is immoral for greens to promote costly inefficient unreliable "renewable" energy to the rest of the world. Not only does it push up your own electricity costs it stops people in

Africa, Asia, South America and India escaping from the grips of poverty and reaching our standard of living. As an absolute minimum, these people need clean potable water and cheap reliable electricity.

China

China has 19% of the world's population. It consumes 53% of the world's cement. Concrete is made from gravel, sand, cement and water and any modern industrial growth is underpinned by concrete because it is a strong, cheap and durable building material. The high *per capita* consumption of cement shows that China is building, modernising and growing. Cement is made from heating limestone and shale to a high temperature with the heat derived from coal. Burning limestone which contains 44% carbon dioxide releases carbon dioxide into the air. China's energy is principally from the burning of coal and it is no wonder that China emits huge quantities of carbon dioxide. By 2020 (or earlier), China will be the world's biggest carbon dioxide emitter. This is good news as it means more and more people in China are escaping poverty and living better lives. We also should thank China for putting so much plant food into the atmosphere.

If China reduces its carbon dioxide emissions, then its economic growth will slow. Notwithstanding, slight changes in the growth rate of China will have no great effect on the world. China will not curtail its growth in response to moralising by unelected wealthy Western greens. They want the same standard of living as enjoyed by city-based Western greens. If Western greens want China to reduce carbon dioxide emissions as a moral imperative, then they are knowingly and hypocritically keeping hundreds of millions in poverty as a result of their ideology. China wisely does not listen to Western greens. Neither should we sensible folk.

Over 400 billion kWh/month of electricity is consumed by China. The United States Energy Information Administration projects that China will bring on over 450 GW of new coal-fired capacity by 2040. The demand for steaming coal in China has doubled from 2011 to 2016. China is reducing domestic steaming coal production and is continuing to close small and inefficient coal mines. Chinese domestic steaming coal adds sulphur gases and particles to the atmosphere thereby providing serious pollution in industrial centres. Massive new non-polluting central power stations, the availability of cheap electricity together with the prohibition of coal use in small businesses and houses is starting to clean up the air in China. The same was done in the UK in the 1950s.

By substituting less polluting and higher energy Australian steaming coals for the high-sulphur high-ash Chinese coals, energy production can increase without a proportional increase in air pollution. In March 2017, natural gas production in China rose 8.2% and coal output rose almost 13% and averages 9.67 million tonnes a day.[16] Australia and Indonesia are the leading suppliers of coking and steaming coal and it is expected that in 2017 both countries will each export more than 450 million tonnes of coal to China.

Coking coal is used as a chemical reductant and source of heat for converting iron ore to steel. There is no suitable replacement. Steaming coal is used for electricity generation. China is also looking to impose local production limits for the special and scarce resource of their own coking coal. With reduced access to domestic coking coal, China's demand for imported coking coal will increase and China already imports 47% of the world's coal production.

China also is the number one producer of wind and solar power generators. They are not silly, they don't use such tech-

---

[16] *Bloomberg*, 17 April 2017.

nology themselves to drive their own industrial revolution. They sell them to the renewable energy fools in the West whose governments subsidise unreliable electricity generation and live off debt. Australia is an example and pays more than $1 billion a month in interest. The West just becomes more inefficient and internationally uncompetitive with unreliable subsidised electricity. The Chinese are laughing all the way to the bank.

In 2016, the increase in China's carbon dioxide emissions was more than 200 million times more than the increase in UK emissions. Whatever carbon dioxide emissions savings the UK, Europe, Patagonia, Iceland, Timbuktu or Australia manage to achieve, it will have not one *iota* of difference to global carbon dioxide emissions by humans. China will make up for the shortfall before you can say coking coal, thermal coal and methane gas. Even though there has been a coal-driven increase in longevity in China, 50,000 cigarettes are consumed each second.

In China there has been a 45% increase in national income, protein sources are now more varied and agricultural productivity has increased. Wine is a measure of consumer affluence. China now owns wineries in France and the New World and, as well as making its own wines, imports wines from many parts of the world. This would have been unheard of 50 years ago. The area of China's corn harvested over the last half century has doubled and each harvested hectare has become more than 4.5 times more productive than 50 years ago. The 120 million hectares of land thus spared from land clearing is twice the size of France. No wonder Chinese forests have expanded more than 30% over the last 50 years. As affluence in China increases, the rate of population increase is falling. In China, the middle class now consumes wagyu beef, quality beef that few Australians have tasted.

China has probably about one trillion tonnes of coal resources and 40 billion barrels of crude oil. The EU bankrolled the building of wind turbines in China and 30% are not even connected to the grid. The Chinese are not mugs. The EU also partially financed the building of dams for hydroelectricity and agriculture with "renewable" energy funds. Green activists would have stopped these dams being built in Europe, UK, North America and Australia. China has fast-tracked its nuclear power generating industry. Despite the large coal imports, China is almost energy-independent and has made big investments around the world to acquire other mineral commodities while preserving local reserves.

While the green activists in the West are forcing wind and solar farms to be built, China is building dams for hydroelectricity and food production while the West is not. China is boosting its coal-fired electricity capacity and using gas for other purposes whereas the West is doing the opposite. Most Western countries have no coal-to-liquids programs and modest nuclear programs whereas China is fast tracking these energy systems. While green activist pressure in the West has led to destroying, ignoring or not using efficient energy systems, China has been racing towards long-term self-sufficient efficient energy. The West is looking increasingly exposed.

India

India's population has doubled since 1960, its national income has increased 15 times. Indians eat more and better than their counterparts in 1960. Since 1960, Indian forests have increased by 15 million hectares, less land is used for agriculture and India now exports food in contrast to former times when it imported food and had famines. India is a success story of chemical, biological and mechanical innovations arising from the Industrial Revolution. The move of Indians and Chinese to cities has actually resulted in a better rural environment because fewer people are now dependent upon the forests for food.

India will build an additional 75 GW of coal-fired electricity in the next five years. Most of the new power stations will be built on the coast because Indian coal mines and existing power stations are inland, many are government-owned and inefficient, mines are small and old and the railway network cannot handle large tonnages of coal. New power stations will require the burning of 2,509 million tonnes of steaming coal, some of which will come from Indian coalfields and most of which will be imported by sea from Australia hence the development of coastal power stations.

Thermal coal demand in India is expected to outpace production by at least 150 million tonnes within five years. The increased number of blackouts in India shows the need for more coal-fired power stations, increased coal imports and an improved power grid. The state-run Coal India Ltd will try to raise production by another 180 million tonnes giving a total production of 615 million tonnes in 2016-2017 compared to 435.5 million tonnes in 2011-2012. Multiple new port projects are underway to enable increased imports of steaming coal. In terms of million tonnes of oil equivalent, in India coal dominates over oil, natural gas, nuclear, hydro and "renewable" energy.

India was poised to increase global market share of coal output from 10.1% in 2016 to 13.1% by 2020. In 2016, India surpassed the USA to become the second largest coal-producing country in the world.

The United States

In late 2016, President Trump was elected. Employment in the US mining industry was at a low point in October 2016. Coal production was at half its peak. America was built on cheap abundant coal yet Hillary Clinton boasted to her green allies that she would finish off every last coal mining job in America despite the US obtaining one third of its electricity from coal.

Since the Trump administration came into power, the mining sector has added 35,000 new jobs[17] which have a multiplier of over four thereby adding another 140,000 jobs. USA possesses more combined coal, oil and natural gas resources than any other nation on Earth. These resources represent trillions of dollars in economic output and countless American jobs, particularly for the poorest Americans.

The same growth rates we see now in China and India happened in the US during the Industrial Revolution. Between 1860 and 2016, the American population grew nine times and the gross domestic product more than 130 times. Corn production rose 17-fold yet more land was planted with corn in 1925 than in 2016. This is due to better farming methods, mechanisation, fertilisers, insecticides, herbicides and genetically-modified crops. The volume of timber able to be left standing in US forests has increased appreciably from 1952 to 2016. None of this has been done by green policies.

At present, the US holds 27.6% of the planet's coal reserves, Russia has 18.2% and China 13.3%. Coal resources are unknown but are orders of magnitude higher. Because coal is present in terrestrial sedimentary rocks that drape over most countries, almost all countries have resources and reserves of coal. Australia, India, Germany, Ukraine, Kazakhstan, South Africa, Serbia, Colombia, Canada, Poland and Indonesia also have substantial reserves and more coalfields are being found every year.

Many nations keep coal in their core energy mix as a national security consideration and restrict its export. Coal also provides more bang for the buck compared with wind, solar, biofuel,

---

[17] *The American Spectator*, 17 April 2017.

wave or tidal energy. Despite green propaganda and wishful thinking, coal is not about to die. Coal will continue to thrive because it does what no other energy source can do. Steaming coal is cheap and there is every possibility of getting cheaper although top quality coking coal is becoming harder to source.

The United Kingdom

The UK was a significant coal producer. A century ago the UK produced 292 million tonnes of coal *per annum*, now it is less than 10 million tonnes. There is no shortage of coal reserves in the UK, but production costs have risen greatly and the EU has put its green bureaucratic nose into the internal affairs of the UK energy industry.

Furthermore, in 1999 the UK was at peak oil production of 2.9 million barrels per day. In 2012, it was 1.9 million barrels per day and oil production continues to decline.

The North Sea oil is almost exhausted yet the UK has huge potential resources of oil and gas that could be won by fracking. The UK now imports almost all the fossil fuel it burns and is greatly exposed to the vicissitudes of the energy market. The lights could very easily go off in the UK. And stay off.

Australia

According to Geoscience Australia,[18] Australia has more than 1,000 years of coal and 30% of the world's known uranium at current production rates. Gas reserves are huge and relatively unexplored, such as in Victoria's Otway and Gippsland Basins. Australia is a large coal producer, it is the second biggest export (after iron ore) yet more than half of Australia's coal mines have production costs above the global average.

With capital costs in Australia two-thirds above the global

---

[18] www.ga.gov.au/scientific-topics/energy/resources/coal-resources

average, Indonesia has now overtaken Australia with steaming coal exports to China. Five years ago, Mozambique and Mongolia were not exporting coking coal. Both countries are now starting to export coal and Botswana is waiting in the starting blocks to be an exporter. US coal exports are set to rise dramatically in the next decade.

In response to noisy political pressure, we might decide to reduce emissions of carbon dioxide. The only effect will be to make our industries uncompetitive and to increase our cost of living. This has already started to happen. Environmental carping about coal and carbon dioxide is unrelated to reality.

As a result of heavy Red/Green pressure, Labor wants 50% of electricity from renewables, the Liberals want 42% as per the Finkel report and the Greens want 100% because the only way to stop using fossil fuels is to have an astronomical increase in electricity prices. Labor has abandoned its traditional base and added huge costs to workers, the Liberals have abandoned free enterprise for subsidies and the Greens have abandoned use of anything above their shoulders. In Australian politics, there is a huge vacuum for a real leader to fill.

If one country for some bizarre reason decides to stop exporting coal to save the world from a speculated carbon dioxide-driven global climate change, another country will take up the slack and the markets. Once markets are lost because of unreliable supply, trust is lost and markets cannot be won back. And nothing would happen to global climate by one country reducing its carbon dioxide emissions. However, these export earnings and jobs would be lost. Forever.

Over the last decade, investment in coal-fired power stations has been made very risky by both Labor and Liberal-National Party governments. Power station owners have trimmed costs and reduced maintenance and capital expenditure so much that the end result is a country that once had cheap electricity has committed economic suicide. It is delusional to think that coal has no future. Today, coal has a better future than it has ever had and it is still the world's leading source of cheap electricity. This will not change for generations.

New South Wales is a good lesson in the importance of coal-fired electricity. Even green left environmental activists need fossil fuels to transport food from rural areas or from abroad to the cities. A huge amount of fresh food, flowers and high unit value materials flit around the world as air cargo.

Without fossil fuels and dams for domestic and industrial electricity in NSW, there would be no cooling, heating, cooking, refrigeration, transport, employment or communications. Nothing, unless you don't mind days of facing the north end of a horse slowly heading south. If that's the world that the green left environmental activists want us to live in, then let's see them lead by example and shout their ideology from the caves on a cold, wet and windy night. I won't be listening.

In NSW in 2014, there was a mix of electricity generation from coal, natural gas, hydro, wind, diesel, bio and coal seam gas.[19] The table shows what we already know. If heavy-duty base load power is needed, then coal (10,760 MW; 59.3%) and hydro (4,510 MW; 24.9%) do the heavy lifting 24/7. Gas (2,144 MW; 11.8%) and hydro can be used to increase electricity generation in peak load times.

Of the 18,138 MW capacity, wind capacity is 550 MW (3.0%) but, because the wind does not blow all the time, the input into the grid is about 110 MW (0.6%). The sugar cane waste (bagasse) only has a 68 MW (0.4%) capacity and electricity generation is seasonal. Diesel (106 MW; 0.6%) is used when required.

---

[19] http://www.aemo.com.au/About-the-industry/Registration/Current-Registration-and-Exemption-lists2465

More than 71% of electricity in NSW is generated from fossil fuels. In Australia, 65% of Australia's electricity needs are produced from coal reflecting the fact that coal provides the main base load demand in Australia. Base load power is needed 24/7.

Calls for moves for clean energy are driven by those who will profit from its adoption. Why on Earth should we pay for expensive clean energy (that is not clean) when we already have cheap clean energy? Boilers at coal-fired power stations are 96% efficient. Exhaust heat is captured and reheats water, the 4% heat loss is through the boiler walls. Some moist air is condensed, precipitators remove 99.99% of small particles derived from coal burning and carbon dioxide escapes up the stack to the atmosphere. Coal-fired power stations generate electricity efficiently with some 8% of the generation cost being the cost of coal.

And what does the carbon dioxide emitted from the stacks do? Fertilises plants. What can be wrong with this? Plants neither know nor care whether the carbon dioxide they use is from human emissions or from natural emissions. Carbon dioxide is carbon dioxide.

At night, power is used in factories, mines, office air conditioning, refrigeration and houses. About 65% of capacity is needed at night. Demand increases as soon as the sun rises and suddenly increases at breakfast, evening meal times and advertisement breaks when people boil water for a cuppa. Demand begins to decrease when people start going to bed. Winter demand is high for heating as is summer demand for air conditioning.

Whatever the green left environmental activists may desire, the reality is that NSW will continue to have a reliable coal-hydro-gas mix for peak- and base-load electricity. Why? Because it works.

Table 1: Major existing power stations in NSW[20]

Power Station	Location	Technology	Capacity (MW)
Appin	Illawarra	Gas[a]	55
Bayswater	Hunter	Coal	2,800
Bendeela	Nowra	Hydro	240
Blowering	Snowy	Hydro	80
Broadwater	North Coast	Sugar cane waste	38
Boco Rock	Nimmitabel	Wind	113
Broken Hill	Broken Hill	Solar	50
Capital	Tarago	Wind	140
Condong	North Coast	Sugar cane waste	30
Colongra	Central Coast	Gas[b]	724
Cullerin	Upper Lachlan	Wind	30
Eraring	Hunter	Coal	3,000
Eraring	Hunter	Diesel	56
Gullen Range	Goulburn	Wind	172
Gunning	Gunning	Wind	47
Guthega	Snowy	Hydro	80
Hume	Snowy	Hydro	70
Hunter	Hunter	Gas[b]	50
Liddell	Hunter	Coal	2,200
Mt Piper	Central West	Coal	1,400
Murray	Snowy	Hydro	1,575
Smithfield	Smithfield	Gas[c]	175
Tallawarra	Illawarra	Gas[c]	435
Tower Point	Illawarra	Gas[a]	41
Tumut	Snowy	Hydro	2,465
Uranquinty	Wagga Wagga	Gas[c]	664
Vales Point	Central Coast	Coal	1,360
Woodlawn	Tarago	Wind	48

a = coal seam methane; b = open cycle gas turbine; c = combined cycle gas turbine

---

20 http://www.resourcesandenergy.nsw.gov.au/investors/projects-in-nsw/electricity-generation

In 2007, Australia had one of the lowest electricity costs in the world. Between 2007 and 2013, Australia had a Labor Federal government that brought in a carbon tax and various Federally-funded renewable energy schemes. In 2011, the electricity prices in Western Australia, Victoria, NSW and South Australia had risen so much that they ranked just behind Denmark and Germany.

By 2017, South Australia had the dubious honour of having the most expensive power in the world after closure of base load coal-fired power stations. South Australia placed its economy in the unreliable puffs of sea breezes and daytime sunbeams. Blackouts became a costly nuisance and embarrassment to the State government. Many households could not pay their electricity bills.

As the joke goes:

Q: *"What did South Australia have before candles".*

A: *"Electricity".*

The message is clear. If you want to build an economy and create jobs in an industrial society, then mine coal for power generation.

If you have a pathological illogical religious hatred of coal, then bad luck.

High energy low emissions power stations (HELE)

A new survey by Urgewald,[21] an environmental group based in Berlin, shows that more than 1,600 coal-fired electricity generating plants are planned or under construction in 62 countries, increasing global coal-fired power capacity by 43 per cent. The bulk of those plants are in the G20 countries, with China, India and Japan leading the pack. The firms building

---

[21] www.thefifthestate.com.au/briefs/new-database...coal-plant.../93130

the plants and supplying their key components are also almost entirely based in the G20 countries, including leading companies such as India's National Thermal Power Corporation (now the largest builder of coal-fired plants worldwide), China's State Power Investment Corporation, Japan's Marubeni and Germany's Siemens. And where the plants are being built in poorer countries, the projects are typically supported by generous taxpayer-funded export assistance in the form of concessional loans and outright grants.

The Minerals Council of Australia[22] estimated that over 1,015 HELE coal-fired generators are currently in construction worldwide and another 1,231 are planned or under construction. These new, ultra-supercritical and advanced ultra-supercritical coal power plants use new technology to burn coal at higher temperatures than in a conventional plant and higher temperature-higher pressure steam is generated resulting in reduced coal consumption and reduced carbon dioxide emissions.

Australia is the only country that doesn't regard these new super-efficient coal plants as clean energy. Australia has not built a new coal-fired power station for eight years. This has resulted in a skyrocketing of electricity costs. Australia is the world's largest coal exporter and this coal goes to HELE clean energy generation in Asia. Coal-fired generators in Australia that are due for retirement could be replaced by HELE generators. Has Australia given itself a problem?

Germany is building new HELE plants as is Japan (45 new plants), China (400) and India (600). In east Asia, a total of 1,250 new HELE plants are being built or planned. This new technology does not rely on the vagaries of the gas market or the weather. HELE is the cheapest and most reliable energy option

---

[22] www.minerals.org.au/news/show/category/coal

and Australia has a good chance of missing out on producing cheap electricity.

That coal-fired generation is growing is unsurprising. According to estimates by the International Energy Agency, its costs are still 20 per cent to 30 per cent below those of natural gas and nuclear, and are even lower compared with intermittent sources such as solar and wind which require costly backup generation, subsidies and expensive new transmission networks.

The G20 should see US President Trump's recommendation that developing countries be encouraged to design and operate fossil-fuel capacity "more cleanly and efficiently" as obvious common sense. Countries such as Egypt, Pakistan and Malawi, which currently have no coal plants, are investing in large-scale coal generation, thus reshaping the long-term politics of climate change negotiations.

What is this filthy black evil demonised substance?

Coal is of vegetable origin. Therefore it is good. It is organic. Therefore it is better. It is natural. Therefore it is the best. Coal is a form of solar power because it is highly concentrated solidified solar energy. Burning of coal recycles carbon dioxide back to the atmosphere. It therefore should be revered by greens.

It contains organic carbonaceous matter, inorganic material (minerals) and fluids such as coal seam gas (methane) and water. The carbon and hydrogen compounds in coal burn to produce heat, light, gas, residual solid ash, water vapour and carbon dioxide. Many coals contain fragments of fossilised wood, leaves, bark, fungus, pollen and spores and roots that go from coal into the underlying rocks.

Plant material dies, accumulates and decomposes and in

favourable settings it can form coal. This decomposition is driven by bacteria and oxygen. If plant material is to accumulate and be preserved, it must be water saturated in a bog or swamp and covered with water to stop oxidation. The material that forms is peat. Bacterial activity almost ceases and the peat accumulates if the swamp is not drained.

There is a popular view that coal swamps formed in steamy Amazonian-type jungles. Nothing could be further from the truth. Peat bogs form at high latitudes in cold climates where bacterial decay of accumulated vegetable material is very slow. Covering of peat with sediment followed by subsidence and compaction convert peat into brown coal and, with further pressure and temperature, converts brown coal into black coal and eventually anthracite, nature's cleanest burning coal.

The process of coal formation takes place at depth and later geological processes uplift the coal seams to at or near the surface. For many technical and economic reasons, coal cannot normally be mined at depths greater than 500 metres. That will change and massive deeper reserves will be available in the future.

This process of increasing the maturity from peat to brown coal to black coal results in the loss of water and an increase in carbon, coal density and the amount of energy that the coal yields. There are two major types of industrial coal, thermal coal and coking coal.

Underground mining of coal can produce dust. Coal dust, like many other dusts (e.g. wheat husk dust), can be explosive. During the mining of coal, waste rock is sometimes mined and coals might have internal seams of waste rock such as sandstone, shale or volcanic ash. Much of this material is removed by coal washing. In this process, coal is mixed with water containing suspended magnetic iron oxide (magnetite) to create a dense liquid. Coal floats and waste rocks sink. The traces of magnetite stuck to the washed coal and waste rock are removed by electromagnets and reused.

Compared to unwashed coal, washed coal produces more heat, reduces transport costs and produces fewer waste products at smelters and coal-fired power stations. The best quality coals have low water, low ash (mineral and rock particles in coal) and low sulphur contents. Some coals on the west coast of the South Island of New Zealand have an extraordinarily low ash content (less than 1%) whereas most coals have an ash content of 10 to 20%. Some of the ash in coal cannot be removed by washing.

Before coking coal can be used, a comprehensive suite of physical and chemical analyses need to be performed. Measurements need to be taken to determine how quickly coal will wear out equipment. Tests of coking coal need to be undertaken to determine how much the coal will swell, when it will soften in the furnace, how strong the coal will be in a furnace, what gases will be emitted upon heating and when gas will be released from the coal.

During the process of conversion of peat to black coal by heat and pressure, gases are produced. The world's peat bogs, peaty tundra, swamps and soils leak methane gas into the atmosphere as do humans, bovines, termites and many other insects. This leakage process continues during conversion of peat to coal. In places, leaked methane (swamp gas) spontaneously ignites (with the help of phosphorus gases and microorganisms) making swamps appear dark evil spooky places at night with flickering lights, dancing phosphorescent glows and flames.[23] In more modern times, the odd dancing lights from swamp gas ignition have been interpreted as UFOs, aliens and will-o'-the-wisps. I guess this depends on your state of sobriety.

---

[23] In Shakespeare's *The Tempest*, Caliban is frightened witless by lights in a swamp from the combustion of methane (swamp gas).

Gases produced by the conversion of peat to coal leak to the surface at variable rates, can be trapped in overlying rocks and are trapped in the fractures in coal. The main gas is methane although carbon monoxide, carbon dioxide, hydrogen sulphide and nitrogen are also common. Accumulation of poisonous carbon monoxide is deadly.

Over the centuries, there have been some shocking multiple fatalities in underground coal mines from explosions of methane and dust. Mixtures of inert gases such as carbon dioxide and nitrogen have asphyxiated miners because of the lack of oxygen. Fall of the roof, walls and an inrush of water have also produced many fatalities. These explosions still take place in jurisdictions where the occupational health and safety laws are not as rigorous as in the Western world.

In order to avoid such gas explosions, modern mines are well ventilated and equipment that may produce a spark is not used. Electrical switching equipment is non-sparking and, where metal meets metal, zinc against zinc contacts are used. As iron can spark with striking silica-rich rocks, it is not used. Areas where tungsten carbide cutting teeth on mining equipment may spark are wet and ventilated. Matches, cigarette lighters and smoking underground are strictly prohibited.

Some coking coal contains a large amount of gas and, as an additional safety procedure, patterns of drill holes from the surface penetrate the coal seam to extract methane before mining as a safety precaution. For nearly a century, this coal seam gas has been bled from the coal and used at the mine site to generate electricity for the operation.

Greens have only just discovered that there is such a thing as coal seam gas. Those in the coal mining industry have known about it for centuries.

So much for the progressive greens.

Dirty coal

Most Northern Hemisphere coals have a high sulphur content. This is because the swamps from which the coal derived were close to the shore. Regular events of slight subsidence or sea level rise meant that peat swamps were sporadically inundated with sulphate-bearing seawater that was chemically reduced by decomposing plant material to sulphides. Some sulphur is also chemically bound onto organic compounds. Most large Northern Hemisphere deposits of coal are Carboniferous in age and the exhaust gases from burning this coal need to have the sulphur compounds removed.

The geological time period named the Carboniferous is a good indication of what was happening at that time. Carbon dioxide was sequestered from the then very carbon dioxide-rich atmosphere into plants. This modified plant material is now coal and, by burning coal, we are recycling sequestered carbon dioxide back into the atmosphere so that it can undergo another lap of natural sequestration.

In the mid 20th century, the burning of these high sulphur coals for household heating and cooking in the UK, Europe and USA created noxious fogs called pea soupers. Tens of thousands died from respiratory problems. This is exactly what is happening in China at present with small industry emitting particulates and sulphur gases creating dreadful smog and respiratory problems. In the Western world in the mid 20th century, this public health problem was solved by having large centralised coal-burning power stations that scrubbed particles and gases from exhaust gases. The domestic use of coal was banned.

Heating and cooking at home was then by reticulated cheap electricity and not by burning sulphur-rich coals as previously. The smogs and respiratory problems disappeared. This shows that pollution can be stopped but only when communities become wealthier and apply simple engineering solutions to problems. This was an environmental decision taken by responsible Western governments 60 years ago. No greens were involved.

The first step to stop massive pollution in China from many small factories burning low quality coals has already been taken. Previously China would import low quality coal with 20% ash and 1% sulphur. Imports of these coals is now banned and so some 50 to 55 million tonnes of poor quality sulphur-rich coals will not be imported and burned. Although sulphur is scrubbed out of waste gases, even a small amount of sulphur oxides released into the air can cause smog, acid rain and respiratory problems.

By contrast, Southern Hemisphere coals and those from India are slightly younger (Permian), formed in a deltaic environment up slope and further from the sea and accordingly have a lower sulphur content because of the lack of inundation by the sea. The large Southern Hemisphere coal deposits of India, South Africa, Australia, South America and Antarctica formed during very cold times. In places, coal seams are interlayered with debris left behind by retreating glaciers.

This environment is also well represented in the modern swamps of the Northern Hemisphere where peat directly overlies glacial debris left behind by retreating glaciers at the start of the current interglacial (about 14,700 years ago).

Coal seam gas

Uplift of coal-bearing sequences results in rock destressing due to unloading, the rocks fracture and gases migrate into fractures in coal. This is a process of natural fracking and is common at relatively shallow depths. Industrial fracking is not used for coal seam gas extraction. With fracking at far greater depths for shale gas and oil, uplift has not naturally fracked the rocks so fractures are artificially produced by immense hydraulic pressures.

Coal seam gas has been drained from coal seams before mining because of the potential risk of gas explosions underground. Australia has drilled holes and produced coal seam gas for generating electricity as part of safety measures in the coalfields for 70 years.

Because coal seams are generally enclosed by impermeable rocks, coal becomes a shallow reservoir for pressurised methane gas. Once the coal is penetrated by a drill hole, the gas drains from high pressure in the coal seam to low pressure where the drill hole has intersected the seam. The gas then naturally rises up the drill hole. The hole is cased, can be plugged at any time and flow can be adjusted with valves in surface facilities.

The footprint of a coal seam gas drill hole surface facility is about the size of the average lounge room. There are millions of holes around the world each year that penetrate the water table and tried-and-proven procedures of casing holes to stop water table contamination are well established. Coal seam gas can be very cheaply extracted from seams that may be too deep or too poor in quality to mine. All that coal seam gas extraction does is accelerate the removal of gas that would naturally leak out of the coal over time.

Underground coal seam mining does not remove all the coal. This is because pillars of coal need to be left for supporting the roof. In old mines, only about 30% of the coal was removed. In modern longwall mines, most of the coal is removed.

Some weird and wonderful things happen in the subterranean world. A biochemical reaction between coal and water is driven by microorganisms and methane is produced. Openings in old coal mines are filled with methane. By drilling into these spaces (called goaf), methane can be extracted. What's more, the methane continues to be produced and the goaf operates like an underground gas storage tank. Again, drilling through the water table uses cased holes. Entering an abandoned coal mine often leads to a Darwin Award.[24]

There have been a number of corporate cowboys trying to rapaciously enter the coal seam gas industry for a quick profit and they have had shocking practices of venting some of the non-flammable gases, such as hydrogen sulphide, to the surface and using cheap plastic casing rather than steel. Every new industry attracts grubs who think they can get rich quick. Landholders have rightfully been incensed. These landholders have actually joined their mortal enemies, the greens, as anti-coal seam gas groups. It is not landholders and green groups that are running the cowboys out of town, it is the serious players in the industry and the regulators. Later the greens will of course run true to form and come back and bite the landowners.

There is also great potential for offshore coal seam gas as many coal basins are both onshore and offshore. One of the biggest basins is in the North Sea where there is an estimated three trillion tonnes of coal underneath it. This is the perfect setting for a massive coal seam gas operation that could power the UK.

Opposition by greens to coal seam gas extraction may be because of a lack of knowledge, because methane is a hydrocarbon gas or because anything associated with coal is off limits.

I suspect that the real reason for green opposition is that coal seam gas is a cheap source of energy and that coal seam

---

[24] Posthumous award for removing oneself from the gene pool by one's own stupidity. For example, a thief's gun jammed while trying to hold up a supermarket, he turned the gun around to look for the problem, pulled the trigger and blew his head off.

gas electricity bypasses the green control of wind and solar power. This opposition has nothing to do with the environment and everything to do with power over the average person by unelected minorities. As a result, you pay more for electricity.

In every case around the world when there has been a government coalition involving greens, unemployment has risen, productive industries have been destroyed, costs have risen, massive amounts of money has been wasted, freedoms have been lost and the government has flirted with totalitarianism. Again, the costs are borne by the average person, as per usual.

Oil shales

Coals dominated by spores and pollen can be used to make petroleum. These are the oil shales, sometimes called cannel coal or sapropelic coal. When lit with a match they burn, they float on water and they were a source of petroleum in the 19th century and World War I (e.g. Newnes, NSW; Joadja, NSW).[25] They are different from shales that hold crude oil and gas that are now being exploited by horizontal drilling followed by fracking.

These mined oil shales were crushed and heated in a retort, petroleum vapour was released and then condensed to form various fractions of liquid hydrocarbons. Although a tonne of oil shale could release up to 100 litres of hydrocarbons, there were monstrous amounts of rock waste produced. This is exacerbated because after crushing, heating and distillation of petroleum, the residue has a volume 30% greater than the unmined oil shales and hence is too voluminous to fill underground or open pit space. Probably the biggest reserve of oil shales are in the Green River Formation of Wyoming, Utah and Colorado (USA).

---

[25] A.J. Kraemer and H.M. Thorne, 1951: Oil-shale operations in New South Wales, Australia, *US Department of the Interior, Bureau of Mines*.

Huge reserves of oil shales exist in many coal-bearing sedimentary basins as well as coal seam gas and deeper shale gas. Petroleum from oil shales is uneconomic unless the crude oil price stays above $US 120 per barrel for a long time. Until then, oil shale sits on the reserves bench of the energy game and is a security against the closure of sea traffic or ultra-high crude oil prices.

Carbon capture and storage

Carbon capture and storage is another way of green activists saying no to coal.

In Australia, the National Geosequestration Laboratory has sequestered $49.4 million annually of taxpayers' funds. This is your money.

Some 95% of Canada's remaining known fossil fuel reserve is bitumen. If the example of fracked oil and gas from basins in the Lower 48 of USA is anything to go by, then Canadian sedimentary basins also have huge volumes of oil and gas to be released by fracking. Only 5% of Canada's bitumen has been exploited to date. The Canadian bitumen industry currently employs 130,000 people, generates $CAN 5.5. billion per year and adds 2% to the Canadian GDP.

Canadians are tying themselves in knots about the release of 70 million tonnes of carbon dioxide annually (0.13% of global emissions) and if all of Alberta's exploitable bitumen were used, 22 billion tonnes of carbon dioxide would enter the atmosphere. If carbon dioxide is the sole variable driving a complex climate system, then models speculate that global temperature would rise by 0.4°C by burning all of the known Alberta bitumen. The Athabasca tar sands contain a heavy oil that needs upgrading by hydrogen addition to convert it from tar to synthetic crude oil for refining into the various fractions of petroleum products. Making hydrogen produces carbon dioxide that is separated from nitrogen through an absorption amine technology process and it is this carbon dioxide that is sequestered.

One of the world's largest carbon capture and storage facilities is the Shell Quest project in Alberta. It can capture and sequester 2 km underground a third of the carbon dioxide produced by the conversion of bitumen to lighter fractions at about one million tonnes *per annum*.[26] Over the 25-year life of the Scotford installation, 27 million tonnes of carbon dioxide will be captured and 50 million tonnes will be vented to the atmosphere. The carbon dioxide is transported by a 64 km pipeline, injected into a porous sandstone underlain by impermeable metamorphic rocks and overlain by impermeable shales and salt formations. Leakage is checked, as is seismic activity and ground deformation. Quest received $CAN 865 million from the Alberta ($CAN 745 million) and Canadian ($CAN 120 million) governments, the total project cost $CAN 1.35 billion, will operate for 10 years and the project has received a two-for-one carbon credit arrangement with Alberta province for each tonne sequestered.[27]

The energy used to compress and inject carbon dioxide deep underground is unknown. It is wonderful to see science and engineering solve problems however I feel that a non-problem has been solved. There has been a boom in science and engineering trying to solve the problems created by carbon dioxide sequestration. It has not been shown that human emissions of carbon dioxide drive global warming and it looks like a huge waste of money to stop plant food being put back into the atmosphere where it belongs.[28]

---

[26] http://bit.ly.2k5ncA]
[27] https://www.globalccsinstitute.com/projects/quest
[28] https://www.albertaoilmagazine.com/2016/02/is-carbon-capture-technology-doomed/

All potential breakthroughs aside, the current technology for carbon capture and storage remains an expensive endeavor for governments and the private sector to solve a non-problem.

Gas

Many rocks are rich in gas.[29] The gas occurs in pores and fractures. Biological material trapped in sediments is decomposed to methane as the sediment is heated under pressure to solid rock. Any black sedimentary rock contains a large amount of decomposed carbon-rich biological material and methane. During the process of conversion of sediment to sedimentary rock, methane gas migrates and can be trapped in overlying rocks or may leak into the atmosphere.

There is a constant leak of methane from rocks into the atmosphere along porous rocks, fractures and faults. I have been underground and heard methane hissing out of rocks and drill holes. This is one of the reasons why there is no smoking allowed in underground mines.

If a river has cut its path along a fault, there will be constant bubbling of methane through the water. Artesian springs are where warm salty water from kilometres depth has been squeezed to the surface together with gases such as methane, carbon dioxide and rotten egg gas. Such springs contain rare traces of oil. Groundwater contains methane and upon depressurisation when the water nears the surface, the methane bubbles out of the water.

During decomposition of trapped biological material in sediments, both oil and gas may form, especially if the sediments formed in a shallow marine environment. The oil and gas form in a source rock, commonly an old coral reef, and later migrate upwards into a trap rock that is porous and permeable. Up to

---

[29] Mainly methane (CH_4).

30% of the trap rock can be space that is filled with oil or gas. If the rocks at depth become too hot, the gas is driven upwards into the atmosphere and the hot oil may form pools on the surface or the lighter hydrocarbons may boil off to leave pitch or tar at or near the surface. The tar sands in the Los Angeles region were a death trap for pre-historic animals which became pickled and preserved.

Gas migrates further than oil, so trap rocks often contain pores filled with gas overlying rocks with pores filled with oil which in turn overlies rocks with pores filled with salty water. Pressurised gas produced from onshore and offshore wells flows to the surface and undergoes processing before being transported along pipelines.

Gas producers argue that because natural gas[30] has only one carbon atom, it is a preferred fossil fuel to coal which has many carbon compounds and emits more carbon dioxide upon burning. However, what the gas producers don't tell us is that natural gas is processed to extract carbon dioxide (which is vented to the atmosphere), rotten egg gas (which is captured), water, oil condensate and sometimes helium.

For some 75 years, oil and gas wells have been fracked in order to increase oil and gas flow and to remove precipitates of minerals and waxes in fractures. Gas will flow under its own pressure. However, some oil may flow under its own pressure and most oil needs to be pumped or displaced by filling the pores in the sedimentary rock trap with water. As oil and gas are produced from a far greater depth than coal seam gas, oil wells have cemented double or triple casing to prevent oil loss and water contamination from ground water or sea water. Oil and gas are produced from well beneath the water table and are isolated from groundwater. Gas produced from onshore

---

[30] Mainly methane (CH_4) with less abundant higher hydrocarbon gases.

and offshore oilfields is known as conventional gas. Both oil and gas also naturally leak from depth and some oil slicks at sea are from natural leakage.

With new technology, vertical wells (drill holes) can now be deflected into horizontal holes of up to 3 km length. This can only take place in highly pressurised rocks at depth. Casing is run along wells during drilling. It is normally triple tube with cement seal and such casing does not permit gas or oil to be lost to enter overlying rocks or overlying water tables. Casing for the horizontal hole is perforated and a high-pressure water-sand mix is pumped at ultra-high pressure into the perforated casing. The grains of sand wedge fractures open, traces of lubricants are used if oil is expected and traces of acid are to dissolve any new minerals formed by depressurising rocks at depth.

After fracking, the water pumped into the fractures takes about a week to rise to the surface and then gas or oil will flow. When all the oil or gas has been bled from the fractures, the process is repeated and fractures are pushed further into the rocks surrounding the horizontal hole. Fracking in the USA has changed the geopolitics of oil, has taken the control of international oil trade out of the hands of OPEC and has led to an economic boom in the USA. The USA is now self-sufficient in oil and gas and should be for hundreds of years.

Extraction of groundwater, gas, oil and rock from depth involves destressing of rocks under pressure. This produces earth tremors, most of which can only be detected with very high precision instruments. These earth tremors are at the same magnitude as tremors produced along the coast from waves and from storms, wind, tides, traffic, loading and unloading of dams with water, underground tunnels for trains and traffic and loading of the substrate with high rise buildings. These tremors are far lower in magnitude than regular natural tremors produced from uplift, compaction and compression of rocks. The crust of the Earth is always bending, breaking and readjusting. This tells us what we already know: Planet Earth is dynamic.

One of the popular scare campaigns with gas extraction is that earthquakes are produced. They are not. The energy released during an earthquake is many orders of magnitude greater than from an earth tremor. There are about 10,000 significant earthquakes and tens of millions of earth tremors each year.

There is great confusion in the community about methane produced from coal at a shallow depth (coal seam gas), gas continually forming in old coalfields from a coal-water-bacteria biochemical reaction in underground spaces (coal seam gas), gas produced from a greater depth in trap rocks in an oilfield (conventional gas) and gas released from sedimentary rocks at an even greater depth by induced fracturing (unconventional gas). Just to make matters even more complex, there are some naturally fracked rocks in oilfields that host gas that has migrated from a deeper conventional gas source.

This appears to have been a little too difficult for some politicians. In Victoria (Australia) there is a blanket moratorium on onshore drilling for gas whether this gas is coal seam gas, conventional gas or unconventional gas. A magnificent own goal has been kicked because offshore conventional gas is processed, sent 2,000 km by pipeline to Queensland for liquefaction and exported as LNG under long term contracts. Meanwhile, Victoria has known onshore gas wells that flow, has two huge basins with a few kilometres thickness of gas-bearing sedimentary rocks and a gas shortage. Local gas is selling for above the international market price and the exports contracts have left very little gas for domestic use. To make matters worse, the onshore gas has very little carbon dioxide compared to the offshore gas, does not need processing and could be tapped into existing pipelines that are only a few kilometres from wells. You couldn't make this up if you tried.

The Northern Territory has also banned onshore gas exploration, despite having the most attractive sedimentary basin in the world for unconventional gas. The basin is traversed by gas pipelines. The Northern Territory could follow in the footsteps of the USA and be an energy giant. Instead, the politicians, in their infinite wisdom, have decided that unconventional gas is politically incorrect and the consequence of this decision is that the Northern Territory has high unemployment and little revenue.

Nuclear

We live in a radioactive world. Many rocks are naturally highly radioactive (e.g. granite, salt) and radioactive materials leak from the ground (e.g. radon gas). Plants can be quite radioactive (e.g. banana), as can many fertilisers. We are constantly bombarded with particles from space. House bricks are slightly radioactive as are house fire alarm systems. Radioactivity is everywhere.

If emissions of carbon dioxide were really a problem for the green left environmental activists in Australia, then they would welcome a long-term program of building nuclear reactors in order to greatly reduce Australia's emissions. If one believes as a matter of environmental religious ideology that coal is old technology, then we should get with numerous other countries and use nuclear energy. Renewable energy will not keep an industrial society functioning. Base-load energy from nuclear and coal and peak-load energy from gas and hydro is the ideal mix for an industrial country.

More than half the world has access to some electricity

generated by nuclear fission. USA generates about 30% of the world nuclear energy. About 5.1% of the planet's total electricity is generated from nuclear reactors. If Australia wants to continue to grow and have large base load electricity for smelting, refining and general industry, then it must either increase the number of coal-fired power stations or go nuclear. To have both would create energy security. Your electric light does not know whether the electrons it uses come from coal, gas, nuclear, hydro or renewables.

A few hundred kilograms of uranium were produced in South Australia in the 1930s (Mount Painter and Radium Hill). Uranium was mined from 1954 to 1971 at Rum Jungle (NT) and Radium Hill (SA), Mary Kathleen (Qld) (1976 to 1982) with Nabarlek (NT) (1979 to 1980), Ranger (NT) (1981 to present), Olympic Dam (SA) (1988 to present), Beverley (SA) (2000 to present) and Honeymoon (SA) (2011 to 2013). There have been numerous other discoveries over the last few decades (e.g. Mulga Rock and Lake Way, WA).

Australia has the largest resources of uranium in the world (31% global total) and is the third largest producer in the world (~8,000 tonnes contained uranium in yellow cake or 12% of global supply) after Kazakhstan and Canada. Australia does not beneficiate yellow cake into fuel rods and does not have the capacity to reprocess used fuel rods. We have lost thousands of new jobs by not adding value to our yellowcake exports.

Some 35% of all energy exported from Australia is as yellow cake. Exports are to Europe (37.8%; mainly Belgium, Finland, France, Germany, Spain, Sweden and UK), USA (33.6%) and Asia (28.6%; mainly Japan, South Korea, China and Taiwan). USA generates 30% of global nuclear power.

Natural uranium ores contain 99.284% uranium 238 and 0.711% uranium 235, the ore is processed to make yellow cake[31] which is sold at the mine gate and is processed elsewhere into a product that is enriched to about 20% uranium 235 for use in a reactor.

The natural decay of uranium is a slow process that produces heat, sub-atomic particles and new isotopes.[32] In some places in the world, highly radioactive granite is covered by a natural blanket of insulating rocks. Driven by green zeal, for several years in outback South Australia, there were experiments whereby very high pressure water was forced from a drill hole into these hot granites so that the rocks fractured and steam rose to drive turbines to produce electricity. Fracking of granite was strongly supported by the environmental movement for

---

[31] Different mine processing plants produce different yellow cakes such as uranyl hydroxide, triuranium octoxide, uranium dioxide, uranium trioxide, uranyl sulphate, sodium-para uranate, uranium peroxide, ammonium diuranate or sodium diuranite are loosely called "yellow cake".

[32] The natural radioactive decay of uranium 238 is ultimately to stable lead 206.

Type of radiation	Nuclide	Half life
α	uranium 238	4.5×10^9 years
β	thorium 234	24.5 days
β	protactinium 234	1.14 minutes
α	uranium 234	2.33×10^5 years
α	thorium 230	8.3×10^4 years
α	radium 226	1590 years
α	radon 222	3.825 days
α	polonium 218	3.05 minutes
β	lead 214	26.8 minutes
β	bismuth 214	19.7 minutes
α	polonium 214	1.5×10^{-4} seconds
β	lead 210	22 years
β	bismuth 210	5 days
α	polonium 210	140 days
	lead 206	Stable

the production of hot dry rock geothermal energy whereas fracking of oil- and gas-bearing shale elsewhere will apparently ruin the planet.

For many technical reasons the process of hot dry rock geothermal energy failed. It was touted as the great hope for clean green energy, even serial alarmist and kangaroo specialist Tim Flannery bought shares in one of the companies assessing this process (Geodynamics) and the Rudd government poured $90 million of your money into this highly speculative venture. This unproven technology in the middle of nowhere would have required building hundreds of kilometres of new power lines if successful. At that time, Australia had tried-and-proven reliable cheap coal-fired electricity in populated areas. The government's hot dry rock geothermal power joins other disastrous government and hence taxpayer-funded climate actions such as the "free" insulation, global warming conferences and wave generators that sank.

For power generation, radioactive decay of uranium is sped up in a nuclear reactor. The heat is used to create steam that drives turbines. By spinning magnets in a coil using this steam energy, electricity is produced. However, the accelerated breakdown of radioactive material produces particles that need to be shielded and a huge amount of heat hence there needs to be efficient cooling systems in the power plants.

Decay of uranium produces radioactive waste. The high-level waste occupies 3% of the volume and has 95% of the radioactive content compared to medium level waste (7% volume, 4% radioactive content) and low-level waste (90% volume, 1% radioactive content).[33] If one person used nuclear power for a year to produce electricity, the waste would be 30 grams. This waste continues to emit heat and particles,

---

[33] www.world.nuclear.org

especially over the first 50 years after use in a reactor. Most waste is uranium and plutonium which can be reprocessed into a mixed oxide fuel for use again. Two isotopes, technetium 99 and iodine 129 have long half-lives (220,000 years and 15.7 million years respectively). About 80% of the isotopes used in nuclear medicine are technetium 99 with smaller amounts of other isotopes.[34]

In Sweden, a country considered to be an environmentally-conscious democracy, Oskarshamm has three nuclear reactors and an interim spent fuel storage facility and Oesthammar has three reactors. The Swedish Nuclear Fuel and Waste Management Company undertook feasibility studies in eight municipalities for a deep waste repository. A poll showed that 79% of Oskarshammar and 75% of Oesthammer residents supported a deep waste repository in their own towns.

By contrast, Australia has 31% of the world's uranium reserves and exports yellowcake yet piously declines to add value by creating well needed jobs and infrastructure in embattled rural areas geologically suitable for waste disposal. Australia is the only G20 country without nuclear energy. Furthermore, Australia exports yellowcake knowing full well that waste products will be generated. The geologically stable ancient rocks of Sweden are no different from those of Australia.

Australia has closed university departments that teach nuclear science and engineering and hence the scientists and engineers who will be needed for the inevitable future nuclear power stations will have to be trained abroad. We have a handful of trained nuclear scientists at Lucas Heights. Ironically, new nuclear scientists and engineers will probably be trained in nuclear power stations that derive their yellowcake from Australia.

---

[34] thallium 201, iridium 192, samarium 153, caesium 137, iodine 131, iodine 125, iodine 123, ruthenium 106, palladium 103, yttrium 90, strontium 89, gallium 67, cobalt 60.

There are 30 countries that have 438 nuclear reactors generating electricity. Another 67 reactors are under construction in 15 countries.[35] The long-term construction jobs and highly-skilled workers required to run reactors creates jobs for generations, as does the additional electricity. Modern nuclear plants have costly but deliberate over-design and environmental lawfare is one of the main costs in building a large nuclear reactor.

The processing of nuclear waste also produces many highly skilled jobs. Australia is ideal for a vertically-integrated industry comprising mining, yellow cake production, construction of fuel rods, leasing of fuel rods to other countries, reprocessing of rods, waste disposal and releasing of processed fuel rods. This would provide generational employment and Australia would become the OPEC of the nuclear world. Inland Australia is ideal because of isolation and lack of threats from earthquakes, volcanoes and tsunamis.

Australia is a dry country. Some 400 Gigalitres of water is lost annually by evaporation in the cooling towers of coal-fired power stations.[36] Seawater, saline underground water, helium or liquid metal would be the best fluid for cooling nuclear reactors thereby releasing fresh water for other purposes. Australia is well blessed with uranium ores and a sparsely populated hinterland if reprocessing needs to be isolated. However, acts of Federal and State parliaments would have to be revoked if Australia were to go nuclear.[37]

---

[35] http://www.nei.org
[36] Most media networks imply that the condensing steam from the cooling tower of a coal-fired power station is an emission of carbon dioxide. The media networks have been reminded many times that this is not carbon dioxide yet they continue to show these images. This is fraud.
[37] Federal Acts: ARPANS Act 1998, Section 10, Prohibition on certain nuclear installations and the EPBC Act 1999, Section 140A. State Acts: Victoria Nuclear Facilities (Prohibitions) Act 1986, Queensland Nuclear Facilities Prohibition Act 2006, NSW Uranium Mining and Nuclear Facilities (Prohibition) Act 1986.

Between 1958 and 2007, the 10 MW High Flux Australian Reactor (HIFAR) operated at Lucas Heights (NSW) and was used for materials research, production of medical isotopes and for irradiation of silicon for the computer industry. In 2006, it was replaced by the 20 MW Open Pool Australian Light water reactor (OPAL) which is in the top league of the world's 240 research reactors. This reactor produces 25% of one of the world's medical isotopes[38] and, because it produces neutrons, requires little cooling.

There are complaints from residents in Sydney about the proximity of the Lucas Heights reactor. Houses were built proximal to the reactor after it was built. When I first went to the HIFAR reactor, it was in the middle of the bush and people knowingly built houses close to a small safe research reactor. The reactor needs to be close to Sydney's airport to allow transport of short-lived medical isotopes to rural hospitals otherwise they would decay to uselessness.

Australia is an ideal place for small portable nuclear reactors to service remote towns and mines off the grid. Most of these sites are currently powered by diesel generators. Small modular reactors have outputs between 1 to 300 MW. Their components can be transported to site by truck, they can be sited at places that might not be suitable for a large reactor, there can be incremental additions, modules can be put together quickly, reactors can be cooled by water, metals or helium and small portable reactors offer a number of economic, operational, non-proliferation and safety advantages.[39] When the job is done at a mine, the reactor can be picked up and moved elsewhere.

A large number of modular reactors are currently in ser-

---

[38] Mo^{99} precursor to Tc^{99}.
[39] https://www.energy.gov/ne/nuclear-reactor.../small-modular-nuclear-reactors

vice in ships, submarines, satellites, remote areas, universities and research institutes. There is a great diversity of small modular reactors available from manufacturers hence a brand new facility doesn't need to be created from the drawing board.[40]

If Australia was really serious about maintaining its standard of living and quality of life, it would add employment-creating nuclear-powered electricity to the grid, use small portable nuclear power stations to run remote mines and towns, and establish a nuclear reprocessing industry to beneficiate spent fuel from other countries. Maybe Australia could launch itself into the 1950s and purchase nuclear-powered submarines. However, Australia has contracted to purchase nuclear submarines from which the nuclear power unit is removed and replaced with a diesel power unit.

I'm sure that green activists with cancer or one of numerous tissue diseases would take the high moral ground and reject treatment because the life-saving medical isotopes came from the nuclear industry.

Hydro

New hydroelectricity plants could increase the proportion of renewables in Australia. The chances of the green left environmental activists and politicians allowing another hydroelectricity plant to be built by damming rivers is zero, despite there being many favourable sites.

In low rainfall countries like Australia, it is only the coastal mountains that are a suitable site for hydroelectricity generation. In the north, there are mountainous areas that receive high rainfall in the wet season. The Great Dividing

---

[40] www.world-nuclear.org/.../nuclear.../nuclear-power-reactors/small-nuclear-power-rea...

Range in eastern Australia catches orographic rainfall and this flows eastwards into the Pacific Ocean.

Elsewhere in the inland of the continent where rainfall is lower, there are neither the rivers nor the topographic elevation difference needed for hydroelectricity. After World War II, there were concerns about security of supply of food in Australia. The Snowy Mountains hydroelectricity scheme was built to divert water into inland Australia for agriculture and electricity generation was always considered a by-product. If the Snowy scheme was primarily for electricity, then it would have been designed totally differently.

The speculated Snowy 2.0 pumped hydro scheme is years away, is uncosted and was deemed uneconomic in the 1980s when it was first being promoted. It is proposed that excess wind and solar power will be used to pump water from a low altitude dam to one at higher altitude. This requires a huge amount of energy and there is frictional energy loss. The Snowy 2.0 scheme does not address the fundamentals. Electricity is horrendously expensive from wind and solar schemes because of subsidies and must-take policies hence a Snowy 2.0 would not lower electricity costs and its only merit is that it would make peak load power more reliable.

Furthermore, we already have a Snowy 2.0. It is Jindabyne. It was designed to use off-peak coal-fired electricity in the wee hours of the morning. At the time it was built, there was no 24/7 air conditioning in city offices, little overnight refrigeration and few factories and mines operated 24/7 hence there was excess electricity from the coalfields. We now don't have that excess.

Because of rainfall and topography, Australia is not a good place for hydroelectricity and we'll leave that for places like Quebec, Ecuador and Ethiopia to use their incessant rainfall and topography to generate cheap reliable electricity.

Energy security

Few Western countries have energy security. The West is extremely exposed because greens object to an efficient mix of coal, gas, oil, nuclear and hydroelectricity and have persuaded politicians to embrace feel-good inefficient ideological election-winning energy that can neither provide base load electricity nor the required energy density. To make matters worse, Western countries are preparing for speculated global warming by trying to reduce carbon dioxide emissions and are not preparing for an inevitable global cooling event. In World War II, the guns protecting Singapore were permanently mounted in one direction to repel an anticipated Japanese advance. The Japanese soldiers came from the other direction and Singapore fell. A lesson?

Between 2005 and 2015, the global oil production has started to decline and the oil price rose about 15% *per annum*. Conventional oil production in the US decreased. And then came unconventional oil from fracking. Oil production in the US has increased massively and the crude oil price has decreased. Just when there is doom and gloom, some inventive technical person solves a problem and the world changes. What have the greens done to change the world?

Fracking for oil and gas production has provided the US with huge reserves and the US is now no longer dependent upon the Middle East or Venezuela for its energy. The US is now self-sufficient (except for a few heavy oils that it sources from Canada).

Some 30 billion barrels of oil and 35 trillion cubic feet of gas have been discovered in the US shale oil and gas revolution. Maybe this has given enough breathing time for a clear-headed US energy policy? Oil consumption in the US has also fallen because people have driven their vehicles more slowly. Road deaths have been lowered and the distance covered by vehicles has decreased.

Coal is absolutely vital for energy security and preservation of oil reserves. The attack on coal by green activists is an attack on the energy security of the nation. In Weimar Germany in the 1920s, coal-to-liquid processes were invented and used in World War II because Germany was isolated from sea transport of oil. During embargoes, South Africa also used coal-to-liquid technology. Australia is a net oil importer and could easily be an exporter.

The Bergius process uses powdered coal and hydrogen catalysed at high temperature and pressure and has a high yield per tonne of coal. The Fischer-Tropsch process has a lower yield, is not as complicated and burns coal in pure oxygen to make carbon monoxide and hydrogen. These are converted to liquid petroleum using a catalyst. The Fischer-Tropsch process can use poor quality and water-saturated coal. Australia has huge reserves of such coals.

These coal-to-liquids processes are viable when the crude oil price is above $US 70 per barrel. Coal-to-liquids is the highest value use of coal resources, creates a very clean-burning fuel and provides the most convenient and lowest capital cost energy source for transport for consumers. Countries such as Australia have widespread coal resources of variable quality that occur in each State and huge volumes of hydrocarbons are currently transported great distances from refineries and ports when we could have numerous regional coal-to-liquids plants.

At $US 120 per barrel, all coal-fired power stations could be closed and replaced by nuclear power. Coking coal would still have to be mined. In the US, about one billion tonnes of coal is burned each year for electricity. Known coal reserves and resources could keep the lights on for 250 years. As petroleum reserves are an order of magnitude lower, if the US wanted to replace the seven million barrels of crude oil used daily, then four million tonnes of coal a day for coal-to-liquids would not only provide the required oil but would create employment and energy efficiency savings. The coal-to-liquids processes create a constant stream of carbon dioxide that could be pumped down exhausted oil wells to lower oil viscosity and recover some 34 billion barrels of residual oil.

The only sensible solution to keep the Western world running for the next few hundred years, based on energy density, coal reserves, efficiency and potential petroleum reserves is to have nuclear power for electricity, coal-to-liquid for vehicle transport and dams for back up hydroelectricity and food security. Coal seam gas, fracked gas and shale oil could be used for the chemicals industry and local energy sources. Coking coal would keep the strategically vital metals industry alive.

However, green activism is attempting to stop the use of nuclear power, dams, coal, gas and hydrocarbons. Do greens have a viable efficient alternative? No. Our children and grandchildren will pay dearly.

Save the whales

In the 18th and 19th centuries, we used whale oil for heating and lighting. This was replaced by fossil fuels in the late 19th century. What would you prefer for your heating and lighting? Energy from coal and oil or energy from whale oil? Australia is very critical of Japanese whaling in the Southern Ocean yet Australia only stopped whaling in late 1978. Other countries such as Norway and Iceland still slaughter whales.

Bob Brown is a former Green Party political leader in Australia. He is also the chairman of Sea Shepherd Australia.

This is the same Bob Brown who fulminated when a grounded Chinese coal ship released some oil on the Great Barrier Reef:

> Studies of previous accidents shows damage to the Reef occurs through physical damage to the coral substructures and toxic pollution from marine anti-fouling paint, as well as impacts from oil spills ... Because of the sway the industry has over the government, the Great Barrier Reef has been turned into a coal highway ... The Greens are calling for a Royal Commission into how this situation could occur. Certainly, the coal industry should be held to account.

In early 2014, the anti-whaling ship *Sea Shepherd* pleaded guilty in the Cairns Magistrates Court for polluting Great Barrier Reef waters with 500 litres of oil.

So, it's OK for an anti-whaling ship to pollute the Great Barrier Reef but not OK for a coal carrier to do the same thing. It seems that there is a pecking order of environmental concerns and that saving whales from Japanese fishing boats is higher priority than dropping oil in Great Barrier Reef waters.

Apart from the *Cairns Post* and a few blogs, there has been no mainstream media mention of *Sea Shepherd* polluting the Great Barrier Reef.

Why wasn't this reported by the ABC who receive more than $1.2 billion of taxpayers' hard-earned money each year?

NEW FRIENDS

Wind

Miguel de Cervantes' *Don Quixote* saw wind mills as evil giants fit only for destruction. I do also but for different reason.

Not one country in the world is powered by wind energy.

Wind power provides too much power one minute and not enough the next.

Windmills may have been first used by the Persians in 500 to 900 AD, they were used by the Chinese from about 1200 AD and the Old World is littered with ruins of windmills. When more reliable sources of water and power were available, wind mills fell into disuse. Wind power reached its zenith 400 years ago. Centuries ago, wind was used for pumping water and grinding grain and these processes did not have to be undertaken all the time. Wind energy was replaced by steam generated from burning coal. Since then, the increasing energy requirements, energy density, inefficiency and unreliability made wind power more and more expensive. Small windmills pump water in isolated areas for stock. They are slowly being replaced by diesel- or solar-driven pumps. For hundreds of years, industrial windmills have been useless. They still are.

Despite massive subsidies for competing energy such as wind and solar, coal producers and coal-fired electricity generators are not subsidised and face punitive regulatory barriers and massive environmental and legal costs just to establish an employment-creating business. It is only by mandating and subsidising increasing proportions of electricity consumption from the highest cost renewable sources, thereby driving up electricity prices, that governments have made otherwise cheap coal-fired energy more expensive and now comparable in price to wind energy. A false market has thus been created and you are paying for it with your higher electricity bill.

The International Energy Agency's 2016 Key Renewables Trends show that wind provided 0.46% of global energy in 2014 with solar and tide combined provided 0.35%. The rest of the energy is solid, gaseous and liquid fuels which do the heavy lifting for heat, transport and industry. If we round the wind contribution of total global energy to the nearest whole number,[41] it is 0%. Why do we even bother?

Wind energy damages the environment, health and prosperity.[42] To summarise the main issues, wind power is:

1. *Unstable and erratic.* This, of course, is the bleeding obvious. It is only in parts of Antarctica that the wind blows constantly. Elsewhere in the world, the wind does not blow constantly. When it blows, it is erratic and it may not blow for days. In periods of high wind, the grid becomes overloaded or the turbines have to be shut down to avoid mechanical damage. German wind turbines put out a miserable 18% of their installed rated capacity. Wind companies claim that their turbines will power thousands of households but fail to state that this can only be achieved when the wind velocity is incessant and at an optimal velocity.[43] The wind companies such as AGL feed the public with fraud.

2. *Wind turbines are expensive.* The return on investment takes many years, even in places where the wind blows often (e.g. coastal areas, offshore, hill tops). Unless there are generous long-term subsidies paid by consumers and taxpayers, wind projects are poor investments. Because wind projects use subsidies and not wind to generate income, generated

---

[41] Matt Ridley 2017: Wind turbines are neither clean nor green and they provide zero global energy. *The Spectator*, 13 May 2017.

[42] Gosselin, Pierre, "Wind energy's 8 serious disadvantages: hurts everything from health to wealth", *NoTricksZone*, 16 July 2017.

[43] The power output of wind is proportional to half the cube of the velocity of the wind. This means that wind velocity does not have to vary very much to have massive swings in power output. Minimum wind velocity for power generation is 9.6 km/hr, full power output is at 34 kph and turbines are shut down at 90 km/hr to prevent damage. Wind power output is non-linear in relation to wind speed above 34 km/hr. If the wind is blowing at 70 km/hr there is no more power being generated that at 34 km/hr.

electricity is expensive. Governments have been totally commercially naïve and electricity prices have skyrocketed because there is a mandatory feed-in of wind energy.

3. *Excess wind power is very difficult to store.* We are a very long way away from being able to store large amounts of electricity. Large batteries are expensive, short-lived, heavy and can be a fire hazard. Pumping water to a higher dam is hopelessly inefficient as is the conversion of water to hydrogen using electricity.

4. *Destruction of the environment.* Turbines are in rural, scenic and forest areas. Plants and animals lose their habitats and there needs to be land clearing for the turbines and new access roads. Environmentalists complain about land clearing yet are happy to destroy the environment for wind industrial complexes. Wind industrial complexes require orders of magnitude more land than coal-fired or nuclear power stations.

5. *Birds and bats are slaughtered.* Each year millions of birds, especially raptors, and bats are killed worldwide by turbine blades. If environmentalists were really concerned about saving the planet, they would first try to save unnecessary slaughter of wildlife. If killing on such a scale was done by another industry, there would be uproar.

6. *Flying ice.* In winter at high latitudes, ice builds up on blades and is later thrown as projectiles. This danger to people and property has been totally ignored by environmentalists.

7. *Aesthetics.* Huge towers, spinning blades and land clearing are a blight on the landscape. Many turbines are on the tops of hills. Environmental objections to many projects are often on aesthetic grounds, except

for bird- and bat-chomping wind towers. The proximity of these ugly monstrosities to urban areas lowers property values.

8. *Health.* Wind turbines produce flickering light and infrasound which can have a profound effect on some people. Infrasound, low frequency sound below the human threshold of hearing, is used as a military weapon because the inner ear pressure pulses are disorienting.[44] Long term exposure damages health. If any other industry besides the wind industry damaged human health, environmentalists would be up in arms. If wind turbines are safe, I see no reason why they can't be in the middle of cities in order to avoid voltage drop with long-distance transmission.

9. *Increased carbon dioxide emissions.* Wind industrial complexes emit monstrous amounts of carbon dioxide in their construction and maintenance. Furthermore, carbon dioxide emitting coal-fired power stations need to be operating for when the wind does not blow.

10. *Toxic pollution.* Wind turbine blades cannot be recycled. Land fill and incineration emit toxic chemicals into the environment. If a factory emitted such toxins, it would be closed.

The mantra given by greens is that wind power is non-polluting and free. Not so. Processing of the rare-earth minerals mined at Bayan Obo in China has left a huge toxic, radioactive waste pile that deleteriously affects the locals[45] but is unseen by those that promote wind power. Rare earth neodymium-samarium magnets are used in the turbines.

---

[44] A.N. Salt and T.E. Hullar, 2010: Responses of the ear to low frequency sounds, infrasound and wind turbines. *Hearing Research* 286, 12-21.

[45] http://www.ressourcenfieber.eu/publications/reports/Rare%20earths%20study_Oeko-Institut_Jan%202011.pdf

Low frequency noise, possibly as low as 8 Hz,[46] derives from spinning blades on a wind power station. This has physical and psychological effects on humans and it is not known what effects low frequency noise has on animals. However, mammals feel distress, confusion and fear from the unheard infrasound in the roar of the big cats.[47] These medical effects of noise from wind turbines at this stage are difficult to quantify as medical research is at an early stage and further work is needed. Reports from residents living near wind turbines suggest serious unresolved problems resulting from low frequency infrasound.[48]

The wind power companies and green activists are only too well aware of the effects low frequency noise has on human health[49]. Green left activists are very vocal about what can't be seen (e.g. radiation, carbon dioxide, modified genes) but are hypocritically silent about what can't be heard. Inaudible sound maybe can't be heard but can be felt.

Wind power provides electricity intermittently regardless of demand and cannot provide electricity when it is needed in peak demand times. A spinning reserve of coal- and gas-fired plants needs to keep operating and releasing carbon dioxide to the atmosphere whether the wind blows or not.

Large wind turbines need to extract energy from the grid to start and when the turbine is not spinning it still requires energy for the controls, lights, communications, sensors, metering, data collection, oil heating, pumps, coolers and gearbox filtering systems. This comes from the grid and is provided by burning

---

[46] *Physikalisch Technische Bundesanstalt EurekAlert*, Public release, 10 July 2015, Dr Christian Koch.
[47] E. von Muggenthaler, 2000: The secret of a tiger's roar. *American Institute of Physics*, www.sciencedaily.com/releases/2000/12/001201152406.htm
[48] *Australian Senate Select Committee on Wind Turbines*, 2015.
[49] A.N. Salt and T.E. Hullar, 2010: Responses of the ear to low frequency sounds, infrasound and wind turbines. *Hearing Research* 286, 12-21.

coal. The bottom line is that wind turbines cannot be built, operated or maintained without using fossil fuels.

Wind generating facilities cannot exist without fossil fuels. The energy density of wind is very low. If all of the electricity requirements of the USA were to be from wind, then an area the size of Italy would be required.[50] The resources that a wind turbine uses and the carbon dioxide emissions from its construction and maintenance as well as the necessity to have a supporting coal-fired power station ticking over all the time just don't seem to get mentioned by the green left promoters.[51]

The land area used by wind industrial complexes is orders of magnitude greater than the land area used for gas generators, coal-fired power stations and nuclear power stations. A single unsubsidised shale gas pad of two hectares would produce as much energy in 25 years as 87 giant wind turbines covering 15 square kilometres (with turbines visible from 30 kilometres away).[52] Furthermore, shale gas used to make electricity is not intermittent and unreliable.

Despite an explosion in installed wind capacity since 1990, wind power has achieved a very small amount of the world's total energy generation.[53] Germany leads the race to the bottom with its share of energy consumption reaching only 2.1% in 2016. Wind power has just not penetrated the market yet it is touted[54] as "clean green energy". But is it clean?

A few simple numbers should have sobered enthusiasm for

---

[50] Robert Bryce, 2014: Smaller, faster, lighter, denser, cheaper. How innovation keeps providing the catastrophists wrong. *Public Affairs.*
[51] Wind energy in the United States and materials required for the land-based turbine industry from 2010 through 2013: *US Geological Survey.*
[52] Christopher Booker, 2015: Why are the greens so keen to destroy the world's wildlife? *Daily Telegraph,* 4 July 2015.
[53] K. Richard, 2017: Nightmare of turbine blade disposal: two new papers expose the environmental nightmare of wind turbine blade disposal. *No Tricks Zone,* 22 June 2017.
[54] US Department of Energy 2015.

renewable energy. If the wind turbine is actually working, then 1MW of wind power requires 103 tonnes of steel, 402 tonnes of concrete, 6.8 tonnes of fibreglass, three tonnes of copper and 20 tonnes of cast iron. Metal requires a long chain[55] of mining, smelting and fabrication. All require fossil fuels for this. Mining alone involves high-density energy such as diesel fuel.

Transporting iron ore from the mine to the steel mill requires diesel for trucks and locomotives and bunker fuel for ships. Many ship fuels are dirty and emit large quantities of sulphurous gases into the atmosphere. To convert iron ore into steel, coking coal or rarely natural gas is used and carbon dioxide is vented into the atmosphere. The fossil fuels for steel manufacture are for both energy and chemical reduction of an oxide to a metal. Chemical reduction cannot be done with wind, solar, hydro or nuclear power.

Concrete is composed of aggregate, sand and cement. The precursor to cement is limestone and shale, these are heated and carbon dioxide is released to the atmosphere. To quarry, transport and crush gravel, sand, limestone and shale requires diesel fuel. The energy to heat limestone to convert it to cement requires coal or natural gas. The resources that a wind turbine uses just don't seem to get mentioned by the green left promoters.[56]

Coal-fired power stations cannot be shut down because wind turbines provide electricity. Coal-fired power stations must be kept running for when the wind does not blow. A wind turbine that produces inefficient and intermittent expensive subsidised electricity results in the emission of more carbon dioxide than if the same amount of electricity was generated from a coal-fired power station.

---

[55] Ian Plimer, *Not for greens.* Connor Court, 2015.
[56] Wind energy in the United States and materials required for the land-based turbine industry from 2010 through 2013: *US Geological Survey.*

Global Wind Day is 15 June. This is a feel-good orchestrated media event celebrated by rent seekers with their snouts in the trough getting subsidies and raising electricity prices in a guaranteed market, landowners receiving compensation indirectly from the consumers and the green left environmental advocates promoting various UN agendas at the expense of sovereignty, freedom, the environment and financial common sense.

Global Wind Day is not celebrated by birds, bats, nearby residents who suffer decreased property values, humans proximal to wind turbines suffering health problems from thumping blades and consumers and industry hurting by paying higher electricity prices. Although aesthetics is subjective, most folk see no scenic enhancement in hilltops covered by wind turbines. Electricity consumers, taxpayers, businesses and true environmentalists only breathe out hot air on Global Wind Day. Maybe there are not enough days in the year to set aside special days to celebrate employment destruction, killing of wildlife, industry destruction, fuel poverty, sovereignty loss, freedom loss and UN hypocrisy.

A typical wind turbine has a tower, a nacelle and three blades. The foundation is made of concrete, the tower of steel or concrete, the nacelle from steel and copper and the blades from composite materials.[57] Spectacular fires in windmill nacelles fill the air with black toxic smoke. Although wind power companies state that such polluting fires are rare, the data shows the exact opposite. The nacelle fires are let burn because fire-fighting equipment cannot reach such heights. Toxic unburned oil contaminates soil beneath a turbine.

In summary, if the aim of wind power were to lower carbon dioxide emissions, then this it has to be marked as a monstrous failure. Wind power industrial complexes raise carbon dioxide

---

[57] https://www.vestas.com/~/media/vestas/about/.../pdfs/2006aruk.pdf

emissions because of the use of coal to make the steel towers, copper wiring and magnets and the carbon dioxide emissions from burning limestone to make the concrete base. Furthermore, a coal-fired power station needs to be constantly operating for when the wind does not blow. The huge amount of diesel burned in transport, construction and maintenance adds to carbon dioxide emissions to the atmosphere.

There is no legislation requiring wind power generators to remove the facility at closure, as with other industries. Why is the wind power industry allowed to leave a monstrous mess behind when other industries cannot? Many wind industrial complexes are not decommissioned and are left as rusting, commonly burnt out, relics in areas of great scenic beauty.

Wind turbine blades can last up to 20 years but are lucky to make it to 15 years and then they are tossed into land fill.[58,59,60] This is, of course, if they have not catastrophically separated from the turbine, broken up or burned as so many do. Blades snap, crackle and drop with alarming frequency. Blades are not recyclable and 43 million tonnes of blade waste will be added to landfill sites over the next few years.[61,62] China will have 40% of the waste, Europe 25%, USA 16% and the rest of the world 19%.[63] There are no established industrial recycling

---

[58] B. Tremeac and F. Meunier 2009: Life cycle analysis of 4.5 MW and 250 W wind turbines. *Renewable and Sustainable Energy Reviews* 13, 2104-2110.

[59] A. Arvesen and E.G. Hertwich, 2011: Environmental implications of large-scale adoption of wind power: a scenario-based life cycle assessment. *Environmental Research Letters* 6, doi:org/10.1088/1748-9326/6/4/045102

[60] B. Guezuraga, R. Zauner and W. Polz, 2012: Life cycle assessment of two different two MW class wind turbines. *Renewable Energy* 37, 12: 37-44.

61 K. Ortegon, L.F. Niles and J.W. Sutherland, 2014: The impact of maintenance and technology change on remanufacturing as a recovery alternative for used wind turbines. *Procedia CIRP* 15, 182-186.

[62] https://alethonews.wordpress.com/2017/06/22/unsustainable-43-million-tonnes-of-wind-turbine-blade-waste-by-2050/

[63] P. Liu and C.Y. Barlow, 2017: Wind turbine blade waste in 2050, *Waste Management* 62, 229-240.

routes for end-of-life concrete and composites[64] and there is an unresolved nightmare of the health risks of blade disposal.[65]

Blade disposal is not a trivial problem. In 2012, there were more than 70,000 turbine blades deployed globally. To achieve the US's 20% wind production goal by 2030, more blades will be manufactured and blades up to 100 metres long are being deployed and produced. The designed life span of twenty years is because of physical degradation or damage beyond repair. Blade damage from striking wild life is minor yet it is fatal for birds and bats. The constant development of more efficient blades with higher power generation capacity is resulting in blade replacement well before the twenty-year life span.

Each year some 3,800 blades break off, are thrown to the four winds and end up in land fill. When this 20% wind production goal in the US is achieved, there will be disposal of 330,000 to 418,000 tonnes of blade composite per year. This is equivalent to the waste generated by four million Americans in 2013. Blades will not decompose in landfill because they are designed with a high resistance to heat, sunlight and moisture. Blades dumped in landfill deliver a toxic cocktail into aquifers and water supplies for centuries. Incineration will produce toxic gases and soot. Worldwide, there are 249,365 tonnes of epoxy resin containing highly toxic bisphenol A in wind turbine blades[66] just waiting to enter aquifers and life as part of the environmentalists' dream to have a wind-powered clean green world.

---

[64] S. Pimenta and S.T. Pinho, 2010: Recycling carbon fibre reinforced polymers for structural applications: Technology review and market outlook. *Waste Management* 31, 378-392.

[65] K. Ramirez-Tejeda, D.A. Turcotte and S. Pike, 2016: Unsustainable wind-turbine blade disposal practices in the United States. A case for policy intervention and technological innovation. *New Solutions* 26, 581-598.

[66] www.epoxyeurope.eu/wp_content/uploads/2015/07/epoxy_erc_bpa_whitepapers_wind_energy-2.pdf

Although it is promoted that blades have a twenty-year life, the life of the whole unit is a little more than a decade. Blades commonly fail. For example, at AGL's Hallett 1 (Brown Hill) wind industrial complex south of Jamestown (South Australia), each and every one of its forty-five turbines failed within its first year of operation, requiring its wholesale replacement. The 2.1 MW Indian-built turbines commenced operations in April 2008 and soon after stress fractures appeared in the 44-metre-long blades.

What happened to the failed blades? They were ground up and mixed with concrete used in the bases of turbines erected later. Did AGL tell the local community that the blades and now the concrete bases contain bisphenol A, a chemical so toxic that it has been banned by the European Union[67] and Canada?[68] Did AGL tell the community that their bisphenol A will leach from concrete? AGL clearly knew what they were doing and knew that they were adding toxins to the environment. If a farmer, heavy industry or coal-fired power station deliberately and knowingly allowed bisphenol A to be leached into the environment, their operations would be closed down by the authorities. Governments have been conned by the likes of AGL who can get away with releasing long-lived toxins into the environment over long periods of time.

We end up with an environmental mess because advocates of wind power have closed their eyes to the environmental and industrial realities of "clean and green" energy.

There is extensive literature to show that wind turbines kill birds (especially rare eagles) and bats and are built in areas of outstanding natural beauty. Wind turbines are now the leading cause of multiple mortality events in bats with 3 to 5 million killed every year. If farmers shot that many bats destroying

---

[67] www.efsa.europa.eu/en/topics/topic/bisphenol
[68] www.canada.ca/en/health-canada/services/home-garden-safety/bisphenol-bpa.html

crops there would be an outcry by environmentalists, political action, fines and/or incarceration and yet more regulations. If just one bat is killed at a mine site, there is an uproar from environmentalists. Are bats part of a necessary sacred sacrifice to the god of environmentalism? Tim Flannery is so concerned about bird deaths due to land use change, plastics and fishing[69] that he just happens to forget bird murder in front of our eyes by his environmentally-friendly wind turbine blades.

In a study of migratory bats in North America, there are some sobering statistics.[70]

> Large numbers of migratory bats are killed every year at wind energy facilities ... Using expert elicitation and population projection models, we show that mortality from wind turbines may drastically reduce population size and increase the risk of extinction. For example, the hoary bat population could decline by as much as 90% in the next 50 years if the initial population is near 2.5 million bats and annual population growth rate is similar to rates estimated for other bats species (l = 1.01). Our results suggest that wind energy development may pose a substantial risk to migratory bats in North America.

These known effects of wind industrial complexes show that environmentalists are not concerned about mass killing of wildlife, devegetation, despoiling the environment, destroying habitats, adding toxins to the environment, energy efficiency, increasing carbon dioxide emissions, destroying the health of their fellow citizens, wasting money, impoverishing the poorest in the

---

[69] *The Weekend Australian*, 2-3 September 2017.
[70] W.F. Frick, E.F. Baerwald, J.F. Pollock, R.M.R. Barclay, J.A. Szymanski, T. J. Weller, A.L. Russell, S.C. Loeb, R.A. Medellin and L.P. McGuire, 2017: Fatalities at wind turbines may threaten population viability of a migratory bat. *Biological Conservation* 209, 172.

community and converting First World electricity systems into Third World systems.

The Rio Declaration of 1992[71] included the statement wherein we first saw the so-called precautionary principle:

> ... where there are threats of serious irreversible damage, lack of full scientific certainty shall not be used as a pretext for postponing cost-effective measures.

Too often when the green left has lost arguments of science (on the basis of weak evidence), economics and logic, they invoke the "precautionary principle".

It is just another way of the green activists not agreeing to anything and showing that they don't care about the environment and their fellow man. Why don't greens apply their precautionary principle to bat- and bird-chomping polluting wind turbines that affect human health?

Wind industrial complexes clearly demonstrate that environmentalism is not about the environment.[72] It is about control of every minute aspect of our lives by unelected thugs.

Wind industrial complexes to generate electricity are demonstrably a failed concept. It's time for the supporters to pay for their fantasies instead of forcing taxpayers and power consumers to write the cheques. The wind industry has been promising to deliver competitively-priced electricity but has never achieved it. Australia has a sad history of funding infant industries and subsidising industries such as textiles, clothing, shoes, cars, chemicals, hot dry rock geothermal power and now wind energy. They all failed.

A small country like Australia cannot afford to back losers

---

[71] www.un.org/documents/ga/confl51/aconfl5126-1annex1.htm
[72] Christopher Booker, "Why are the greens so keen to destroy the world's wildlife?", *Daily Telegraph*, 4 July 2015.

and subsidise high capital cost technology even if well-funded lobby groups frighten the public, pressurise politicians and have a supportive media who have not the skills to ask searching questions. Much of the funding for green activism comes from outside Australia. Such support only raises regulatory costs, damages the economy, jobs and the budget and increases debt. The Australian government's wasteful green bank, the Clean Energy Finance Corporation, receives subsidies from consumers through the Renewable Energy Target.

Wind is not free. The green left environmental activists can't do sums. Wind power is horrendously expensive, damages the environment and is unreliable.

The South Australian blackouts[73] are the warning to steer well clear of wind power. It is ruinable energy.

USA

The US Senate Finance Committee[74] has proposed a 2.3 c/kWh production tax credit for wind energy. This is the seventh time since 1992 that a subsidy has been granted to help "the industry compete in the marketplace" and there have been other "temporary" federal subsidies since 1978. One would have thought that after more than 30 years, wind power would have been able to compete. Unless, of course the industry is so hopeless that it cannot survive without massive subsidies.

The US Energy Information Administration[75] stated that the 2013 production tax credit for wind was $US 5.9 billion

---

[73] On 13 June 2015 between 9am and 3pm, the whole South Australian grid collapsed from 750 MW to 50 MW, a drop of 94%. It's been known for a long time that feed-in solar and wind power leads to uncontrolled power surges and blackouts and attempts to increase the proportion of renewables to the grid result in more blackouts and no savings of carbon dioxide emissions.
[74] http://finance.senate.gov
[75] http://eia.gov

and for solar $US 5.3 billion. In 2013, wind and solar in the US provided less than 5% of the total electricity and yet received 50 times more subsidy than coal and gas combined.[76] These additional costs are borne by the taxpayer, thanks to renewable energy mandates in 29 States and the District of Columbia that are designed to guarantee a market share no matter what the production costs for wind and solar might be.

Households in New York State[77] now pay $US 400 *per annum* more than the national average for electricity.[78] Statewide this 53% extra cost over the national average is $US 3.2 billion a year. The 15 wind industrial complexes operating in 2010 produced an output of 2.4 million MW hours. The same amount of electricity could be generated by a small 450 MW gas-fired generating plant operating at 60% capacity and with a capital cost of 25% of the wind turbines. The wind turbines need replacing every 10-13 years at a capital cost of $US 2 billion. Some States have woken up (Ohio, Virginia) and have frozen or stopped renewable energy mandates.

The EIA argues that onshore wind is one of the cheapest forms of electricity and is cheaper than nuclear,[79] coal, hydro and solar. A report from Utah State University[80] shows that the EIA's true costs are around 48% more expensive than claimed. States have enacted Renewable Portfolio Standards which require utilities to purchase electricity produced from renewable sources at a high cost for consumers Wind turbines are often a long way from transmission lines, expansion of the grid is expensive and costs were passed on to taxpayers and

---

[76] http://instituteforenergyresearch.org

[77] http://www.nyiso.com

[78] http://newsmax.com/LarryBell/Climate-Change-Global-Warming/2015/008/03/id/665118/

[79] www.eia.gov

[80] http://www.usu.edu/ipe/wp-content/uploads/2015/04/Renewable-Portfolio-Standards-Colorado.pdf

consumers. Conventional electricity generation needs to be available as backup 24/7 as wind is unable to meet demand. This further drives up the cost of electricity. The Production Tax Credit alone amounts to $US 5 billion per year subsidy to wind producers. And the real costs of wind energy don't include the environmental damage and purported harm to human health.

Europe

No country went as hard and fast as **Germany** to spear their countryside with wind turbines. South Australia came close but it only has 1.6 million people as compared with Germany's 80 million. Germans were told that with more and more wind power, coal and nuclear power could be made redundant by the wonders of sea breezes and the Sun. Reality and ideology are streets apart.

German wind power has averaged 18% of installed capacity since 1990, a small fraction of its capacity. A new study by German power engineers[81] of 18 European countries and their wind power capacity concludes that 100% backup by coal/gas/nuclear is needed for 100% of the time and the more wind capacity is installed, the greater the volatility of the grid. Wind power is an expensive flop.

The Germans are ruing the day they decided to save the world and to quickly convert to wind power. The German energy policy has been a disaster. There is a silent catastrophe taking place in Germany. Subsidies are colossal, the electricity market is now chaotic, electricity grids have problems coping with the sporadic input from wind turbines[82] and carbon dioxide emissions are rising quickly.

---

[81] T. Linneman and G.S. Vallana, 2017: Windenergie in Deutschland und Europa. *VGB Power Tech* 6: 63-73.
[82] Grid systems have tolerances of a few volts and Hertz hence any excessive variation results in a shutdown.

German industry has decamped to other jurisdictions and more than 800,000 German homes have had their power cut off because they can't afford to pay the increased cost of electricity. In just one year, 330,000 households had their electricity disconnected, 6.2 million threats to disconnect power were made and 44,000 had their gas turned off because of the high cost of electricity.[83] The green dream of creating thousands of new jobs has turned into a cold black nightmare. Costs have been driven up because of the legally mandated feeding in of wind and solar power.

The costs of Germany's dash to green energy are rising and the number of green energy jobs are falling. The costs of Germany's subsidised wind and solar power will rise by €14 billion by 2025 when the costs will be €77 billion per year.[84]

US electricity prices are about 33% of those of Germany yet Obama wanted to head down the same path as Germany. Conversely, China is banning wind power projects in six regions.[85]

Even the most committed environmentalists in Germany are asking why old growth forests are felled for wind turbines and whether cutting down forests will save the planet from global warming.

Some Germans are facing such energy poverty that they have taken to the forests to collect wood for heating and cooking. God knows how Germany will be able to provide electricity in winter for the more than a million illegal immigrants who arrived on their doorstep. Germans must also tread carefully on the energy road because they are dependent upon Russia

---

[83] DPA German Press Agency, 2 March 2017, t-online.de, Bundesnetzagentur, *NoTricksZone*, 3 March 2017.
[84] Daniel Wetzel, Die Energiewende droht zum ökonomischen Desaster zu werden, *Die Welt*, 6 March 2017.
[85] *China Daily*, 23 February 2017, "China bans wind power projects in regions".

for gas along the 1,244 km pipeline under the Baltic Sea from Vyborg (Russia) to Lubmin (Germany).[86,87] The Russians have often turned off the gas in various spats with their neighbours.

Germans are generally law-abiding. They have had enough. German farmers have blockaded construction sites, destroyed wind measuring masts and have expressed concern about the crony capitalism of wind complexes.[88] *Energiewende* (transition to renewable energies) has split the community because city-based people are not affected by wind complexes whereas rural Germans suffer health problems and the living area is being ruined for generations.

Denmark has the most expensive electricity in Europe. It has now been overtaken by South Australia who proudly advertise that they are the wind capital of the world. About 50% of Denmark's potential electricity is from wind. When the wind does not blow, Denmark buys coal-fired electricity from Germany, nuclear electricity from France and Sweden and hydropower from Norway. We saw a screaming headline[89] telling us that *"Wind power generates 140% of Denmark's electricity demand."* However, this 140% was for only a brief moment on a windy night at 3 am, which is the time when demand is the lowest. If wind power was efficient and reliable, the media would not need to be misleading and deceptive. Call up the Danish Electricity Authority site[90] and look at the map of inputs and outputs of electricity.

There is far more electricity imported by Denmark from conventional sources than is exported from wind. Denmark

---

[86] Agnia Grigas, "Europe needn't worry that proposed American sanctions on Russia will hit its gas supply", *Scroll India*, 21 July 2017.
[87] "EU divided over Russian gas dependence and sanctions", *Stratfor*, 25 July 2017.
[88] *Die Welt*, 24 June 2016.
[89] http://www.theguardian.com/environment/2015/jul/10/denmark-wind-windfarm-power-exceeded-electricity-demand
[90] http://energinet/dk/EN/EI/Sider/Elsystermet-lige-nu.aspx

is not self-sufficient in electricity, as wind does not provide enough for consumers and employment-generating industry. Wind-power is inefficient costly ideology. It's not hard to see that misguided energy policies have exacerbated the economic decline of Europe.

Denmark's state-owned energy company Dong Energy has stopped building onshore wind industrial complexes because of the public outcry. At times, Denmark's wind energy cannot be used in Denmark. The end result is that about 7% of Denmark's energy is from wind. The rest is imported.

Now to compare Australia with Denmark. Australia has a population of 24 million spread over an area 7,692,024 square kilometres with an electricity demand of more than six times that of Denmark. Denmark has 5.7 million crowded into 43,094 square kilometres and no major electricity-consuming industries such as smelters providing metals to the rest of the world. Australia is an island and we can't rely on neighbouring countries to provide us with electricity if our ideological energy schemes fail. Like Denmark, our electricity consumption is low at 3 am and our pricing is highest at peak morning and evening times. South Australia, which has a higher proportion of wind power, is higher cost than the other eastern States. What a surprise. And who pays?

In Spain, at least 18 million birds are slaughtered annually by wind turbine blades. Bird deaths in Germany are more than 300 per turbine and in Sweden almost 900 per turbine. German turbines kill more than 200,000 bats per year and in the US turbines kill some 2.8 million bats. Not to worry. Greens feel morally superior because they think that wind industrial complexes emit less carbon dioxide into the atmosphere and hence are saving the planet. They are certainly saving the planet from birds and bats yet climate will continue to do what it always does: Change.

If a nuclear- or coal-fired electricity generator damaged the environment as much as wind generators, there would be an outcry yet there is no outcry from environmentalists about the environmental damage by wind industrial complexes. Why not? Is it that environmentalists have no interest in the environment and just want to take control of everyone's life without the bother of having to face an electorate?

Ireland's electricity prices have risen 20% and Ireland has joined the ranks of countries with high electricity costs and employment destruction. Relative European electricity prices are Denmark > Germany > Italy > Ireland > Spain > Portugal > Belgium > Austria > Sweden > Great Britain > Netherlands > Greece > Norway > France > Finland > Poland > Croatia > Romania > Czech Republic > Hungary > Serbia. Electricity is the cheapest in countries generating electricity by coal, hydro and nuclear and the most expensive have renewable energy.[91] Wind companies in Ireland, like in Australia, have production mandated and underwritten by government. The end result is that energy-intensive industries die, jobs are lost and electricity bills rise. The Irish cement industry has shrunk by 70%.

United Kingdom

In the UK wind turbines are subsidised. Owners are even paid for electricity that is not used and are compensated because consumers are not using wind power. This was the UK's attempt to "decarbonise" the economy. What drongo in the bureaucracy agreed to such gouging of the average electricity consumer in the UK? The UK National Grid has now given out contracts for standby diesel generating plants which can be fired up as soon as wind-supplied electricity output drops. This has been done because of intermittent and unreliable wind power. Is using

---

[91] Barry O'Halloran, "Government feels the heat as cost of supporting renewable energy jumps", *The Irish Times*, 5 August 2016.

diesel "decarbonising" the economy? Use of diesel generators certainly does not reduce UK carbon dioxide emissions.

The UK National Grid[92] warned that coal-fired power stations are being closed down so quickly that the spare capacity of 16% in 2012 was only 1.2% in 2015. Most will be closed by 2023 because of EU rules intended to curb "carbon" emissions.[93] What will another cold winter show? The UK is bordering on a self-induced disaster that even the costly emergency back-up diesel generators will not solve.

During winter, warmer periods are often windier while colder periods are more calm. As winter temperatures fall and electricity demand increases, the average wind speed reduces. Neighbouring countries also struggle to provide additional capacity to the UK when the UK's own demand is high and wind power low.[94] Just when power is needed, it cannot be generated.

In the UK, renewable energy costs, principally from wind, create fuel poverty for 2.4 million folk. In the 2012-2013 UK winter, there were an additional 35,000 deaths. This translates as six sick, elderly or vulnerable people killed every year for each installed wind turbine. Not only does wind energy kill wildlife, it also kills the people who subsidise wind energy.

Australia

Renewable zealots don't want to know that a single 1,000 MW wind industrial complex produces at least seven million tonnes of carbon dioxide in component construction and concrete.

---

[92] http://www.ft.com/intl/cms/s/0/f3d1b352-fef4-11e4-84b2-00144feabdc0.html#axzz3ilYqjyfx

[93] http://www.bloomberg.com/news/articles/2014-07-09/most-uk-power-plants-seen-shut-by-2023-on-climate-rules.html

[94] http://www.metoffice.gov.uk/news/releases/2017/study-shows-potential-for-wind-energy-on-coldest-days

Thousands of truckloads of concrete are required just for the footings. Maintenance by diesel-powered vehicles only adds to emissions. Wind industrial complexes need 24/7 backup from carbon dioxide emitting coal-fired power stations. Wind industrial complexes actually increase human emissions of carbon dioxide yet the story touted is that wind industrial complexes save carbon dioxide emissions.

An older wind industrial complex requires 7,600 generators at 20% efficiency to produce 1,000 MW. At $2,000 per kilowatt installation, this would cost $10 billion. This is ten times the cost of a 1,000 MW reliable clean coal-fired generator and more than twice the cost of a reliable nuclear 1,000-megawatt generator. Newer generators are 3,600 kW and the costs and subsidies are even higher.

Wind turbines are increasingly littering rural Australia in an effort to satisfy the Federal Government's 2020 Renewable Energy Target. Every turbine is issued with 8,000 to 10,000 Renewable Energy Certificates each year worth $665,000 to $855,000.[95]

A new 3.8 MW turbine operating at 30% capacity, on average, yields 9,986 RECs.[96] At the current RET price of $85, that single 3.8 MW turbine collects $848,810 in subsidies, the cost of which is added to the retail cost of power and collected via retail power bills.

That subsidy isn't a one off. This continues for the life of the RET, until 2031. A turbine that started operating in January 2017 will collect $11,883,340 in subsidies until 2031. This is daylight robbery and you pay for it with your electricity bills. The Prime Minister is disingenuously telling consumers to

---
[95] http://lgc.mercari.com.au
[96] 24 x 365 = 8760 hrs x 3.8 MW = 33,288 MWh if operating 100% of the time. At 30% capacity, 0.3 x 33,288 = 9,986 MWh dispatched providing and entitlement to collect 9,986 RECs.

shop around for the best electricity deal rather than addressing the heart of the problem. The consumer is being robbed by RETs. Get rid of them and, if there is a contractual problem, make them worth only $0.01. Rome burns yet the Greens want 100% renewables, Labor wants 50% and the Coalition wants 42% and you can look forward to your bills being even higher. If renewables were not subsidised, your electricity bill would decrease to what it was years ago.

This is money for jam for the power generators but the consumers are forced to take electricity at up to four times the cost of hydro-, coal- or gas-fired power. This is the Federal Government's tax on power consumers, retailers are forced to pay exorbitant prices for wind power which is added to electricity bills. Consumers are therefore forced to subsidise inefficient unreliable electricity.

When the Renewable Energy Certificates rise to $90, wind power operators will be issued with $700,000 to $900,000 worth of Renewable Energy Certificates *gratis* until 2031 for every turbine and $52 billion of Renewable Energy Certificates will be issued before 2031. Each wind turbine will earn $12 million just for being a blight on the horizon above and beyond the electricity generated that retailers must buy.

Wind industrial complexes have the life of a parasite. They freeload themselves onto existing grids paid by conventional efficient energy, need subsidies and drain electricity from the grid when not operating. Wind turbines don't run on wind, they run on subsidies. From the taxpaying consumer. These subsidies make your electricity bill prohibitively expensive. Without subsidies, we would save huge amounts of money and would have guaranteed tried-and-proven reliable base load power from conventional sources that have served us well for a century.

Crafty wind power generators have out-foxed naïve green bureaucrats and ideological politicians, monopoly money is printed at the expense of the consumer, electricity prices skyrocket, businesses close and create unemployment, fuel poverty increases and more and more people can't afford electricity in a wealthy 21st century Western country.

The government has created a business opportunity to make money out of air. Wind generators have signed long-term contracts with governments to lock in cash flow for decades even if the wind does not blow and they sleep well to the gentle ring of the cash register. To make matters worse, once a wind or solar power station generates electricity, there is a legal mandate for the grid to take this electricity in place of cheap reliable coal-fired electricity.

The wind generators knew it was a scam hence locked in governments to long-term and must-take contracts. You can't blame the wind companies making money out of community hysteria and government stupidity.

Wind power, touted as renewable energy, has destroyed what was a cheap reliable electricity system. The message is loud and clear from South Australia which has 38% of its generating capacity from wind, all of which attracts the Renewable Energy Certificate subsidy, all of which is unreliable and all of which has placed South Australia in first place. This is not first place for being green, it is first in the world for the most expensive electricity.

South Australia contains about 30% of the world's uranium which could be used to run tried-and-proven reactors yet it has chosen to run an economy on the weather.

What reasonable decision-maker would support policies that favour something of no economic benefit? The central claim that wind power reduces carbon dioxide emissions is a falsehood. Billions in subsidies directed to wind power. This looks like lunacy. It might be graft and corruption (i.e. crony capitalism).

Wind power is exploited by subsidised companies who supply the subsidised electricity industry and the whole premise is based on subsidised junk science. Your electricity bills show that wind power is a con.

Time to stop.

Solar

Throughout most of human history, there has been Sun worship. We have known for a very long time that without the Sun there is no life on Earth. Some 2,000 years ago, Archimedes dreamed that solar energy could be harnessed. Dreamers, Sun worshippers, hippies, the unbalanced and politicians seeking the green left environmental activist vote think that solar will provide the electricity needed for a modern industrial society.

Solar cells (photovoltaics) were invented in 1839. One would have thought that if solar cells were to be a low-cost efficient competitive energy dense system to assist humanity, then 175 years of refinements and improvements would have been enough to make it cheap and efficient. Apparently not. After all, this time was enough to make the steam engine efficient and we still use steam today for the generation of electricity from coal and nuclear fission. During this time, the internal combustion engine was invented and, for the last 120 years, has undergone significant improvements such that hyper-efficiency is gained by modern small capacity turbo-charged diesel engines.

In 1839, a silicon solar cell was 10% efficient and this was excluding light reflection and current leakage. It still is. Why? A silicon solar cell does not create energy, it converts just one wavelength of the whole spectrum of solar energy into electrical energy and the rest of the spectrum creates heat energy. Just one wavelength in the infrared spectrum (1,130 nm) excites an electron to jump to a higher energy state, when the electron falls back it gives out a small amount of electricity (1.1 eV).

We are told that solar power is clean, free, renewable and will go on forever. This may seem true until one looks at the uncomfortable fundamentals such as the environmental and monetary costs. Solar power exists only because of subsidies and the mistaken belief that it reduces carbon dioxide emissions. It does not. Solar power, like wind power, requires infrastructure construction that actually adds to human emissions of carbon dioxide.

Solar power is uneconomic and is subsidised. This is done via taxes and high electricity costs. As soon as subsidies disappear, so does an inefficient costly industry. Japan has cut subsidies and up to 100 companies could go bankrupt in 2017. This is the same number that went bankrupt in all of 2016.[97]

If green left environmental activists want solar energy to be more efficient, then they had better invent a few new laws of physics and persuade electrons to respond to the spectrum of wavelengths rather than just one specific wavelength in the infrared part of the spectrum. Don't wait up.

The only reason Western countries have solar electricity is that it is subsidised. In remote areas with small power needs and where regular maintenance is prohibitively expensive, solar power is sensibly used for lighting, telecommunications, navigation beacons, recording equipment, marine buoys, electric fences, pumps at bores and satellites. This is the market talking. Farmers I talk to also tell me that solar panels for fences, dams and bores only last about five years before they need to be replaced. The only reason Western countries have

---

[97] *PV Tech*, 18 July 2017.

large scale solar electricity is that it is generously subsidised. With your money.

The US Department of Energy concluded that solar electric systems couldn't meet the energy demands of an urban community or industry. Large scale solar arrays, whether photovoltaic or solar thermo-electric are far too variable, unreliable, expensive and ecologically damaging. Solar can only make a minor contribution to the national power requirements. Solar power is not very efficient and the optimal figure used for incident radiation is 10 watts per square metre with an overall system efficiency of 5%.

Other factors affect the efficiency of solar radiation such as latitude, time of year, time of day and aerosols. There are also long-term weather fluctuations due to cycles of cloud coverage that can change the efficiency by up to 4%. Aerosols can reduce the efficiency by almost 30%. Furthermore, in remote areas, lack of regular cleaning off of dust and plant spores from the glass surface covering the photovoltaic cells can result in reductions of efficiency by up to 50%. It has become political fashion to promote renewables. But there are some realities.

In 2015, NSW had no solar power generation capacity whatsoever and, although there are now five wind industrial plants, the total wind power input into the grid was 0.6%. And at what cost? No green left environmental activist in NSW could possibly survive on wind or solar power from the grid.

In 2016, NSW took a great step (and more than likely economically backwards) in building a solar power plant at Broken Hill. At capacity, the solar plant provides a glorious 2.9% or at normal capacity 0.29% of electricity for NSW. The 53 MW capacity plant is privately owned and the capital cost was $166.7 million. The NSW government and Federal government's Australian Renewable Energy Agency provided

$64.9 million of the capital cost.[98] Consumers will be forced to pay extra for this renewable energy. This goes to the bottom line in your electricity bill.

Given the well-known inefficiencies of solar plants, the reality is that on average about 5-6 MW will be available from this new solar plant. Some 35 MW is needed to keep the mines operating at Broken Hill. And if there are no mines, then there are no jobs. I often go underground in the zinc-lead-silver mines at Broken Hill. Do I rely on solar-generated electricity when underground at night to keep the safety systems and pumps operating? No. Will the processing plant be able to operate 24/7 on solar energy. No.

Some green left environmental activists argue that we should invest in solar power as the breakthroughs are just around the corner. What breakthroughs? There are still no cost-effective batteries to store solar energy for use at night and we have been waiting for such breakthroughs for more than a century. The only somewhat sensible suggestion is to use solar energy to pump water uphill into dams and generate electricity when needed at peak times or at night. Dams are off the menu for green left environmentalists.

Surely new developments with metals, metalloids and super conductors are just around the corner and these can be used to make more efficient solar cells? Well ... yes and no. Cells of higher efficiency require the use of exotic, rare and poisonous elements such as germanium, gallium, indium and cadmium. There are no germanium, gallium, indium and cadmium mines in the world and these metals are by-products from the zinc, aluminium and tin smelting and refining industries. To produce 1% of the US electricity requirements from a germanium or gallium solar cell would require three times the planet's annual

---

[98] http://www.agl.com.au/about_agl/how-we-source-energy/renewable-energy/broken-hill-solar-plant

production of germanium and twenty times the world's annual production of gallium.

The zinc (for germanium) and aluminium (for gallium) deposits are yet to be discovered and, if they were, it would not be economic to produce massive excess amounts of zinc and aluminium just to provide germanium and gallium. Zinc and aluminium are the two metals with the highest amounts of embedded energy and, to create small volumes of cheap solar power using other materials, astronomical amounts of cheap conventional base load energy would be required.

If more gallium were to be produced for solar cells, the already marginal aluminium industry would have to greatly increase production, flood the market with massive quantities of unwanted aluminium and use stupendous amounts of electricity. Furthermore, these elements are far more expensive to produce than silicon (the second most abundant element on Earth). Production of these germanium or gallium solar cells would require 17% of the US annual cement production. To make cement, limestone needs to be burned and carbon dioxide is released to the atmosphere.

For every kilogram of aluminium made, 15 kWh of electricity is needed. Some 10% of all of Australia's electricity production is embedded in aluminium (27 TWh[99]) made in Australia and exported. Non-ferrous metals smelted and refined in Australia use 43 TWh of the 80 TWh used in Australian industry in 2012. If Australia were to become a germanium and gallium producer, then it would need a far greater generating capacity and the electricity produced would have to be reliable and cheap. This just cannot be provided by the solar or wind industries.

Efficiency is a banned word from the green left environmental activist lexicon. If a solar panel is to generate maximum electricity, conditions must be ideal (i.e. middle of the day, clear sky,

[99] 1 TWh = 1 billion kWh.

low latitude). The maximum incident solar radiation value is 1,000 watts per square metre. It is claimed that an off-the-shelf solar panel will produce 110 watts for a one metre square panel. This is an efficiency of only 11%. Long-term measurements show that the average incident solar radiation is 125 to 375 watts per square metre.

Solar panels can be flat or can use concentrating collectors and solar trackers to give maximum incident solar radiation for a longer time. But, in high winds, these concentrating collectors and trackers have to be closed down.

Assuming a very generous optimistic efficiency for the average solar panel of 15%, an average output of 19 to 56 watts per square metre would provide only 0.46 to 1.35 kilowatt hours per metre per day at an average of 0.61 kilowatt hours per square metre per day. The glass cover that protects the solar cells reduces the efficiency to 13% and further system losses of 7% can be expected due to localised conditions and the conversion of direct current (DC) to alternating current (AC) that we use in our domestic life.

The US Department of Energy calculated that solar panels have a 10.27% efficiency that equates to 4.25 kilowatt hours per square metre output per day. Solar power advocates quote the maximum radiation value of 1,000 watts per square metre and don't really worry about those trivial little inefficiencies. But then again, the world of the green left environmental activist is all about inefficiency with costs passed on to the consumer.

There are numerous poisonous, flammable and hazardous chemicals used in the manufacture of a silicon solar panel. For example, arsenic, cadmium and lead are used in solders and a cleaning fluid for silicon manufacture is sulphur hexafluoride, a greenhouse gas that is 25,000 times more powerful than carbon dioxide. A recent calculation showed that to manufacture a

2 gram silicon chip, 72 grams of chemicals, 1.6 kilograms of oil or coal equivalent and 3.2 tonnes of water are used.

China hosts the largest number of solar panels in the world (80 GW capacity), twice the amount of the USA. Solar panels are short-lived and contain a great diversity of toxins. By 2034, China will have to retire some 70 GW of solar panels. How will they dispose of the panels? Most solar panels are in remote inland areas such as the Gobi in Inner Mongolia whereas recycling industries are along the Pacific coast.[100] My guess is that the 20 million tonnes of solar panels will be left to rot in the countryside of Inner Mongolia.

Solar power has a low capacity factor. For example, in Germany, it is about 10%. Hence 10,000 MW of solar power capacity is needed to generate the same amount of electricity as a 1,000 MW thermal coal or nuclear power station. Furthermore, when the 10,000 MW solar power generator is producing its maximum of 10,000 MW, the grid system cannot cope and hence huge yet-to-be-invented energy storage systems are needed or the solar power station needs to be shut down. Now that's efficiency.

Free markets are highly competitive and if there was a better way of creating solar panels, a competitor would fill the gap. Costs, efficiency and productivity are dirty words for the green left environmental activists so let's just whisper dirty for a while.

One square metre of a solar panel costs $750. Installation doubles this cost. For a 1,000 MW plant to provide electricity in the depths of winter, 3,230,000 panels are required at a bargain basement price of $4.83 billion. This is only for peak production of 1,000 MW at the optimal time of day. If the solar power station were to compete with a conventional coal-fired thermal power station providing 1,000 MW constantly with a

---

[100] "China's ageing solar panels are going to be a big environmental problem", *China Science*, 30 July 2017.

load factor of 70%, capital costs for a solar power station would be in the order of $100 billion.

The latest 1,000 MW coal-fired thermal power station built in Australia cost $1 billion at current costs and over its 20 year life would consume $2 billion worth of coal. This, of course, assumes that a solar power station would last 20 years at peak efficiency. An optimist would give it five years, at best 10 years because wafers of silicon quickly reconstitute and become even less efficient.

There are claims that solar panels are becoming cheaper and cheaper. From any perspective solar power is too expensive, too environmentally damaging and cannot provide large-scale energy to an electrical grid system. The greens hope against all knowledge that a large-scale low cost method of storing electricity for days, months or years is just around the corner. Such technology does not exist and is not even on the horizon. Whatever the cost of panels, solar power cannot compete without massive subsidies. A recent study in Germany showed that solar power is four times as expensive as power from a prototype nuclear reactor being built in Finland, and which is a particularly expensive design anyway.

In the UK, solar panels generate barely 1% of all electricity at only 10% of their 6.5 GW rated capacity. On 11th April 2015, when solar-generated electricity rose from zero to 3.7 GW, before falling back again to nothing within a few hours, the National Grid had to pay out £500.000. Most of this was to compensate wind facilities for switching off 2.5 GW of wind-generated power. And who really paid? The consumer. Early closures of solar subsidies would save between £40 and £100 million in 2020-2021 in the UK.[101] The same procedure happens in Australia.

---

[101] https://www.goc.uk.government/upload/systems/upload/attachment_data/file/447323/ Solar-PV-within-the-RO-consultation_-_Impact-Assessment.pdf

Electricity

A 1,000 MW nuclear or coal-fired power station occupies an area of 30 to 60 hectares (75 to 150 acres). An alternative 1,000 MW solar power station would have to produce enough energy for an 8-hour day plus provide for storage for the remaining 16 hours. The area required is 55.5 square kilometres. To do this with an efficiency of 10.27%, the area of solar panels, the space between panels to prevent shading and the area of maintenance roads would have to be 128 square kilometres (50 square miles; 12,800 hectares). All plants (and hence animals) would be removed from this 128 square kilometre area just to produce ideological inefficient unreliable electricity. This is green left environmental activism at its very best.

To build a 1,000 MW solar power station, it is not only the solar cells that are needed, there are structural supporting materials, concrete foundations, transmission systems, access roadways and a dispersed area of collectors and converters. The amount of material required is huge. Furthermore, massive earth works would have to be undertaken by diesel machinery emitting carbon dioxide during site preparation. Construction of a solar power station results in massive emissions of carbon dioxide into the atmosphere.[102]

---

[102] To produce the 35,000 tonnes of aluminium for structural support 12,777,950,000 kWh of electricity is needed which is now embedded energy and 735,000 tonnes of carbon dioxide is released during the process of smelting and refining to produce the 35,000 tonnes of aluminium. Some 75,000 tonnes of glass is required to cover solar panels and the manufacture of this glass releases 260,000 tonnes of carbon dioxide into the atmosphere. The embedded energy in the glass is 661,425,000 kWh of electricity. At least 600,000 tonnes of steel are required for the 1,000 MW solar industrial plant and electricity used for the manufacture of this steel would release 3,901,576 tonnes of carbon dioxide and the blast furnace would release 1,218,000 tonnes of carbon dioxide into the atmosphere. The embedded energy in the steel is 9,070,000,000 kWh of electricity. For wiring, some 7,500 tonnes of copper would be required. It has embedded energy of 529,507 kWh. To make this copper, 13,500 tonnes of carbon dioxide would be released from the smelter and the electricity used to run the smelter would release 494,200 tonnes of carbon dioxide into the atmosphere. The two million tonnes of concrete for footings would release 2,706,885 tonnes of carbon dioxide for the electricity used and 360,000 tonnes of carbon dioxide from burning limestone for

Unless solar power generation is very heavily subsidised, it is clearly uneconomic. Why should the average worker pay for environmentally devastating uneconomic unreliable subsidised electricity? We've been conned.

These are minimum figures because carbon dioxide emissions from road building, road and shipping transport, site machinery, manufacture of other metals (e.g. silicon in solar cells, tin/silver at electrical contacts), packaging, office activities etc have not been calculated. Nor have the use of vehicles for maintenance. Thousands of truck-loads of concrete would have been delivered by diesel-powered trucks emitting carbon dioxide and particulates and creating dust.

The real environmental cost for the manufacture and decommissioning of solar cells is not known but it is not pretty. And all this for a short-life solar cell. When solar power was all the rage, numerous Chinese companies were established to manufacture and sell solar cells to credulous Western green-contaminated countries. These companies are now disappearing at a very rapid rate, as are the Chinese poisoned by the pollutants used to make solar cells. Presumably for the greens, saving the planet is worth the human cost of killing Chinese workers.

The use of huge amounts of energy and the release of at least 10 million tonnes of carbon dioxide into the atmosphere just to construct a 1,000 MW solar powered generator in order to save the planet from increased carbon dioxide into the atmosphere does not look like a good idea, especially as at least 128 square kilometres of plant and animal habitats would be destroyed and the energy produced would be inefficient, unreliable and costly.

---

cement manufacture. The embedded energy in the concrete is 6,249,000,000 kWh of electricity. Just for these components alone, some 9,688,661 tonnes of carbon dioxide would have to be released to the atmosphere and the 1,000 MW generator would have to work at an efficiency of 10.27 for over 24 years just to pay back the 20,804,705 kWh of embedded energy.

Another great way of killing wildlife has been invented for environmentally friendly energy generation. It is solar thermal. Sunlight is reflected, focused and is so hot that it melts salt or metals that boil water to create steam to generate electricity. However, birds fly to the bright light and are incinerated midair. This is environmentalism. What we don't hear is about the performance of a solar plant. Crescent Dunes (Nevada, USA) has a 52% capacity, stopped from October 2016 to July 2017 due to a leak in a container of highly corrosive molten salt and the plant managed to operate for only five days in July 2017. Solana (Arizona) has six hours a day of generating electricity at 25% capacity and the Ivanpah plant (California) runs on water and not salt hence has no storage capacity with a 27.4% capacity. Crescent Dunes produces power at $170/MWh despite subsidies through loan guarantees and capital subsidies as tax concessions.[103]

And what does South Australia want to do? Spent $650 million on a Crescent Dunes look-alike for Port Augusta. This is a huge amount of money for a bankrupt State to spend on unproven bird-cooking technology. According to the unctuous press releases, it is claimed that power will cost $78/MWh. The South Australian government has form when it comes to desalinated water and renewables and the bookmakers have very short odds on another expensive white elephant.

Fuel poverty

Government policies on renewable energy targets have resulted in massive never-ending subsidies (paid by the consumer) that have been scooped up by opportunistic businesses signing long-term contracts. Because governments cannot abandon a renewable energy target policy that appeases a few percent of

---
[103] Mark Lawson, "Hype is renewable", *The Spectator Australia*, 26 August 2017.

the electorate, consumers are left with expensive unreliable electricity, a chaotic distribution scheme unable to handle surging and no energy security.

Electricity prices in Australia have risen so much that many people suffer from fuel poverty in one of the most energy rich countries in the world.

We pay for global warming schemes such as the carbon tax (hidden as renewable subsidies), solar rebates and wind farm subsidies. This is on top of price gouging by the renewable electricity generators and the schemes concocted by governments and their bureaucrats. Every month these clowns dream up a new scheme supposedly to stop the planet frying.

Leading IPCC scientists such as Ben Santer and Michael Mann after decades of trying to scam the world about global warming have now joined the crowd of contrarians and stated that temperature has not risen over the last two decades as predicted. The government's chief scientist Alan Finkel admitted in June 2017 that if Australia stopped all carbon dioxide emissions, we'd make no difference to global climate.

We are paying huge amounts for electricity as a non-solution to a non-problem. People are losing their jobs and poorer people and pensioners are shivering in the dark as a result of one of the biggest political cons of all time. This con is underpinned by scientific fraud.

It is simple. People can't pay their bills. In Victoria, the number of disconnections has risen by 140%. Energy rich Australia has energy poverty. To make matters worse, Victoria is copying South Australia. Dan Andrews (Premier, Victoria) has announced that he will be raising the wind and solar share of the State's electricity supply from its current 8% to 40%. Victorians should buy more winter clothes, fill the piggy bank to pay for higher cost and more unreliable electricity and be prepared to lose their jobs.

During winter in Western countries, many unnecessary deaths of older people are caused, or hastened, by fuel poverty. Far more people die in winter from the cold than in summer from the heat. Fuel poverty has been worsened by environmental policies resulting in massive increases in energy costs in order to subsidise unreliable, uneconomic and intermittent electricity sources such as wind and solar.

The UK Department of Energy and Climate Change shows frightening statistics on fuel poverty in the UK.[104] A 2011 YouGov poll showed that up to 24% of UK households suffer fuel poverty and a Confused.com survey showed that 82% of the UK population are concerned that they may not be able to pay their energy bills during winter. This just should not happen in a developed country and is the end result of poor energy policy. Anti-fossil fuel activists have blood on their hands. There is no virtue in supporting renewable energy when it makes people poorer and hastens the demise of the weakest in society.

You are now starting to suffer from fuel poverty. Your electricity prices have skyrocketed.

Expensive energy increases poverty, as welfare agencies attest. Poverty is the biggest killer worldwide, far outranking pollution, war or disease. The poorest countries in the world have the lowest life expectancy. Despite China's air pollution, and this is not carbon dioxide, its massive economic growth has lifted nearly a billion people out of poverty. Life expectancy has increased from 68.3 in 1990 to 75.4 years in 2016. This economic growth was driven by coal. It is claimed that Australia's privatised electricity grid and expensive "renewables" have already killed untold numbers of people. As an example, during the January 2009 heatwave, Victoria suffered its worst power blackout in history. The Victorian Government's Department

---

[104] https://www.gov.uk/government/collections/fuel-poverty-statistics

of Human Services estimated that 374 excess deaths occurred during the first week of that heatwave.

In the industrialised world, energy poverty is defined as households in which 10% or more of family income is spent on natural gas and electricity costs. Energy poverty does not result from having a huge number of appliances running incessantly, it is directly due to policies that compel utilities to provide expensive, politically-preferred green energy. It is a regressive tax that disproportionately affects low and fixed income families which have little money to spend beyond energy, food, clothing, rent and other basics. Each time the electricity price increases, they are hammered even harder. However, now that social democratic parties representing the poor, the workers and the disenfranchised have devolved into urban socialist parties, these folk are the forgotten people. They were called the deplorables in the USA.

There are billions of people who do not enjoy our living standards. Many don't have electricity and, if they do, they will get it a few hours a week at random times. They burn wood, twigs, leaves and dung for cooking and heating and spend hours every day collecting fuel and hauling filthy water from miles away. In Africa, India and other poor regions, more than two billion people still burn firewood, charcoal and dung for cooking. Millions die from lung infections caused by pollution from these open fires, millions more from curable intestinal diseases caused by bacterially-infested food and water and more millions because medicines are spoiled by a lack of refrigeration because clinics don't have electricity, refrigeration or window insect screens.

In many parts of the world, desperately poor people hunt and cook anything that walks, crawls, flies, swims or has a pulse. It matters not whether today's food is an endangered animal or not. Forests, scrub and grasslands have been cleared for firewood and charcoal and, in many poor countries, charcoal is the mainstay of cooking. Poverty is undeniably the worst environmental pollutant. Corrupt, incompetent governments and constant pressure from murderous environmental groups in rich countries all too often perpetuate the misery, joblessness, disease, starvation and early death in poor countries.

In fact, it is worse. The wealthy and increasingly radical environmental movement is not concerned about addressing the obvious pollution problems in poor countries, protecting the environment or being concerned about the health and welfare of their fellow human. UN officials have proudly proclaimed that environmentalism is really about ending fossil fuel use and capitalism, redistributing the world's wealth and controlling people's livelihoods, living standards and freedoms.

In the USA, the green energy policies affect the poorest households three times more severely than the richest households. Rising electricity prices affect all goods and services and for all electricity users in homes, offices, hospitals, schools, malls, farms, factories, mines and especially smelters. There are now 37 million American families earning less than $US 24,000 *per annum* after tax and 22 million households taking home less than $US 16,000 *per annum* after tax. It should be obvious to even the most ethically-challenged environmentalist that wind and solar mandates are unfair, unsustainable and inhumane. In Australia, we are suffering similar ethical problems. This is why your electricity bill is so high.

California has mild weather compared with States clinging onto the hem of Canada. According to the Manhattan Institute, one million households in California now have energy poverty. What was once the USA's "Golden State" now has the highest poverty rate, thanks largely to government requirements that one-third of the State's electricity must come from "renewable" sources by 2020 and one half by 2030. California's rising electricity rates are already nearly double those in Kentucky and other States that use coal and gas to generate electricity. Those attempting to claim the high moral ground in California buy Tesla electric cars. They are wealthy greens and celebrities with a symbol that states "look at me."

The capital cost of electric cars is extraordinarily high. There are free charging stations, access to special lanes on roads and up to $US 10,000 in combined tax rebates. It is only the wealthy who can afford to buy electric cars and they are subsidised by the poor. The economics does not look persuasive as the batteries cost $US 325/kWh which is equivalent to oil selling at $US 350 per barrel, some seven times the sale price of oil in 2017. If the Californians could persuade the rest of the USA to join the race to the bottom, then the cost of everything would rise. Jobs would disappear, living standards would decline as would health, longevity and happiness. The collapse of economies is a well-trodden path with Argentina, Greece, and Venezuela being recent examples.

However, if you thought that California was bad, try Europe. The electricity prices in Europe are double those of California (i.e. 30-45 c/kWh) and green energy policy is destroying employment, industry, health, longevity and economies. In Germany, 330,000 families had their electricity cut off in 2015 because they could not afford to pay the increasing bills. In Bulgaria, 50% of the average household income is spent on energy. In the protected forests of Greece, trees are being cut down because households cannot afford home heating. One-tenth of all families in the EU are now suffering energy poverty. Europe is lining up to kick an own goal.

In the UK, the story is the same. British families pay 54% more for electricity than the average American. Nearly 40% of UK households are cutting back on food and other essentials to pay for electricity. One in three UK families struggles to pay gas and electricity bills. Up to 24,000 elderly British die from illness and hypothermia each winter because they cannot afford proper heating. Many are forced to choose between heating and food and others spend the day in buses, shopping centres and public buildings to stay warm.

Those in the NSW Environment Office live in the comparative warmth of Sydney. Their offices are air conditioned 24/7. They do not experience the cold winters of inland NSW. Many country folk rely on red gum firewood and, in some areas, 75% of the household heating is from red gum. The NSW Environment office has deemed from the comfort of their air-conditioned offices that no longer should red gum be used; pensioners can shut the curtains and wear more clothes in an effort to reduce power costs and save more dead red gums.

While it is +15°C outside in Sydney, it is -5°C in many parts of rural NSW. The NSW Environment Office is funded by taxpayers who are pushed further into energy poverty by those they fund. This disconnect between the cities and rural Australia has the potential for civil unrest.

Maybe the do-gooders could get electricity to Africa. The International Energy Agency states that more than 75% of people in Madagascar, Ethiopia, Somalia, South Sudan, Sudan, Kenya, Tanzania, Uganda, Rwanda, Burundi, Congo, Central Africa Republic, Chad, Niger, Burkino Faso, Liberia, Sierra Leone, Guinea Bissau, Guinea and Mauritania have no electricity.

In most of the rest of sub-Saharan African countries only 50 to 75% of people have electricity. There are 150 million people in Ethiopia, Congo and Sudan that have no electricity. And in countries where there may be electricity, the wires may be live for two hours every second day. Hospitals, schools and productive businesses cannot be run with irregular electricity. More than 100 coal-fired power generating plants are in various stages of planning and development in Africa, excluding South Africa. Africa's embrace of power is in part the result of its acute shortage of power.

Bio power

Australia has a small biofuels industry. The burning of sugar cane waste (bagasse) to create steam to drive a turbine produces a very small amount of seasonal electricity in northern NSW.[105] Bacterial decay of waste produces methane. Capping of garbage transported to old open pit mines (e.g. Woodlawn, NSW) allows methane to be harvested for the driving of electricity-producing gas turbines.[106] Excess heat is also used. Many farms have small methane generators using gas developed from waste in anaerobic digesters. There has been great concern in other parts of the world about releasing toxins from anaerobic digesters into river systems so there are problems with this energy source also.[107]

The UK is committed by law to a radical shift to green energy. By 2020, the proportion of electricity generated from "renewable" sources is supposed to triple to 30% and with about 10% of total electricity generation from biomass. The only way to produce so much electricity from biomass burning is to do what people have been doing for thousands of years. Chopping down forests. And why? Because the EU rules deem that burning wood is "carbon-neutral".

The extreme of green madness is the conversion of UK's Drax power station in Yorkshire from coal to wood. Drax was the largest coal-fired power station in Europe and it could generate

---

[105] http://www.resourcesandenergy.nsw.gov.au/energy-consumers/sustainable-energy/bioenergy
[106] http://www.veolia.com/anz/our-services/services/municipal-residential/recovering-resources-waste/woodlawn-bioreactor
[107] http://www.sswm.info/content/biogas-electricity-small-scale

up to 3,960 MW of electricity. This required 36,000 tonnes of coal per day. But, Europe wants to reduce carbon dioxide emissions from coal burning. In order to reduce emissions from Drax, burning 70,000 tonnes of wood per day wood for nearly 4,000 MW of electricity is needed. This wood is to come from the US (North Carolina) and will be shipped as pellets over 5,000 kilometres across the Atlantic Ocean from the purpose-built Chesapeake Port in Virginia.

To convert the Drax power station from coal to wood pellets, the UK taxpayers will have to pay £700 million and the new wood-generated electricity will triple the cost of electricity. The Drax Group plc will be subsidised over £1 billion *per annum* by the British taxpayer for this green miracle. Last year, Drax received £62.5 million in green energy subsidies and this figure is set to triple as the amount of biomass burning increases. The UK government has decreed that electricity customers will pay £105/MWh for Drax's biomass electricity which is £10 more than onshore wind energy and £15 more than electricity from a new nuclear power station to be built at Hinkley Point, Somerset. The current market electricity price is £50/MWh.

To harvest some 70,000 tonnes per day from the other side of the world is no mean feat. After clear felling North Carolina's forest of maples, sweet gums and oak in the swamp lands, the wood is converted to pellets in giant energy-hungry factories using large amounts of energy derived from coal- and nuclear-fired electricity generators in the US. For the 20-year life of the Drax power station, 511 million tonnes of wood will be harvested using diesel equipment from trees in the US to provide expensive subsidised renewable electricity in the UK.

Trees just cannot grow fast enough to feed wood-fired electrical generators and "peak wood" will be reached very quickly because the trees harvested take 60 to 100 years to re-grow. This North Carolina wood will be shipped to the UK using oil-fired ships, unloaded using diesel equipment and transported using diesel trains. The equivalent of 46% of the energy generated by the wood-fired Drax generator will be used to transport the wood. Because wood has a lower energy density than coal, Synapse Energy Economics estimated that wood burning emits 50 to 85% more carbon dioxide than burning coal for the same energy output. The harvesting, transport and pelletising will produce huge amounts of carbon dioxide outside the EU jurisdiction so this does not come into the EU equation.

Where is the environmental impact statement for the Drax power station? A UK wood-fired power station has just passed on its environmental impact to the US for short-term gain. Harvesting forests results in increased soil carbon dioxide emissions. With such massive harvesting of wood, plants and animals would suffer massive habitat destruction. US environmental groups claim that these forests comprise some of the most biologically important forests in North America and that there are risks to wildlife survival and biodiversity (especially birds). And what about the otters and pileated woodpeckers that inhabit these swamps? They don't really matter as the EU and UK greens are saving the planet. It is habitat destruction that drives species extinction, not climate change. This is green environmentalism at its best. Destroy the forests and their animals in another country for the sake of feeling good at home.

Green activists used to demonstrate against the Drax power station burning coal. It was Europe's single largest carbon dioxide emitter. The company now boasts of its "environmental leadership position" and state that they are the biggest "renewable" energy plant in the world. Demonstrations by green activists have ceased. The passive start to the environmental movement in the 1970s was against harvesting forests. Now, in order to keep the ideological home fires burning, green policies have led to the clear felling of 70,000 tonnes of trees per day in North Carolina for burning in the UK.

The UK government estimates that by 2020, 11% of the UK's electricity generation will be from wood. Will greens have massive protests against biomass harvesting and burning? No. Air pollution will be worsened, UK electricity will be dependent upon a non-EU country and far more carbon dioxide will be emitted to the atmosphere than if gas was burned. And who will pay for these inefficiencies and subsidies? The poor UK worker. Not that there are many UK workers left now as a large proportion of employment age people in the UK, especially in the north, live off welfare. And who pays for the welfare? The poor UK worker. Again.

In the UK, wood provided about 33% of energy in the times of Queen Elizabeth I. In the times of Queen Victoria it provided 0.1% of the energy. Why? The Industrial Revolution needed far more energy than could be provided by the fastest growing trees and so coal was used. If wood had been used to energise the Industrial Revolution, then a land area one and a half times the UK's land area would have been needed. Coal is plentiful and has a high energy density. English forests started to grow back and in 2000, the forest area was three times that of 1900. The same is true for all European countries. It was the use of coal that saved the forests, not the greens or environmentalism.

In the US, a 40 MW wood-fired electricity generator in Cassville (Wisconsin, USA) burns 1,000 tonnes of wood each day provided by 30 different suppliers. Eventually the wood will be harder to harvest, transport distances will increase and costs will rise. And all this for a mere 40 MW. The 100 MW Picway coal-fired generator in Ohio looked at converting to wood but could not find a reliable long-term supply of wood and closed in 2015 when more stringent Environmental Protection Agency emission regulations took effect. Jobs were lost.

In Virginia (USA), taxpayers paid $US 165 million to convert the Altavista Power Station from coal to biomass. The power station is owned by a private company Dominion Virginia Power. Why taxpayer funds should go to a private corporation is beyond me, especially as Virginians will pay a higher cost for electricity. Politicians promoted the conversion as a method to "help meet Virginia's renewable energy goal". The claim was made that the Altavista station would be using biomass that would otherwise go to landfill. However, the Department of Energy has shown that 65% of biomass-generated electricity actually comes from wood and the rest is from waste.

The EU Council of Ministers has refused to cap the use of biofuels. Originally they wanted their 10% "renewable" energy target for transport to come from biofuels. This was then reduced to 7% but there was no agreement about this cap so Europe is left with 10%. This will cost European taxpayers €13.8 billion per year for a reduction in emissions of nine million tonnes. In November 2010, even Al Gore claimed that his advocacy of corn ethanol, which uses around 40% of all corn produced in the USA (or 15% of the world's corn), failed to feed the hungry and is wastefully used in engines. What Gore did not say is that for every 1°C average increase in temperature in the US Corn Belt, productivity increases by 10%. For every 1°C average decrease in temperature, the latitude for growing corn shifts nearly 150 km southwards. A cooling event would greatly reduce the available land for growing food. It's happened before, it will happen again.

In the EU, crop biofuels (e.g. vegetable oils, sugar beet, canola, harvest waste, wood chips, ethanol) have replaced 5%

of fuel used in transport. If biofuels were burned solely for transport, then carbon dioxide emissions would drop by 59 million tonnes by 2020. However, the International Institute for Sustainable Development showed that deforestation, fertilisers and fossil fuels used to produce the required biofuels emit about 54 million tonnes of carbon dioxide hence only five million tonnes of carbon dioxide emissions are saved. This is 0.1% of the total EU emissions.

Furthermore, the carbon dioxide saved is just emitted elsewhere with the net effect leading to an increase in global carbon dioxide emissions. If standard unbelievable climate models used by climate "scientists" are run, EU biofuel use will postpone a modelled slight temperature rise of 0.00025°C by 2100 by 58 hours. Even climate "scientists' must be able to see that use of biofuels is pointless. An area as big as Belgium is used for growing biofuels and a similar-sized area is used for European imports. Biofuel farmland uses as much water as flows down the Seine and Elbe rivers combined. All this for a saving of five million tonnes of plant food. European farmers now use fast growing trees such as poplar, willow and *Eucalyptus* for biofuels. These plants emit the toxin isoprene. A Lancaster University study suggested that the EU's 10% target will cause an extra 1,400 deaths at a cost of £1 billion annually from isoprene poisoning.

There is an additional cost for biofuel electricity to UK taxpayers of £6 billion a year. Each tonne of carbon dioxide that is emitted to the atmosphere costs the British taxpayer £1,200. By contrast, the EU's cap-and-trade system costs about £4 per tonne and the British are paying 300 times as much for their carbon dioxide emissions than people on the continent. To make matters worse, economic estimates show that to cut emissions of carbon dioxide by one tonne, there is about £4 in environmental costs. All these nonsensical costs should not be been imposed on the taxpayer as it has yet to demonstrated that carbon dioxide harms the environment or drives climate change.

One would have thought that the greens would encourage the emissions of carbon dioxide because it is plant food. By emitting carbon dioxide, we are fertilising the planet and helping green plants grow. There is now good satellite evidence to show that the recent small increase in atmospheric carbon dioxide has resulted in a slight greening of the Earth. In the atmosphere, carbon dioxide is a trace gas, the dominant greenhouse gas is water vapour and, without these two greenhouse gases, there would be no life on Earth. Light, water and carbon dioxide create plant material, but this process of photosynthesis appears to be unknown to greens.

There is a moral argument. Land is being used to grow fuel and not food in a world where one billion people are hungry. Food prices have been driven up by heavily subsidised biofuel farms taking the place of food farms. However, although environmental, economic and moral arguments can be aired against the biofuels industry, it is a huge business. Big green vested interests are living off subsidies and tax concessions. The costs of climate policies to stop the alleged global warming are now globally about £1 billion a day. Wind turbines cost about 10 times the estimated benefits and solar power costs about 100 times the benefits.

What happens if carbon dioxide does not drive global warming as I show? All this money has been wasted rather than used in lifting people from poverty, creating employment or preparing for a real disaster.

The biofuels green dream is an environmentally damaging immoral nightmare. Do you ever get the feeling sometimes that we've all gone mad?

Batteries

Not one country, state or city in the world relies on batteries for electricity or even electricity backup. We already have far better storage of energy than short-lived batteries. It is called coal, gas, oil and uranium.

Renewables have failed to keep jobs in South Australia, businesses are going broke and the average person struggles to pay their electricity bills. Rather than the government stating that they got policy wrong, they are going further down the renewable track and have floated the idea of battery back-up. The pain will continue in South Australia. Australia is on a downhill slide of addiction to wind and solar and denial of responsibility.

The fast money white shoe brigade was in to see South Australia's Premier Jay Weatherill before you can say excessively high electricity bill. The same mob that pushed renewable energy are now pushing for batteries. As soon as there is a quid to be made from the stupid, then they are surrounded by sellers. Meanwhile, those at GetUp!, who prove that the education system has been dumbed down, cannot do simple calculations, imply that batteries generate electricity, and that power engineers got it wrong, suggest that batteries are the greatest thing since sliced bread and will solve South Australia's self-inflicted problems, but never seem to mention words like cost, longevity, discharge rate and high electricity bills.[108] On a recent trip to Australia, Al Gore stated:

> I have a lot of admiration for South Australia because it's now leading the world – the largest battery ever, and it will be the first of many.

A recommendation or endorsement by Al Gore is the kiss of death.

---
[108] www.getup.org.au/sa-power?t=53t1t1OQ

South Australia will spend $111.5 million of a $500 million go-it-alone energy plan. Nine hybrid diesel generators that consume 80,000 litres an hour for 276 MW should be in place before the State election. They will be leased with an option to purchase. Will they be in place if there is a long hot summer? After two years, they will be converted to gas generators. By using diesel generators, the argument about carbon dioxide emissions has been shown to be vacuous. If the Port Augusta power station had been kept open, nearly $100 million would have been saved. The South Australian premier stated:[109]

> Stable renewable policy energy settings have created the right environment for clean energy investment, which has helped make Australia and particularly South Australia global leaders in renewable energy generation.

It certainly is the right environment for "clean" energy investment if dopey governments can be tied up for decades with contracts for wind, solar and now batteries. No one really cares if Australia or South Australia are leaders in renewable energy generation. People want to be able to open their quarterly electricity bill and not keel over in shock and worry about ways to pay exorbitant rip off bills. South Australia has given us a glimpse of the clean energy future and it's not pretty.

The Minerals Council of Australia estimates the added cost of battery storage could push the cost of wind power from $92/MWh to between $304 and $727/MWh.[110] As Matt Ridley wrote in *The Australian* in 2016, you would need 160 million Tesla Powerwalls to cover one day's electricity consumption in just the UK or 3.3 billion for one week's consumption if all the UK's heating and transport was electrified.

---

[109] Chris Kenny, Headlong rush into green energy without a sensible plan is no way to operate, *The Weekend Australian*, 12-13 August 2017.

[110] Andrew White: Energy giant flags 20 pc rise in power bills, *The Australian*, 11 June 2017.

In South Australia, the proposed 129 MWh battery will require about five million cells. Lithium cells are prone to spontaneous ignition from flaws in manufacture. Tesla cars with a lithium battery often explode and burn.[111] If any of these cells has a meltdown the whole module will burn up and if adjacent modules are not sufficiently isolated they too will be ignited. A further risk is also likely to exist in the form of short circuits igniting other modules if one fails. Lightning strikes on the battery complex or on nearby power lines also present high risk.

Graeme Lloyd in *The Australian* summed up the battery madness.[112] Tesla's Elon Musk claims that he can build the world's biggest battery to stabilise the grid in South Australia, a failing State. As with wind factory deals, there is no transparency and no costings for scrutiny. The battery is far too small to cover the vast amounts of electricity that go missing when the wind does not blow.

There are key unanswered questions. How many times can the batteries be charged before they are useless? What is the replacement cost of batteries? Is there sufficient wind power turbine capacity now to maintain the battery charged against expected fluctuating load? Will more turbines be required and at what cost? How is the battery protected against spontaneous combustion or explosions like those experienced at other installations? Will properly designed battery storage be considered an essential component of solar and wind power and be included in any assessment of energy from these renewables?

It appears that Musk lives off taxpayers' money and you can be sure that your taxes and electricity bills will keep him in

---

[111] electrek.co/2017/03/31/tesla-model-s-fire-manchester-crash/
[112] "Can South Australia battery power Elon Musk's Tesla dreams?", *The Australian*, 15 July 2017.

the manner to which he is accustomed. He's got form. The *Los Angeles Times* reported that in 2015 Musk's companies received about $US 4.9 billion in government subsidies and incentives.[113] Nice if you can get it. Tesla Motors also collected more than $US 517 million in 2015 from competing car manufacturers by selling environmental credits.

If Musk's horrendously expensive batteries are purchased using South Australian taxpayers' hard earned cash, will they solve the headache that South Australia gave itself?

If there is again a total blackout in South Australia, fully charged batteries will provide enough electricity for households, businesses, factories, mines and smelters for four minutes.

You couldn't make this up if you tried.

Energy history

In the long ago

Up until the 17th century, feet and wood were used to produce energy. People and animals died like flies and land was completely deforested for the making of iron and glass. In the 18th century, wood was the main source of energy and forests were destroyed at a very rapid rate. In fact, the area of forests now is far greater than it was in the 18th century. By the 19th century, coal was the main form of energy. This was a period of great innovation. It was the high energy density of coal that drove this innovation and the Industrial Revolution. In the 20th century, it was oil that drove industrialisation, transport and trade while coal was used for smelting and in coal-fired power stations.

Once there was abundant cheap coal-fired electricity in the 20th century, economic growth accelerated. Electricity is the

---
[113] www.latimes.com/business/la-fi-hy-musk-subsidies-20150531-story.html

multiplier of economic growth. The 21st century is the century of gas. Industry, domestic heating and cooking, transport, power generation and some smelting uses gas. And what will the future centuries bring? Supplies of gas as methane hydrates are huge, cheaper methods of creating hydrogen are constantly being invented and we may finally enter a nuclear age with fission and perhaps fusion. Nuclear fission is where a large atom breaks apart whereas nuclear fusion is where small atoms combine to make large atoms. Both processes give out energy. Energy from the Sun is from nuclear fusion. For all of my life, nuclear fusion has been only 20 years away. It still is.

The first flush of renewables

Spain is the best example of what not to do.

A 2009 testimony about Spanish renewable energy to the US House Select Committee on Energy Independence and Global Warming,[114] showed that for every green energy job financed by Spanish taxpayers, 2.2 jobs were lost. Only one out of ten green jobs was in maintenance and operation of already installed "alternative" energy plants and the rest of the jobs were only possible because of high subsidies. Each green job in Spain cost the taxpayer $750,000 and green programs led to the destruction of 110,500 jobs. Each green energy megawatt installed destroyed 5.39 jobs elsewhere in the economy. I'm sure those pushed into unemployment by green activism feel that they have made a sacrifice for a higher cause.

On 16 January 2009, during a visit to an Ohio wind turbine component manufacturing business President Obama stated:

> And think of what's happening in countries like Spain, Germany and Japan, they're making real investments in

---

[114] http://www.markey.senate.gov/GlobalWarming/index.html

renewable energy. They're surging ahead of us, poised to take the lead in these new industries.

Sunny Spain was touted as the perfect place for solar power generation. Spain spent a fortune on constructing solar and wind power generators with generous subsidies. Spain became so clever at generating solar electricity that it even managed to do it at night. Generating solar electricity at night? No, it is not the new physics.

Night generation of solar power was because subsidies were so incredibly high that solar power companies could make money by illuminating solar panels with floodlights at night. The floodlights were powered by diesel generators. No wonder Spain went broke.

President Obama did a pretty good job of increasing unemployment by promoting renewable energy. If he had followed the Spanish example he might have done even better. Spain went broke due to the extraordinarily high cost of electricity and subsidies.

Energy in Australia

Barely 25 years ago, Australia was regarded as an energy superpower with abundant coal and gas reserves that ensured reliable, secure and cheap electricity. Business then was about taking that electricity and using it to create goods and services. In that environment, commercial business flourished. Now, the green activist environmental business is all about generating erratic, unreliable and ludicrously expensive wind and solar power. Our present power system is neither cheap nor reliable and white shoe brigade lobbying by the likes of Gore and Musk will only make it worse.

In Australia, electricity is generated from coal (73%), natural gas (13%), hydropower (7%), wind (4%), solar (2%)

and bioenergy (1%).[115] In 2013, Australia generated 247 TWh of electricity of which 234 TWh was sent to customers.[116] The difference of 13 TWh was used by the power stations themselves and 12.6 TWh was lost during transmission leaving some 221 TWh for final consumption.[117] For Victoria, demand varies between 3,900 and 10,000 MW and in NSW between 5,800 to 15,000 MW.

Australia's high carbon dioxide emissions

Australia has high *per capita* carbon dioxide emissions. This might not be because we're being naughty but because we are internationally recognised as a good place to do business. This is changing because of our high electricity costs.

The 24 million people in Australia generate 1.5% of annual global human-induced carbon dioxide emissions. USA emits 14 times and China emits 26 times more carbon dioxide than Australia. Australia has 0.33% of the global population.

We have a high standard of living, a landmass of 7,692,024 square kilometres with a sparse inland population and greenhouse gas-emitting livestock. We need to transport of livestock, food and mined products long distances to cities and ports and we export of ores, concentrate, coal, metals and food sufficient for 80 million people. This all results in high *per capita* carbon dioxide emissions. Australia's exports of coal, iron ore and gas contributes to increasing the standard of living, longevity and health of billions of people in Asia.

If Australia emits 1.5% of global annual carbon dioxide emissions, and as 3% of the total annual global emissions are

---

[115] www.originenergy.com.au
[116] http://www.world-nuclear.org/info/Country-Profiles/Countries-A-F/Appendices/Australia-s-Electricity/
[117] http://www.world-nuclear.org/info/Country-Profiles/Countries175-A-F/Appendices/Australia-s-Electricity/435

estimated to be anthropogenic and as the atmosphere contains 400 parts per million by volume of carbon dioxide, then one molecule in 5.6 million molecules in the atmosphere is carbon dioxide emitted from humans in Australia. This molecule has an atmospheric life of no more than seven years before it is removed from the atmosphere by natural sequestration into the oceans, life and limey sediments.

Australia has far greater economic priorities than to change a whole economy, to increase energy costs, to decrease employment and to decrease international competiveness because of one poor lonely molecule of plant food in 5.6 million other atmospheric molecules. It cannot be shown that this one molecule of plant food in 5.6 million other atmospheric molecules derived from Australia has any measurable effect whatsoever on global climate.

Australia exports a significant share of the global refined aluminium, zinc, lead, copper and gold and hence takes a *per capita* emissions hit for countries that import and use Australia's metals because smelting and refining in Australia result in carbon dioxide emissions. Neither smelting nor refining of the metals for other countries could take place without burning fossil fuels. Australia is heavily dependent upon coal for electricity and nearly three-quarters of Australia's electricity is from coal-fired power generators.

Historically, by world standards, Australia's electricity is low cost. This is why aluminium smelters were built in Tasmania, Victoria, NSW and Queensland and most of the product is exported. About 11,000 kWh *per capita* is used for the Australian aluminium industry. Zinc also has a huge amount of embedded electrical energy. Australia exports zinc metal from smelters at Risdon (Tas.), Port Pirie (SA) and Townsville (Qld). Because Australia has long-term political stability and once had long-term cheap electricity, smelters were established and, as a result, Australia has a higher *per capita* carbon dioxide emissions than other countries that do not smelt.

Manufacture of aluminium is energy intensive at 27 TWh/year which is 10% of Australia's gross electricity production. Zinc, lead, copper and nickel manufacture at 14 TWh/yr are also energy intensive. Most copper and nickel concentrates are exported because there are only copper smelters are at Olympic Dam (SA) and Townsville (Qld) and there is one nickel smelter at Kambalda (WA). Copper, lead and zinc concentrates are exported from Derby (WA), Fremantle (WA), Port Pirie (SA), Adelaide (SA) Darwin (NT), Port Kembla (NSW), Newcastle (NSW) and Townsville (Qld). Over half the Australian generating output total of 80 TWh is exported as embedded energy in refined metals and concentrates. Most of the growth in value-adding manufacturing in the past 20 years has come from industries which are energy-intensive.

Growth has occurred in Australia because of relatively low electricity prices coupled with high reliability of supply and the proximity of natural resources such as bauxite (NT, WA and Queensland), zinc ores (Macarthur River, N.T.; Century and Mount Isa, Qld; Broken Hill, and Cobar, NSW; Rosebery, Tas.), lead ores (Macarthur River, NT; Mount Isa and century, Qld; Broken Hill and Cobar, NSW; Rosebery, Tas), copper ores (Mount Isa, Qld; Cobar and Parkes, NSW; Olympic Dam and Prominent Hill, SA; Rosebery, Tas) and nickel ores (Kambalda, Forrestania, Mount Keith, Leinster and Murrin Murrin, WA).

We should be proud of our high emissions. It shows that we are good place for business which means that there are more jobs. A steel mill uses coal to reduce iron oxide into iron metal and the carbon in coal is oxidised to carbon dioxide. A modern economy cannot rely on sea breezes and sunbeams to make steel, aluminium, zinc, lead or copper or generate base load electricity for industry. A decarbonised economy would be a deindustrialised economy.

Annual Australian *per capita* carbon dioxide emissions are in the order of 20 tonnes per person. There are 30 hectares of forest and 74 hectares of grassland for every Australian and each hectare annually sequesters about one tonne of carbon dioxide by photosynthesis. Forests and grasslands in SE Australia fix about 30% of all of Australia's carbon dioxide emissions. As per normal, it is rural Australia that carries the population centres of Australia. On the continental Australian landmass, Australians are removing by natural sequestration more than three times the amount of carbon dioxide they emit. Crops remove even more carbon dioxide from the atmosphere. Australia's net contribution to atmospheric carbon dioxide is negative and this is confirmed by the net carbon dioxide flux estimates from the IBUKI satellite[118] carbon dioxide data set.

Australia's continental shelf is 2,500,000 square kilometres in area. Australia has a huge coastline with currents bringing warm water from the north to the cooler southern waters (East Australian Current, Leeuwin Current) and through the Great Australian Bight (Zeehan Current). Carbon dioxide dissolves in Australian waters and living organisms extract dissolved carbon dioxide from seawater to build corals and shells. This natural marine sequestration locks away even more Australian emissions of carbon dioxide and adds to the negative contribution of atmospheric carbon dioxide made by Australia.

Using the thinking of the IPCC, UN and activist green groups, Australia should be very generously financially rewarded with money from populous, desert and landlocked countries for re-

---
[118] global.jaxa.jp/projects/sat/gosat/topics.html

moving from the atmosphere its own emitted carbon dioxide and the carbon dioxide emissions from many other nations. By this method, wealthy Australia can take money from poor countries. This is, of course, normal for the green industry.

Satellite measurements show that there has been a greening of the planet over the last few decades, thanks to a slight increase in traces of plant food in the atmosphere. Without carbon dioxide, there would be no plants and without plants, there would be no animals. Geology shows that atmospheric carbon dioxide has not driven global warming since planet Earth formed. Why should it now?

Dangerous global warming did not occur in the past when the atmospheric carbon dioxide content was hundreds of times higher than now. Each of the major ice ages was initiated at a time when there was more carbon dioxide in the atmosphere than now. The planet has not warmed for the last two decades despite a massive increase in carbon dioxide emissions during the industrialisation of Asia.

It has never been shown that human emission of carbon dioxide drives global warming and the Australian contribution of one lonely Australian molecule in 5.6 million is not worth expenditure of a single penny.

It is because Australia does not change governments with guns and has a stable democratic political administration that large long-term capital-intensive investments such as smelters and refineries can operate. This perception of stability is fast changing with a judiciary making political judgments on environmental and regulatory matters, a politicised bureaucracy, increasingly ineffective politicians, anarchistic unions and green left environmental activist movements (some of which are supported by the unions and foreign foundations).

If emissions of carbon dioxide are a worry to activists,

then Australia does more than its fair share to reduce carbon dioxide from the atmosphere. Do we ever hear this from green left environmental activists? All we hear is constant doomsday stories by activists constantly crying wolf.

Media networks are only too happy to broadcast that Australia is the worst polluter in the world but don't mention that carbon dioxide is not a pollutant and, if some scientifically misguided activist lawyer makes laws to call carbon dioxide a pollutant (as the EPA has done in the USA), then Australia is therefore adsorbing and fixing carbon dioxide "pollution" from elsewhere on the planet. However, "worst polluters" is a great headline whereas the facts are boring. For those that espouse that human emissions of carbon dioxide are the major cause of climate change, then Australia should be viewed as the hero and not the villain.

Facts tend to pour cold water on hysteria.

Emissions targets

The Labor Party realised that there were billions to be made by drinking from the renewables energy trough. The Labor/Green alliance increased the Federal RET to 45,000 GWh in 2010 driven by ex-unionist environment minister Greg Combet who, along with his old union mate Gary Weaven, used the union superannuation money they controlled to invest in wind power outfits like Pacific Hydro. The Labor Party is beholden to the wind industry.

Once upon a time, the Labor Party would have come to the rescue of the poorest and most vulnerable in the community. Now they are a business to scam the poor and vulnerable. We hear nothing from Labor about the horrendously high household electricity bills. Labor is pushing for a 50% RET, maybe to gain votes in marginal city electorates that would otherwise go to the socialist green left or maybe to increase income for their businesses. Whatever the reason, more of the poor and vulnerable will have their electricity disconnected. The 2017 Finkel report wants a renewable energy component of 42.5% of total electricity yet the system cannot cope with half that, as South Australia has shown.

Hot air comes from the green left environmental activists and from misguided politicians about increasing the proportion of renewables. The large amounts of renewable energy that would be needed to keep people employed and homes on the grid can't be stored efficiently. The more renewable energy put into the grid, the more coal- or gas-fired back up would be needed. Rather than close coal-fired power stations, they would have to be kept running and/or replaced with modern coal-fired power stations as the electricity demand increases because wind power is unreliable and intermittent. The wind just does not decide to blow at times of peak power consumption.

In Australia, fossil fuels contribute to more than 80% of the total energy mix. Do the green left environmental activists have a better energy mix? How would food be delivered from farms to cities if fossil fuels were not used for semi-trailers? Currently wind and solar in Australia provide a very small amount of the primary energy production and the capital costs, recurrent costs[119] and environmental damage to expand this sector of the industry are horrendous.[120]

In a television interview, we were told by Labor's Kevin Rudd a decade ago that renewable energy would only cost $1 a year.

Oh, good grief Kerry. Our manageable renewable en-

---

[119] A. Moran, Submission to the Renewable Energy Target Review Panel, 2014.
[120] Ian Plimer, *Not for Greens*, Connor Court, 2014.

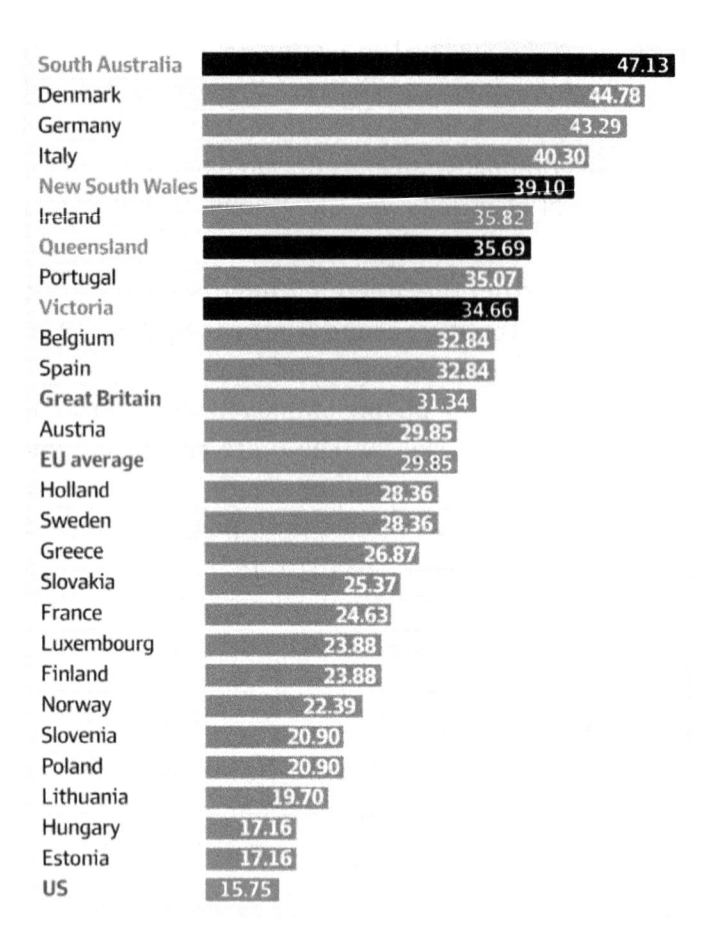

Climate Change Delusion and the Great Electricity Rip-off

Region	c/kWh
South Australia	47.13
Denmark	44.78
Germany	43.29
Italy	40.30
New South Wales	39.10
Ireland	35.82
Queensland	35.69
Portugal	35.07
Victoria	34.66
Belgium	32.84
Spain	32.84
Great Britain	31.34
Austria	29.85
EU average	29.85
Holland	28.36
Sweden	28.36
Greece	26.87
Slovakia	25.37
France	24.63
Luxembourg	23.88
Finland	23.88
Norway	22.39
Slovenia	20.90
Poland	20.90
Lithuania	19.70
Hungary	17.16
Estonia	17.16
US	15.75

Figure 13: Retail electricity prices (in c/kWh) for Eastern Australian states compared with international prices (Source: US Energy Information Administration).

ergy target policy, 20% by 2020, we've been working on that for ages ...

Our position on this is consistent and clear cut. One, unlike the Howard Government, we accept a robust target of 60% reduction in greenhouse gas emissions by 2050 against 2000 levels ...

We have taken this position of a new ambitious but responsible nonetheless renewable energy target of 20% by 2020 based on modelling which has been done by MMA, a modelling firm which is in conjunction with Monash University, looked at all the variables which go into the economic impact of such a renewable energy target ...

That is that they calculate that between now and about 2045 that you'd be looking at a total impact on the economy of somewhere between $600 and $800 million or something in the vicinity of $45 per person over that period of time or something like $1 per person per year.

Maybe Rudd, blinded by green ideology, was totally uncritical of what he was fed and maybe was stupid enough to believe that a restructuring of the nation's energy system would only cost taxpayers $1 per year. Maybe he lied? Maybe he was fraudulent? Maybe he is a total twit? Probably all three. Rudd wanted people to believe that it would only cost $1 per year to feel good and green. The end result was the pointless destruction of Australia's cheap and reliable electricity system which will take decades to rectify. An uncritical fawning interview with the ABC's resident green activist Kerry O'Brien was certainly not in the service of the Australian taxpayer. That same taxpayer funds the ABC.

We had calamitous warmings in Australia about imminent human-induced global warming due to our emissions of carbon dioxide. This was used by the renewable rent-seekers and their political enablers to set up Australians to pay nigh on $80 billion in subsidies, soft loans, ludicrously generous guaranteed feed-in tariffs for roof solar and punitive fines. This has resulted in crippling and ever-increasing power bills and placing south-eastern Australia's grid on the brink of collapse.

National Energy Market

Once the National Energy Market was established, electricity prices escalated.

Much of the dilemma rests on how various levels of government have mistaken their responsibilities. Who really is responsible for the security of supply to the national grid? Under the Constitution energy policy falls to the States but national climate change policy, particularly the Renewable Energy Target, has overlaid this and had a direct impact, as has the National Electricity Market (NEM), which is effectively legislated and controlled by the States under Federal guidance. Before the nation has even had a chance to bed down these overlapping responsibilities and ensure consumers and industry have access to secure and affordable energy, the States again have signalled an intent to play out of their division.

Whatever the relative role and merits of renewable energy, storage, peaking generation and interconnection, it is clear that South Australia's headlong and unilateral rush to a 50% renewable energy share seriously weakened its network, hastening the demise of coal-fired generation, pricing out gas-fired generation and, without adequate storage, increasing reliance on interconnectors with Victoria.

Rather than revise downwards their State-based renewable targets the Labor governments of Victoria, Queensland, SA and the ACT have signed a proclamation, with former US vice-president Al Gore no less, promising zero net emissions by 2050. This potentially commits them to energy policy overreach.

In reality, it is the NEM via interconnection of electricity and gas supplies across borders that has disrupted supply to the States and there needs to be a co-ordination of electricity generation and supply policy. COAG is the appropriate forum to win a nationally co-operative approach. That is why threats by the States to go further and "go it alone" on clean energy targets are disturbing, especially when the States clearly have not planned wisely for their energy needs so far. Bans on gas exploration and exploitation in Victoria, NT, NSW and Queensland have worsened supply and increased price pressures which have fed directly into electricity costs. Even federal Labor leader Bill Shorten has conceded there is a need to act in this area.

It is worth noting that the US has achieved energy independence in the past decade by vastly increasing oil and gas use largely on the back of fracking and associated technologies. This is crucial because as we rely increasingly on intermittent renewable energy the nation will need to greatly expand peak and back-up generation capacity from gas. While Canberra did nothing under the RET to ensure renewable generation was spread across the national market, South Australia should have realised that rushing to a 50% renewable share in one State without adequate storage or back-up was reckless.

There are fears Victoria is heading in the same direction after the closure of the 1.6 GW Hazelwood coal-fired generator. Reassurances from Mr Gore that South Australia and Victoria are looking to lead the world, and from Elon Musk that his battery plan carries some "risk" as a world first, do nothing to provide comfort. We need a national approach to provide investment certainty. Many in the energy industry and politics believe this demands bipartisanship so that we know any policy setting will last beyond a single electoral cycle. "A comprehensive and truly national policy framework will not be achieved by just some jurisdictions 'going it alone'," is the warning from the Business Council of Australia. It is right.

In 2016, the Australian Federal government panicked and commissioned the Finkel report from the Federal chief scientist. This report says we are facing a crisis and the way out is to have more renewable energy. The Finkel report had no consideration of past climate changes. It relied upon climate models based on theory and assumptions to predict the future. The main assumption was the climate is driven by human emissions of carbon dioxide.

But as the States, Territories and Federal governments adopt all Finkel recommendations save for the Clean Energy Target, there must be some concern that the desire for cooperation and bipartisanship could lead to the adoption of second-rate policy. A Clean Energy Target must be left for the federal government to set and manage; and it won't be easy to balance environmental, parliamentary, Coalition and economic objectives.

All governments and parties must adopt a clear hierarchy on Malcolm Turnbull's "trilemma" of priorities: affordability, security and emissions reductions. This cannot be achieved with a Clean Energy Target and playing around the edges trying to bully electricity companies to lower prices or suggesting to consumers that they shop around for the cheapest electricity does not address the heart of the problem. Given our economic needs and the trifling impact our emissions have on global trends, security and affordability must come first with the aspiration of meeting emissions reductions targets coming a distant third. The Coalition faces no more difficult or vital task.

The Australian power market calamity was totally unnecessary, avoidable and predictable. Taxpayer subsidies to meet State and Federal renewable energy targets reached $3 billion in the 2015-2016 financial year.[121] And who pays? The taxpayer and electricity consumer. The overall cost of subsidising renewable energy generation has nearly doubled since 2011 and the Renewable Energy Target will just not go away and die. Irre-

---
[121] Geoff Chambers, "Clean-coal cheaper option than renewables", *The Australian*, 3 July 2017.

spective of how efficient conventional coal generators are, as long as the Large Scale Renewable Energy Target remains in place, their ability to dispatch cheap electricity to the grid is dictated by the weather which determines the output from wind and solar sources.

In mid 2017, AGL announced a 16% rise for residential electricity in NSW. In the ACT, ActewAGL increased the price by 18.95%. The South Australian rise is 18% and Queensland 3.5% after the government owned power stations were ordered by the Premier to cut network costs.[122] At the time of writing, other energy producers were yet to announce price rises.

The Finkel report warns that if power companies don't respond to high prices by installing fresh capacity, the reliability of the network could be compromised over the longer term. Victoria and South Australia are at risk of damaging summer blackouts because of the closure of the 1.6 GW Hazelwood brown coal power station.[123] Hazelwood had been generating more electricity than all of Australia's wind generators combined. Its electricity was cheaper and more reliable. The South Australian solution is simple. Spend a fortune of taxpayer's money on diesel generator back up. These will use 80,000 litres per hour of diesel fuel and emit monstrous amounts of carbon dioxide to the atmosphere. But wait a minute, didn't South Australia kill its economy by changing from tried-and-proven coal-sourced electricity to wind power because carbon dioxide emissions are harmful and lead to human-induced global warming? Stupidity? Crony capitalism?

The Finkel Report suggests that there is a "once in a generation" opportunity to develop a reliable, low emissions energy

---

[122] Sid Maher, "Electricity consumers facing big hikes despite Finkel report", *The Australian*, 10 June 2017.
[123] David Uren, "Finkel sounds alarm on power; states in danger of blackouts", *The Australian*, 10 June 2017.

system. Rubbish. We had one before. However, it ignores the only zero emissions baseload energy source available – nuclear power. That's despite the fact that more than half the world's population live in countries that have access to nuclear power.

Finkel purports to resolve Australia's power market chaos, caused by the erratic and chaotic supply generated by wind and solar generators. A solution offered was to back up wind and solar with coast-to-coast mega-batteries. There is no country in the world with industrial scale electricity systems using batteries. Wind power champion Denmark? No. Wind and solar power champion, Germany? No. The planet's high-tech leader, USA? No. Such batteries do not exist.[124]

The Finkel report makes a solution to Australia's self-inflicted energy wound even more difficult. Households just do not believe the voodoo models that try to persuade consumers that with more wind and solar power, electricity prices could actually fall. They've heard it all before and continue to face electricity price hikes of up to 20 to 30 per cent each year. Renewables have a 16% penetration and the system has fallen apart. What will happen if Renewable Energy Targets of 50% are achieved? Finkel has renamed the Renewable Energy Target (RET) to the Clean Energy Target (CET). As a result, the future looks even bleaker. More renewables will just make the system worse and gas contribution in 2020 of 6% will fall in 2050 to 5%.

Judith Sloan[125] hit the nail on the head:

> Finkel's Soviet-like command that coal-fired electricity plants must give at least three years' warning of closure, even if these plants are haemorrhaging money in the meantime. He can't be serious.

---

[124] Andrew Follett, "*New York Times* admits batteries necessary for green energy don't exist", *The Daily Caller*, 6 June 2017.

[125] Judith Sloan, "The end to the climate wars? Not on your nelly", *The Australian*, 10 June 2017.

The report ignores an important point of company law and corporate governance. Coal-fired power stations must close when they are unprofitable and cannot be mandated to stay in business losing money for three years. A company director who allows his company to trade while insolvent faces personal liability for the debts incurred from the time at which the company was deemed insolvent.[126] Companies are a going concern if they can pay their debts as and when they fall due. If the company is unprofitable, the directors place it in voluntary administration and set in train the process for winding up the company. If legislation to give three years' notice was presented to parliament, directors of every coal-fired power station would close down all coal-fired power stations rather than operate a potentially insolvent company.

The Finkel report did not justify a clean energy target and did not explain how and why coal-fired generators are being driven from our electricity system. The report uses the words clean and unclean (50 times) hence is a not the document of a scientist but that of an advocate.

During the four decades when State governments were independently responsible for producing and distributing their own electricity within their own State without Federal government meddling, real electricity prices fell by about 45% giving great benefits to industry and consumers. State electricity monopolies competed against each other to attract industry. Once the National Electricity Market started, there was more than a 90% increase in electricity prices and prices are still heading north. Finkel argues for "shared accountability" with more institutions taking control. Apart from "shared accountability" in reality being used to avoid accountability and avoid stronger governance, the history of electricity generation

---
[126] Section 588G of the Corporations Act 2001.

in Australia shows that when there are fewer organisations involved, the costs are lower.

The Finkel report has included skewed modelling and contains a patchwork of contradictory recommendations.[127] What Finkel did not address was that if the wind was not blowing or the Sun was not shining, then electricity could not be purchased at any price because there is no electricity.

Time is fast running out for Australia to renew its ageing energy infrastructure before major employment-generating industry move offshore and even more households succumb to skyrocketing power bills.

Table 2: The Finkel report[128] projections of electricity costs for 2020.

Generating system	Cost per MWh
Super critical coal	$ 76
Ultra-super critical coal	$ 81
Closed cycle gas	$ 83
Large-scale photo voltaic	$ 91
Wind	$ 92
Open cycle gas	$123
Large scale photo voltaic with storage	$138
Solar thermal with storage	$172

Why were the possibilities of nuclear and hydro not examined? Furthermore, while these are only projections it is blindingly clear that the world's biggest coal exporter should be powering as much of the country on its own coal. However, Finkel kills off the possibility of using the cheapest energy source by suggesting a Clean Energy Target of 600 kg/MWh should be the boundary between clean and dirty energy. Burning black coal produces 800 kg carbon dioxide/MWh and

---
[127] Brett Hogan, "What will they Finkel next?" *Institute of Public Affairs*, 20 June 2017.
[128] p. 201.

brown coal 900 kg carbon dioxide/MWh. This number of 600 kg/MWh assumes that human emissions of carbon dioxide drive global warming and that carbon dioxide is a pollutant. It was deliberately calculated to put cheap coal-fired electricity out of business.

AGL and other renewable generators are playing a dangerous game by constantly advertising that wind power is cheaper than coal-fired power. Taken at their word, there would be no need for subsidies such as renewable energy certificates, mandated targets or mandated fines imposed on retailers for not taking wind power. They might get what they fear the most: Real competition rather than subsidies.

The review should have been guaranteeing long-term energy security at lowest cost for Australian employers and families. As coal has provided and continues to provide the lion's share of Australia's electricity for the past century, the effective use of Australia's huge coal reserves should have been the centre piece of the review's findings. Instead, it recommended an energy future without coal to meet Australia's commitments to the Paris climate agreement. Whoever signed the Paris agreement certainly did not do so on behalf of Australian employers and families.

Finkel can never be accused of under-selling his 42.5% renewable energy target, reboxed and rebadged as a Clean Energy Target. Finkel forecasts power price cuts based on work by Jacobs (formerly SKM). In 2014, SKM and ACIL Allen told us that Australia's wholesale electricity price would fall to around $50/MWh in 2017. It is rising from the current $130 to $150/MWh. It is courageous of Finkel to rely on the economic models of Jacobs, who have form. Finkel tells us that electricity prices would fall under his favoured Clean Energy Target. If you believe that, then I have a perfectly good almost new Sydney Harbour Bridge that I can sell you.

The Finkel report reads like a communist manifesto: Dictating the terms on which commercial enterprises must supply power and whether or not it's profitable to do so.

Finkel estimated the capital costs of building new coal-fired power stations in NSW and Queensland that would create electricity at below $48/MWh, a figure that includes finance and fuel costs. What was not stated is that weak government adds costs because they would not stand up to environmental anarchists Hell bent on stopping employers and families gaining cheap and reliable electricity. We have a lot more pain to suffer in Australia before we have cheap reliable base-load electricity.

The Finkel report tells us what we already knew.

> The past few years has seen the retirement of significant coal-fired capacity from the National Electricity Market while there has been no corresponding reinvestment in new dispatchable energy.

and

> If new dispatchable capacity is not bought forward soon, the reliability of National Electricity Market will be compromised."

This is bureaucratic speak for: We have a huge problem unless we build more coal-fired power stations. Now. Neither coal-fired nor nuclear power stations are being built in Australia. Matters can only get a lot worse before it gets better for Australian electricity consumers. This depends upon both State and Federal governments making decisions on behalf of consumers and not greens and subsidised carpetbaggers.

Finkel's suggestion that we keep coal-fired power stations operating, despite being rendered unprofitable and bleeding cash by heavily subsidised wind and solar power, is as ridiculous as his suggestion that adding more wind and solar power will lead to lower price rises. Irrespective of their age, Australia's coal-fired plants deliver reliable and affordable electricity, whether it's dark or sunny and whatever the weather.

Australia's energy future does not look too good. Everyone loves renewables, until they can't deliver the electricity as and when required and while the costs are modest. Over the next 20 years, some 8,500 MW of cheap reliable coal-fired electricity will be retired in NSW. To replace this with the equivalent solar generation would require around 29,200 MW of nameplate solar power capacity employing 387 million solar panels spread over 71,450 hectares. This is 30% of the area of the ACT.

Energy will not be produced at night so any 24/7 industry such as refrigeration of food needs back up coal-fired electricity. To this must be added the extra capital costs for transmission of electricity from its inland area of generation to the regional and coastal areas of high consumption. There would be no excess electricity in NSW to feed other States or the ACT. The Finkel report did not even consider that there is an identified disaster if coal-fired power stations are not replaced with newer coal-fired generators.

Australia's energy regulator figures reveal almost 60,000 households are on electricity hardship payments and another 151,862 customers are on electricity payment plans. There was a spike in power disconnection after the introduction of the carbon tax, with NSW experiencing a 38% increase between 2011-2012 and in 2014-2015, there were 32,930 electricity disconnections. Victoria's disconnections rose 45% over the same period and peaked at 34,496 in 2013-2014 while Queensland's rose by 25% and peaked at 29,692 in 2014-2015.

Let the States go it alone the way the system was originally established. First dismantle the interconnectors so the States are completely responsible again for providing the power to their own individual State. While there is the NEM in the mix we get the same blame shifting we have with hospitals, schools and roads, and the States will blame the Federal Government and the Federal Government will blame the States.

South Australia's power crises are due to an obsession with renewables. In South Australia, we are living the green dream. South Australia gets 44% more from the GST pot which means that it can afford to reject common sense, build oodles of wind farms, ban GM crops, impose a bank tax and ban the nuclear industry. South Australia now has the winning combination of skyrocketing prices, load shedding, blackouts, subsidised unreliable wind power, business closures and job losses.

Carpetbaggers

Governments created the perfect opportunity through its subsidies and must-take mandates for carpetbaggers. If government legislation removes the Renewable Energy Target, subsidies and must-take mandates, renewable energy investors could hardly complain because their due diligence would have shown that:

(a) human emissions of carbon dioxide do not drive climate change,

(b) populist politics on one issue has a short life, and

(c) subsidies and protective mandates don't go on forever, especially if the mood of the electorate changes.

The mood has changed. We are being robbed blind with high electricity costs.

AGL built a large proportion of South Australia's wind generators as well as Australia's greatest wind farm disaster at Macarthur in Victoria. They have started an advertising campaign about how they want to keep the renewables subsidy gravy train rolling on and used a fuzzy freak as the key character who advised that AGL is "getting out of coal", that they are all about delivering "sustainable" wind and solar power to your door and that it all comes "with no compromise to you." These words mean absolutely nothing. The Annual Report of AGL shows that more than 80% of the profits are from coal-fired electricity[129] and there would be a shareholder revolution if AGL abandoned its core business for ideological reasons. AGL profits increased 14% in 2017 and the profit projection is for an increase of 25% in 2018.

They don't seem to understand that the community is a great supporter of renewables until they become unreliable and expensive, that AGL is a huge generator of electricity from coal, that wind power is unsustainable and that AGL make super profits at the expense of fuel poverty and of unreliable and excessively expensive electricity. Maybe they are feeling that their rorting of the system is coming to an end.

In an attempt to bolster its green credentials, AGL claims it has "plans to get out of coal as smoothly as possible – embracing cleaner, more sustainable sources of energy like solar, wind and hydro." This process will begin about 2020 and finish in 2050. This is despite 90% of AGL's actual energy generation being from coal, mainly Bayswater (NSW), Liddell (NSW) and Loy Yang A (Vic). AGL has financially benefited from the closure of the Northern (SA) and Hazelwood (Vic) power stations.

Although these closures represented 4% of the national market capacity, they constituted a far greater share of baseload supply. At the current year's spot price, AGL will see an increase in revenue at no extra cost of $3 billion. Origin, China Light and Power and the Queensland government will gain $1 billion extra. Generators are estimated to increase their revenues, at no extra

---

[129] AGL Annual Report for 2016, 10 August 2017; the hirsute freak with his synthetic family features throughout the report.

price of production, from a little over $7 billion in 2015 to more than $22 billion in the 2016-2017 financial year. The generators know that they are gaming the market, gouging huge profits from households and businesses with no additional investment and creating hardship for the community. If renewable energy can accept subsidies, then they can also bear super profit tax. If ever there was a good argument for super profit tax, then the wind power generators have shown it.

The business model of the wind industry is simple. Replace cheap and efficient coal-fired electricity designed to run 24/7 for 365 days a year with open circuit gas turbines. Gas prices have gone through the roof in a couple of years and there are political moves to prevent gas being exported and reserving it for domestic use. Never mind about long term contracts which were obtained and then used to guarantee funding of infrastructure. The war on gas is not about gas supply. It is to prop up wind and solar electricity which depend upon open cycle gas turbines that stop the grid collapsing when the wind does not blow and the Sun does not shine. The wind and solar outputs collapse on a daily basis. Open cycle gas turbines can be fired up in a few minutes to produce peak electricity at four to fives times the cost of coal-fired electricity. These turbines are rather like a jet engine and can run on gas, kerosene and diesel.

When wind power collapses, the rort starts. The much-needed electricity can be sold at $2,000/MWh all the way up to the highest price allowed by regulation of $14,000/MWh. By contrast a coal-fired power station can profitably deliver electricity for $50/MWh. The capital outlay for an open cycle gas turbine is relatively low and operators need upwards of $300-$400/MWh before they'll get out of bed. By not having abundant domestic gas, the gouging of consumers is possible by wind and open cycle gas turbine operators.

To make matters worse, governments of both colours in Victoria have imposed bans on drilling for onshore conventional gas. The gas shortage would easily be solved by drilling wells on numerous well known targets in the Otway and Gippsland Basins close to gas processing plants and pipelines. Without the obstacle of gas availability and a reasonable price system, the grid managers are able to extract obscene electricity prices when the wind stops blowing or the Sun stops shining.

This is a tried and proven energy gaming system is not unique. Enron gamed the Californian power market by deliberately shutting off power plants it controlled, waiting for the grid to reach the brink of collapse and then offering power at 1,000 times the going rate. Power companies are interested in profits, not providing cheap electricity to create employment in other businesses or the environment.

South Australia, the land of sea breezes and sunbeams

The governing Labor party in South Australia was warned by power engineers in 2006 that the grid would fail with the introduction of wind energy and again in 2009 that wind power would destabilise the grid.[130] However, the green ideologues in the bureaucracy and government knew better than power engineers. Had reason prevailed, there would have been no electricity crisis in South Australia.

South Australia introduced wind power to lower carbon dioxide emissions, apparently a dangerous harmful "pollutant" which, when generated by humans supposedly will produce dangerous global warming and kill the planet. Or so the story goes. Just how the miniscule emissions from South Australia would do this has never been shown. South Australia's global contribution of carbon dioxide to the atmosphere is just

[130] Sheradyn Holderhead, "Labor warned wind farms would destabilise energy grid in 2009", *The Advertiser*, 20 March 2017.

one molecule in 84 million molecules of other gases in the atmosphere. Nothing South Australia does will change the planet's atmosphere. All it will do is bankrupt South Australia.

South Australia believes it can run an industrial economy on sea breezes and sunshine yet blackouts and load shedding, business closures and unemployment show the government made a wrong and costly decision. Maybe the market should dictate rather than cynical inept politicians and their green bureaucrats.

South Australia is pushing for 50% renewable electricity. Not far behind on this suicidal aspiration is Queensland (50% by 2050), Victoria (40% by 2025) and the Federal government (23% by 2020). How this will actually be achieved has never been explained.

There has been no due diligence on these renewable energy targets and wind turbines are regarded by governments as perpetual motion machines with no costs.

The South Australian government allowed efficient cheap coal-fired powered stations to close on a matter of ideology. Alinta's Northern power station tried to negotiate a deal with the South Australian government for a $25 million subsidy. The small subsidy to Alinta would have solved South Australia's electricity problem and would have produced cheap reliable base load power. Some 70% of it would have been spent on the 250 km railway from the Leigh Creek coal mine to the 520 MW Northern power station at Port Augusta. The rest would be spent on the Leigh Creek mine and Leigh Creek town, airport and water supply. Royalty relief was also sought.[131] This offer was kept secret by the South Australian government while they promoted the spending of $550 million on batteries, gas-fired

---

[131] Michael Owen and Meredith Booth, "Jay Weatherill rejected $25 million deal to save Northern power station", *The Australian*, 30 March 2017.

generators and diesel generators to try to fix the problem they created.

The subsidised wind power in South Australia can't keep the lights on, an election is coming and so the government has become creative in order to stay in power. It will use diesel generators. Each year, 30 million litres of diesel will be burned to yield a total of 79,000 tonnes of additional carbon dioxide each year. Because of wind turbines, South Australia will now result in the emission of more carbon dioxide to the atmosphere at the bargain basement price of $100 million for the fleet of diesel generators.[132]

The South Australian government will immediately throw $550 million of taxpayers' money at diesel generation; a gas-fired peak load generating plant, either open cycle gas turbines or perhaps gas fuelled piston-engine generators; and a $150 million 100 MW battery that will power South Australia for a whole four minutes when systems again fail. No one knows the fire risk of such batteries, how many times they can be recharged and their life.

South Australia is not alone in having to emit even more carbon dioxide to make up for the shortfalls in renewables output. Tasmania spent $64 million on leasing, site establishment and operational costs for 220 MW of diesel generation when a combination of Basslink carrying electricity from the mainland failed and a drought depleted the hydroelectricity generation.

The *Wall Street Journal* called the electricity disaster in South Australia as the energy shortage that no one saw coming.[133] Many of us saw it coming but *The Wall Street Journal* was happy to believe its own warmist green ideology. To avoid a massive State-wide blackout again, during the January 2017 heat wave,

---

[132] Michael Owen, "Interim power generation to cost SA taxpayers $100m a year", *The Australian*, 30 June 2017.
[133] *Wall Street Journal*, 16 July 2017.

regulators shut off electricity to 90,000 homes. It had happened before.

In the previous free market for electricity, the system was run by engineers. With the tightening hand of the state and its bureaucrats, government intervention had destroyed the electricity system in Australia and elsewhere. Until 2006, Australia has a good system. The wholesale cost of electricity was in the order of $25 to $40/MWh. It then cost three cents an hour to burn ten 100-watt lightbulbs.

Follow the money

Renewable Energy Certificates are money for jam. Queensland and NSW were powered by black coal, Victoria and South Australia were powered by brown coal, Tasmania by hydroelectricity, Western Australia by black coal and gas. The Renewable Energy Target (RET) was introduced by the Howard government in 2001. The Federal parliament legislated to force electricity retailers to buy renewable electricity. Under the RET wind and solar generators were paid an exorbitant amount for their output and gained Renewable Energy Certificates (REC) of about $80/tonne of carbon dioxide "saved". The problem is that human emissions of carbon dioxide don't drive climate change so the whole exercise is useless tokenism.

Whatever the price of electricity, the sure, safe and large income was from REC. No wonder the number of wind turbines has increased. The subsidies (ultimately paid by the consumer), the RET and REC rorts attracted carpetbaggers like AGL who, despite their propaganda, have no concern for the environment. Just visit one of their subsidised wind industrial complexes. They are quite rightfully concerned about the bottom line. The coal fired power stations that produce 75% of the country's electricity stopped investing, other than essential maintenance, and started to close because they could not compete in the artificial market.

South Australia is now a net importer of energy. In Victoria in summer, consumers will be dependent upon Tasmania's hydroelectricity if it rains in Tasmania and if the Basslink cable does not break again. This is the end result of green politics which is trying to ban new gas fields, new pipelines, new coal mines, new coal-fired power stations, new hydroelectricity dams and nuclear power. The green mantra is to increase renewables to 40%. South Australia has shown that this is not possible and is using yet-to-be dismantled gas generators and diesel generators. All major political parties have fallen for this tosh. It can only end in tears.

How to kill off efficient industry

Australia is blessed with abundant coal, gas and uranium unequalled on a *per capita* basis in the world. The use of wind and solar to power a Western modern industrial society is one of the truly bad ideas over the last 100 years. However, Australia is now paying the highest prices for electricity amongst its trading competitors. Why? In South Australia, consumers are lucky to get electricity at any price and, when they do, it is almost double the price paid by consumers in Victoria.

When the wind blows for a few hours at a stretch in South Australia, its last remaining base load power station[134] is unable to dispatch power to the grid and, accordingly, receives no revenue. However, it continues to incur fuel, maintenance, wage and other costs. It's that combination that destroyed the profitability of South Australia's last coal-fired power plant at Port Augusta and helped destroy the viability of Victoria's Hazelwood plant. Subsidies paid in the form of Renewable Energy Certificates enable wind power plants to flood the market, either giving away power or even paying the grid

---

[134] AGL's Torrens Island, 1,280 MW gas-steam plant.

manager to take it, leading to the closing of reliable base load power stations.

South Australia established an industrial base under the leadership of Sir Thomas Playford. The State generated its own electricity which was reliable and cheap and this attracted businesses to South Australia. Since the advent of renewable energy, South Australia has now the highest household electricity prices in the world[135] and relies on electricity from Victoria to keep the lights on 24/7. Businesses are closing or moving.

For example, a family-owned 38-year old specialist plastics recycling company Plastic Granulating Services in South Australia was forced to close, with a loss of 35 jobs, after its electricity bills soared from $80,000 per month to $180,000 per month over the last 18 months.[136] The closure led to waste companies closing with further job losses and now there is no plastic recycling plant to service six councils in South Australia.[137]

This is green politics at its best. Create unreliable ideological energy based on intermittent sea breezes, close down tried and proven cheap electricity generating systems, create expensive electricity that cripples the poor and bankrupts industry, create unemployment and destroy recycling, something dear to the heart of environmentalists. You couldn't make this up if you tried.

Those that vote for the Greens and think that they are doing something for the environment should think again. They should think that, because there has been no warming for 20

---

[135] Michael Owen and Meredith Booth, "State leads the world – for power prices", *The Australian*, 28 June 2017.
[136] Jade Gailberger, "Recycling firm to shut, 35 jobs lost, as SA government ignores plea for help over soaring power bills", *The Advertiser*, 27 June 2017.
[137] http://joannenova.com.au/2017/06/in-sa-recycling-businesses-going-broke-due-to-electricity-cost

years, then maybe renewables are not necessary. They should be aware that it has never been shown that human emissions of carbon dioxide drive global warming hence any emissions reduction scheme is flawed.

If the total renewable energy is to increase to 50% as Bill Shorten, the Federal Opposition Leader in Australia wants, then wind, solar, hydro and biomass will have to increase from 7% to 50%. Mr Shorten stated:

> If we do not get serious about tackling climate change, if we don't get serious about investing in renewables, then we cannot say we are serious about economic reform.

How this will be done was not stated and he is certainly serious about economic reform because such actions will drive the economy backwards. Neither Shorten nor anyone else has shown that human emissions drive global warming hence the need for renewables is not necessary. The renewable energy myth depends on the untruth that renewable energy reduces carbon dioxide emissions and this emitted carbon dioxide changes climate. Furthermore, the above statement demonstrates that economic reform is a joke. Investing in renewable energy actually means subsidising even more inefficient unreliable renewable energy. If Mr Shorten can't show me that human emissions of carbon dioxide drive global warming, then he should pay my excessive electricity bills.

When Paul Kelly (*The Australian*) asked:

> How, as shadow treasurer, could you allow the leader to commit the party to the 50% renewable energy target without any framework of analysis, without work done on the impact of the scheme, the costs, any analysis whatsoever?

The answer from the Labor shadow Federal Treasurer Chris Bowen was: "Well, because I believe in renewable energy."

This deflective moronic throw-away line is an admission that the whole global warming gravy train is a matter of religious faith. God help the taxpaying workers of Australia if Labor become a government and again enact what is considered popular policy made on the run. Sacrificing jobs and increasing costs to families for no climate gain is not good policy. God help those of us who are struggling to pay our power bills. Why didn't Kelly nail Bowen? Is it because Kelly is a warmist or doesn't he understand that people are bleeding to death with high electricity bills based on a false premise?

For a senior Australian politician to argue for an increase in renewables ignores some fundamentals. Australia's emissions amount to almost 1.5% of global emissions yet the Australian Labor Party wants renewable energy at twice the cost than that paid for by the world's largest emitters. The renewable energy mandate guarantees that the costliest capacity remains on stream whereas the cheapest capacity is prematurely scrapped.

By contrast, coal has accounted for 90% of the 4,500 MW that has been shutdown or whose shutdown has been announced. With the renewable target nearly doubled as the Labor Party wants, the costs of electricity will rise by $86 billion which will add another $600 *per annum* to the average family on top of the already high electricity bill of $1,600 *per annum*. This will shut down industry and drive even more industry jobs offshore.

The cost of building wind and solar plants is horrendous, they damage the environment and they increase carbon dioxide emissions. Will increased taxation and electricity costs be used to pay for such white elephants or will the debt be increased to leave a problem for future generations? It's all very well to claim that one is worried about the future environment that our grandchildren may inherit. What about the debt that they inherit from overspending for our glorious and wonderful ideological inefficient white elephants such as unwarranted desalination plants, wind turbines and solar energy facilities? With 50% renewables in the mix, the end result will be to raise the price of electricity, to make electricity supply unreliable and to drive away employment-generating industry.

This hits the poor the hardest and some struggling employment-generating business will just close and move elsewhere. The policies of the political party that claims it is the workers' party will just increase unemployment. It's clear that the Labor Federal Opposition is looking at the populist vote in a forthcoming election, has not thought through the implications of Australia having 50% of its electricity from renewables and just does not care about the social, employment, personal and economic havoc their policy will create. Whatever it takes.

How would the green left environmental activists keep the lights on and keep people employed if there were a great reduction in the use of coal, oil and gas? Before subsidies can be given, there has to be a vibrant economy. No alternative energy company or investor is in the business to save the world. They are in the business to make money, an anathema to the green left environmental activists.

The myth of green dreamtime

There is never a good time for bad public policy.

For half a century, the green left environmental gurus have been telling us that renewable energy would soon be as cheap and as reliable as coal, gas, oil and nuclear energy. It isn't. The turning point and great new discoveries are just around the corner and what is required is more public funding and tax breaks before there is abundant, cheap, clean, reliable energy that will just materialise out of thin air with a sprinkle of pixie dust.

We have waited for 50 years, the dreamtime is over and there are far too many examples showing that renewable energy is not environmentally friendly. Huge areas of wild lands and wildlife habitats are now industrial wastelands of wind and solar energy facilities, food-growing areas have been replaced with biofuel crops and there has been genocide of birds and bats. Apart from the high environmental costs, the economic costs have forced a rethink. In most jurisdictions, it has been a case of legislate in haste (to win votes) and repent in leisure (to win votes).

Electricity must be reliable, secure and affordable. It isn't. If you want to wreck a country, plug into wind and solar energy.

3
Transport

DIESEL VEHICLES

Diesel or die

Diesel trucks keep Australia alive. Forget singing *Kumbaya*, basket weaving, organic food and bicycles. Noisy, smelly, large trucks keep the cities fed. Without diesel trucks, we are dead.

Bicycles, electric cars and shanks pony are the preserve of city-based greens who don't get out into the real world and have no understanding of how basic commodities such as food, water and electricity are produced and transported.

All mines use diesel machinery. Mines in isolated areas use diesel generator sets. Many towns rely on diesel engines to generate electricity. Numerous businesses, hospitals and institutions have diesel gen set emergency backup power. No hospital can have a patient in the operating theatre without a backup electricity-generating system, especially in South Australia.

Crop production uses diesel station vehicles, tractors, trucks, pumps and generators. Without diesel fuel, we cannot produce food. Australia is self-sufficient and produces food for about 80 million people in the world. Almost all food, building materials, general freight and bulk commodities travel by road in diesel trucks. The rest is carried by diesel trains and bunker fuel-fired ships. Food distribution and storage is on a just-in-time basis. Most Western countries have low stockpiles of diesel fuel used for food transport from rural to city areas and we are always two weeks away from food shortages and civil unrest.

In Australia, transport of food to cities is by diesel vehicles. Most of Australia's small inefficient old crude oil refineries have been closed and refined liquid hydrocarbons now travel to Australia by ship. These are stored and distributed all over the country. If a neighbouring country stopped refined fuels from entering Australia from refineries in Singapore as retaliation because Australia denied live beef exports, there is only two weeks of diesel stockpiled.

Most Western countries do not have national food or energy disaster plans. Quick decisions in the absence of food and energy disaster plans will need to be made in a time of crisis. Such quick draconian decisions have only ever been made in times of war. Western politicians most commonly are lawyers and decisions take a long time. China has led the way with long-term energy, resources and food security strategies. China is run by engineers and scientists.

I'm sure that green activists are aware that there are only a couple of weeks between a meal and hunger.

Without fossil fuel, we are dead.

The UK disaster

Our obsession with cutting carbon dioxide emissions has had terrible consequences. In the UK, 11 million people own diesel cars. A decade ago, the UK government advised that diesel cars would help the country lower its carbon dioxide emissions.

The push to diesel cars was driven hard by Professor Sir David King, Prime Minister Blair's personal scientific advisor.[138]

---

[138] Victoria Allen, "Hoodwinked by a green zealot: The scientist behind the dash for diesel called CO_2 "worse than terror", *Daily Mail*, 5 April 2017.

King was quoted as claiming that carbon dioxide was worse than terrorism. King has now stated that he was wrong.

Now the UK government is telling diesel car owners that their cars release particulates and nitrous oxides and they will be penalised.[139] Car manufacturers are making a fortune out of the great green swindle. Those who experienced the winter pea-souper fogs of England[140] in the 1950s and 1960s question whether children are currently being exposed to illegal levels of toxic air, as London's mayor Sadiq Khan claimed when recently declaring a public health emergency. He claims that 9,000 Londoners are dying prematurely. The Great Smog of December 1952 killed 4,000 people. These were actual deaths. Can London's cleaner air now be killing 10 to 15 times as many people? Well, no.

Khan is quoting estimates derived from modelling that is unrelated to real deaths. Oh dear, here we are misled again by models and not real validated data. In the 1950s, soot and sulphur dioxide were inhaled from millions of households and businesses burning dirty sulphurous coal. Today, people are inhaling particulates and nitrous oxides. Why is this? Because the Labour government's green incentives stimulated the substitution of petrol cars by diesel cars.[141] However, the good news is that diesel cars have become far more efficient and particulates and nitrogen oxides emissions have been falling steadily since 1970.

The home of diesel

Germany has tried to increase the proportion of diesel cars on the road in an attempt to reduce carbon dioxide emissions.

---

[139] *The Times*, 26 April 2017.
[140] Michael Fitzpatrick, "Is our air worse than in the Fifties?", *Daily Telegraph*, 17 April 2017.
[141] http://www.transport-watch.co.uk/topic-34-great-dirty-diesel-scare-0

The modern German diesel car is as lively as a petrol car, and is far more economical and cheaper to maintain. All of a sudden, Germany decided that diesel cars were too polluting for the environment and is now in a blind rush to electric cars.

Older diesel cars are now no longer permitted in cities and towns. Policy makers who wrote the new legislation thought that by killing off cars younger than 30 years in the cities and towns they would get rid of all diesel cars because winter road salting and the constant purchasing of new cars meant that there would be no ancient diesel cars on the roads. No one drives 30-year-old cars in Germany, the home of modern quality cars. Or so they thought.

The folly of social engineering is demonstrated by my mate Max in Germany. He has two diesel Mercedes cars. One turns 30 in 2019. As an engineer, he has meticulously maintained these vehicles in pristine condition himself. From 2019, he will legally be able to drive his old polluting car in towns and cities where a modern less polluting Mercedes less than 30 years old will be banned. He is waiting to be pulled over by the Bundespolizei. What will they do?

ELECTRIC CARS

Is the solution electric cars? Solution to what? Electric cars are being marketed as the automotive equivalent of the iPhone. The first electric car was invented in 1828[142] and the first battery-powered car was that of Robert Davidson (Aberdeen, Scotland) in 1837. If electric cars were to conquer the world, they would have done it already.

Enormous hope rests on electric cars as the solution by the motor industry to carbon dioxide emissions. Bearing in mind that it has not yet been shown that human emissions of carbon

---

[142] Invented by the Hungarian Anyos Jedlik.

dioxide drive global warming, one wonders what all the fuss is about. What an opportunity for the vehicle industry to persuade every consumer that they need a new car. Urgently. Now.

Marketing for the electric car market is intense. This does not mean that they are a reliable useful product. City-based greens are heavily promoting electric cars. They may use an electric car in a flat city for short distances and feel better for it but they are totally useless in hills and out of town. They have no idea that those in rural areas need a vehicle that is reliable, can travel great distances and can be filled with fuel quickly in many places.

Electric cars did not substitute for petrol and diesel vehicles in the 19th and 20th centuries for the same reasons as today. They are too heavy and expensive, they have a very small range, need a long charge time and the battery wears out very quickly. However, compared to a petrol vehicle, they are quiet and accelerate like a bat out of Hell. If electric cars catch on, pedestrian deaths will increase – especially those crossing the road while playing with a mobile phone or listening to music.

A light bulb moment

Tesla is planning to build a giga-factory to make lithium ion batteries, but it may find itself with obsolete technology if suddenly a new technology appears. It has happened before and will happen again.

For example, governments mandated to remove tungsten filament incandescent lights in favour of compact fluorescent lights to help tackle "climate change" and lower household electricity bills. The UK government sent "free" compact fluorescent lights to every residence at a cost of £3 billion.

These new globes were slow to warm up, flickered, created eye strain, had no ambience, had a shorter life than expected and created dangerous disposal problems. These globes were less popular with consumers than producers who had tooled up to produce them.

As soon as the price of LED lights fell, consumers abandoned the compact fluorescent globes for the cheap, more efficient, safer and more compact LED lights. The government backed the wrong technology and this cost the UK taxpayer that £3 billion. If governments mandated the wrong battery technology for electric cars, the costs to the taxpayer could be orders of magnitude higher. If a government backs lithium ion battery technology, do they really know that emerging technology such as a vanadium battery will not displace the lithium ion battery? Governments should let the free market do the heavy lifting.

The lithium ion battery may be marketed as the greatest thing since sliced bread. However, all is not that rosy. Ask the owner of a Samsung Galaxy 7 telephone. Why do airlines make repeated safety announcements about lithium ion battery telephones? The batteries of electric cars are certainly not environmentally friendly and are an explosion and fire risk. Batteries, although touted as lithium batteries, are heavy with most of the weight as copper. This makes electric cars inefficient.

We rely on science, technology and engineering for our existence yet few people know anything about these fields, especially politicians. Society is at great risk by allowing technological decisions to be made as a result of populist community pressure on politicians because most politicians are self-interested lawyers who are eminently inexperienced and unemployable outside the legislative chambers and only want contact with the electorate when an election looms.

The best example of such risk is electricity. In Australia, the electricity system used to be controlled by crusty experienced power engineers who understood how systems operate. They have been replaced by green activist ideologues, politicians and their sycophants and carpetbaggers who have no idea how electricity is generated, transmitted and used. The end result was inevitable and you now see it in your massive household electricity bill.

Electric car emissions, costs and battery life

Several tonnes of carbon dioxide are released even before the battery leaves the factory. Over the lifetime of a Tesla Model 3 car, 27.1 tonnes of carbon dioxide are emitted whereas a petrol-powered BMW 320i car releases 22.8 tonnes.[143]

The UK government is at it again. The May 2017 Conservative manifesto stated: "We want almost every car and van to be zero emission by 2050, and will invest £600 million by 2020 to help achieve it."

Sounds like a huge amount of money is going to be wasted again, as was done with light bulbs. Ask any London black cab driver about this wonderful dream if you want a forthright answer. For more than 150 years, scientists and engineers have been trying to make a better battery for electric cars. There is nothing yet on the horizon for battery cars in 2020.

The Nissan Leaf is the world's best-selling electric car. In the UK it sells for £16,680. Without batteries. If you want batteries they can be leased for £80-90 a month. If you actually want to buy your own batteries, you'll have to fork out another £5,000. The total cost is £26,180 compared to the petrol Nissan Micra (£11,995). The Leaf will do 200 km before it needs a recharge which will take at least four hours from special recharging sockets or 12 hours from an ordinary electric socket at home.[144]

---

[143] Johan Kristensson, "New study: Large CO_2 emissions from batteries of electric cars", *New Technology*, 29 May 2017.
[144] Ross Clark, "Road to Nowhere: Electric cars won't get us very far", *The Spectator*, 5 August 2017.

If I drive from Broken Hill to Sydney in my diesel car, it takes 14 hours with no fuel stop. If I drove in a Nissan Leaf it would take at least 84 hours with fuel stops. The distances Broken Hill-Wilcannia and Wilcannia-Cobar are so great that the Nissan Leaf would run out of power somewhere out in the donga. It is a totally useless car for Australian conditions with our long distances, sparse population and the lack of charging facilities.

A 2017 road test[145] of the Leaf on real conditions with hills, corners, braking and acceleration showed that the fuel range was really 110-120 km. *Top Gear*[146] tested the same car with a claimed range of 200 km. Under test conditions, it was 96 km. Nissan has contracted to spend $US 5.6 billion developing the Leaf which will be constructed in Sunderland, an area in the UK crying out for jobs. Maybe UK car buyers will be incentivised to buy the home-made Leaf at the expense of other brands. Who knows what deals have been done?

Testing of other brands was also disappointing. The Tesla 3 in the UK showed that in warm weather the range on flat ground was 320 km. There was no air conditioning. In winter, the range was about 160 km and there was no heating.[147] And if you want to drive a distance in undulating country in winter, forget it. You'll be able to get far enough away from home that you can't walk back when the car dies.

A report on a test drive of a Chevrolet Volt, a petrol-electric hybrid, showed:

> For four days in a row, the fully charged battery lasted only 25 miles (40 km) before the Volt switched to the reserve gasoline engine. Eric calculated the car got 30 mpg (7.8 litres/100 km) including the 25 miles (40 ki-

---

[145] https://www.whatcar.com/nissan/leaf/.../review/ʾ
[146] https://www.topgear.com/car-reviews/nissan/leaf
[147] https://www.driving.co.uk/...reviews/first-drive-review-2018-tesla-model-3/ʾ

lometres) it ran on the battery. So, the range including the 9-gallon gas tank (34 litres) and the 16 kwh battery is approximately 270 miles (434 km).

In a Volt, it would take nearly five hours to drive the 434 km at 100 km/hr and then add 10 hours to charge the battery and you have a total trip time of 15 hours. In a typical road trip your average speed (including charging time) would be 32 km/hr.

According to General Motors, the Volt battery holds 16 kWh of electricity. It takes a full 10 hours to charge a drained battery. Depending upon the cost of electricity, the Volt costs at least seven times as much per kilometre as a petrol car and the initial purchase price is twice that of a typical sedan petrol car.

General Motors produced the EV1 which had a range of 160 to 225 km. It was adored by celebrities and wealthy environmentalists. GM did not sell the cars. They leased them. This car was so successful that GM recalled all of them and turned them into scrap.

A few little problems are hiding in the wings. Charging a Tesla requires a 75 amp service. The average house in the UK has a 100 amp service. In an average street of 25 houses, only three houses can be concurrently charging their Tesla cars otherwise the system collapses. If we are all to use electric cars, we need a new electricity system. The National Grid pointed out that as domestic electric circuits are currently configured, putting the kettle on while charging an electric car would blow fuses due to circuit overload.

It is well known that lithium batteries explode when the electrolyte violently reacts with water. Mistreatment during charging or discharging can cause outgassing. In the UK, if electric cars become ubiquitous, there will be at least another 18 GW of electricity needed on top of the existing 60 GW. The UK now has 3.9 GW of backup electricity provided by diesel and gas hence the arguments about reducing carbon dioxide emissions are meaningless.

A cordless electric motor such as in a car needs a battery. Batteries are heavy, bulky and contain large amounts of metals. They are slow to charge and can explode. Recharging a battery creates emissions from coal-fired power stations that provide the electricity. The carbon dioxide emitted from building an electric car is far more than from the building and use of a conventional petrol or diesel car. Manufacture of just an electric car battery creates more carbon dioxide than a petrol car's emissions over eight years.

In the UK, cars travel 250 billion kilometres each year. If the smallest electric car[148] was used for these 250 billion kilometres, then that mileage would add an extra 16% demand on the existing electricity grid.[149] Where will this extra electricity come from? If it was from wind, then there would need to be an extra 10,000 onshore and 5,000 offshore wind turbines requiring a subsidy of at least £2 billion *per annum*. What is not considered with electric vehicles is that 40% of road transport fuel is used by heavy vehicles.

Once a vehicle has done a few miles and has been charged many times, the batteries are far less effective. After 84,000 km on the clock, the range of the Leaf was reduced to 72 km and after 145,000 km, the range was 56 km. This is less than a person could travel by horse 200 years ago. Even for a city based person, an older electric car is useless.

On 5 July 2017, Volvo announced it would become the first major car manufacturer to go all electric saying that every new car in its range would have an electric power system available from 2019. The company said the announcement marked "the

---
[148] Nissan Leaf.

[149] Matt Ridley, "How the electric car revolution could backfire", *The Times*, 10 July 2017.

historic end" of cars solely powered by petrol or diesel and "places electrification at the core of its future business". Volvo will launch five fully electric cars across its range between 2019 and 2021. Other leading carmakers are heading in the same direction. Toyota is developing next-generation models with low or zero carbon dioxide emissions. Volkswagen aims to sell one million pure electric cars a year by 2025 and Nissan currently offers two 100 per cent electric vehicles, including a big selling vehicle in Britain. Making a car produces large carbon dioxide emissions. This is not mentioned in the sales hype.

World sales of the Tesla in the second quarter of 2017 were 22,000. Despite a high share price, Tesla has never made a profit. Sales of Tesla cars in Hong Kong in April 2017 were zero after authorities removed a tax break and vehicle prices rose from $US 75,000 to $US 130,000. When Denmark's electric vehicle incentive program expired last year, new car registrations for electric vehicles of all brands fell 70 per cent. Tesla's car sales in Denmark dropped by 94 per cent.

The California State Assembly passed a $US 3 billion subsidy program for electric vehicles, dwarfing the existing program. This appears to be a Tesla bailout. Tesla will soon hit the limit of the Federal tax rebates, which are good for the first 200,000 electric cars sold in the US per year beginning in December 2009. In the second quarter after the manufacturer hits the limit, the subsidy gets cut in half, from $US 7,500 to $US 3,750. Two quarters later, it gets cut to $US 1,875. Another two quarters later, it goes to zero. Losing a $US 7,500 subsidy on a $US 35,000 car is a huge deal. The Tesla Model 3 would be hard to sell without the Federal subsidy of $US 7,500. This new bill would push Californian taxpayers into filling the void. It would be a godsend for Tesla.[150]

---

[150] "Green Cronyism Gone Wild: It Looks Like The State Of California Is Bailing Out Tesla", *Business Insider*, 17 July 2017.

Tesla lost $US 773 million in 2016, an improvement compared with the $US 888.6 million it lost in 2015. General Motors, in stark contrast, made a $US 9.4 billion profit in 2016. GM sold 10 million vehicles worldwide in 2016, more than 130 times Tesla's 76,230 vehicles for 2016.

It is a high-risk business that is reliant on government largesse. If the state pushes one type of technology too hard, it risks shutting down the creation of a far more efficient alternative. The introduction of hot-house growth risks entrenching immature technology at the expense of newer and better technology.

This is exactly what we saw with light bulbs in the UK and it is happening again in the UK where the government is being pressurised to ban petrol and diesel engines and replace them with electric cars. The elephant in the room is never mentioned. Diesel trucks are absolutely essential for the transport of food and general freight. If diesel trucks are also banned, then there is a huge problem for city people to eat. Diesel cars consume very little fuel, it is the trucks carrying goods into and around cities that consume most of the diesel and emit most of the gases and particles.

If the problems of range and battery cost can be solved by some yet undiscovered technology, then banning diesel and petrol cars may not be such a disastrous policy. Don't wait up.

In the meantime, long car trips may become a pleasure of the past.

Solar cars, trams, trains and planes

Solar cars

Solar cars provide a bit of a giggle. Races across Australia with solar cars, one driver, no luggage and a fleet of fossil fuel burning back-up vehicles are a regular spectacle. Trips take longer, are more expensive and certainly more uncomfortable than in a conventional vehicle. The aerodynamics, friction and electronics of solar vehicles decide winners and losers, not the solar panels. Silicon solar cells utilise the 1,130 nm wavelength to produce 1.1 eV and no amount of breast beating about solar cells can change the laws of physics.

Solar energy is rightly used in the outback for bore pumps, telephones, railway signals and electric fences. No one in their right mind would use a solar vehicle in rural Australia, unless one wishes to be left high-and-dry in an isolated area after batteries are exhausted or it's dark.

Solar trams

Al Gore came to Melbourne in July 2017 to promote his new film *An Inconvenient Sequel*. He was photographed with fawning fools trying to show their environmental credentials. Why anybody would want to be photographed with one of the biggest environmental scamsters in the world is beyond me. After riding on a Melbourne tram, the Victorian Minister for Energy, Environment and Climate Change Lily D'Ambrosio had a rush of blood and announced that Melbourne will have trams that run on solar energy. What did Gore put in her drink? She probably then got taken home in a large chauffeured car. Does she not think before flapping her lips or did she consult some really very clever people?

How will people get to and from work on a cloudy day? What happens if people want to go home after dinner in the city? Imagine 100,000 people at the Melbourne Cricket Ground at a night football game. Will the government provide magic carpets so they all can get home or shall they be provided solar almanacs rather than tram timetables? And if the trams are to have batteries, will the tram system cope with the extra weight, the inordinately long charging times and the fire risk. The Melbourne tram system uses 35 MW of power. Where will such a huge amount of solar power come from? Even South Australia's proposed giant battery could only run the Melbourne tram system for about three hours.

Solar train

In Byron Bay (NSW), a plan for a diesel train was hijacked by green councillors. Local government attracts some pretty low calibre characters, especially in coastal and hippie areas where there is a long history of body and brain abuse. The council now wants a solar train with diesel backup. Despite Byron Bay's latitude and strange people, I can see the solar train being a diesel train with wind-resisting panels on the roof.

There are serious questions about the quality of our politicians at local, State and Federal levels.

Solar plane

Solar Impulse 2 flew around the world. It was a solar-powered propellor plane that managed to circumnavigate the world (42,000 km) without fossil fuel. The aim was to promote clean technology. There was no mention of the monstrous amount of embedded energy in the solar panels, fuselage and four engines (which needed fossil fuel lubricants) and the huge amount of energy used with backup and ground facilities. The first test flights with Solar Impulse 1 were in 2009.

On 9 March 2015, Solar Impulse 2 started its circumnavigation of the planet with two pilots, no passengers and no freight. I hope the pilots were good long-distance swimmers. It took 16 months to circumnavigate the planet, the plane had to wait at some places for weeks for favourable winds before flying long sectors. It had to wait months for repairs to overheated batteries. To circumnavigate the world took 250 times longer than by a conventional jet aircraft. I'm sure that Boeing, Airbus, Lockheed, Bombardier, Embraer and other aircraft manufacturers are quivering in their flying boots as they can see their businesses going broke.

The small Australian superannuation company Future Super is run by Simon Sheikh. He previously worked for GetUp!, a socialist organisation funded by communists, foreign foundations and unions that aims to put Australians out of work. Future Super had invested in the company producing solar panels for the plane and proudly advertised:

> As a member of Future Super, your super could be helping to power projects like Solar Impulse, the world's first zero-fuel solar plane to fly around the world.

I think I'd prefer my super to give me a good financial return as required by law rather than chase ideological dreams.

4
CLIMATE CHANGE

The take home messages in this chapter are that:

(a) no one has ever shown that the human emissions of carbon dioxide drive climate change,

(b) the scientific process has been corrupted,

(c) peer review is controlled by a small cabal of climate catastrophists so that contrary views don't get published,

(d) climate models have failed,

(e) when contrary scientific interpretations are aired, green activists resort to violence, and

(f) surveys on a consensus opinion of "scientists" have been cooked.

THE KEY QUESTION

Show me that human emissions of carbon dioxide drive global warming?

No climate "scientist" has ever been able to answer this question. The whole scare story of human-emissions driving global warming is based on this false assumption. As a result, you are paying through the nose for unreliable electricity.

The whole basis for human-induced global warming can be found in just one chapter (Chapter 9) of the IPCC Report AR4. This chapter states that we dreadful humans are responsible for global warming and a terrible fate awaits us. How do we know this? From a computer! But wait, it gets better. The computer model, code and data for these frightening predictions are secret.

Only one media person has chased down the key question with a lead reviewer of the IPCC Report AR4 Chapter 9. It was Alan Jones (Radio 2GB, Sydney, 25 May 2011) in an interview with climate activist David Karoly.

Jones: Is there any empirical evidence proving human production of carbon dioxide – as distinct from nature's production – caused global warming? Is there? In these reports? Yes or no?

Karoly: Yes.

Jones: Now where would I find that in Chapter 9, that's your chapter.

Karoly: Sure. You would find that evidence in the peer-reviewed scientific studies and in the data ...

Jones: But where in Chapter 9?

Karoly: So...

Jones: Where in Chapter 9? Where can I open Chapter 9, because I looked at it, where if I open Chapter 9 is that evidence? Where is it?

Karoly: It's ... I can't tell you the page number because I don't ...

Jones: No, no. It's not there. It's not there.

Karoly: What. No, Alan.

Jones: It's not there. You, the Chapter review editor. It's not there. That's why you can't tell me the page number. The evidence is not there.

Karoly: That's not true, Alan.

Jones: Well I've got scientists on stand-by who are going to listen to all of this so your reputation's on the line when you say that. I'm telling you, Chapter 9 is your Chapter. You in fact were the Chapter's review editor and you can't tell me where the evidence is.

Karoly: Yeah, I can. Would you like me to tell you where the evidence is? The evidence is in the spatial patterns and the time variations of temperature changes in the observations...

Jones: Whoa, whoa, whoa. Chapter 9, Chapter 9, David, is the Chapter. It was originally Chapter 12 in the 2001 report. In the 2007 report you were the review editor of this chapter on the direction ... on the detection of climate change. It's now called "Understanding and attributing climate change". Now, to understand climate change you need to know what evidence there was for all of this. In Chapter 9 it's not there.

Karoly: No Alan, it is there. So would you like me to tell you which figure in particular in Chapter 9 shows that evidence? It looks at the pattern of climate variation over the last 50 and the last 100 years and what it does is it makes an evaluation or assessment. It talks about how climate has changed compared with what we'd expect from greenhouse gas variations, it also looks at other factors. Factors like changes in sunlight from the Sun, changes in the effect of volcanoes, natural variations like El Niños, natural variations like the Pacific Decadal Oscillation and what it shows, what it clearly shows, is that the patterns of change are outside natural variability, aren't due to changes in sunlight from the Sun and we can see that sunlight from the Sun would cause more warming in the daytime when the Sun's really important but we've actually observed more warming at night. We've seen changes in the temperatures in the lower atmosphere and the upper atmosphere which clearly show that the changes are due to the increases in greenhouse gases and aren't due to natural variability and aren't due to other factors. And we've ...

Jones: Chapter 9, Chapter 9 doesn't contain any of that detail. Can I go on?

Karoly: Yes, it does.

Karoly gives the game away. There is absolutely no evidence in the key IPCC chapter to show that human emissions of carbon dioxide drive global warming. You heard it from the review editor of the IPCC's Chapter 9.

We've been conned. It has cost Australia billions and every day you pay exorbitant electricity prices because of climate activism.

SCIENTIFIC METHOD

Climate "science" is not science, because it ignores the scientific method. This is why many genuine scientists and non-scientists reject the theory of human-induced global warming.

Science is married to evidence and bathes in modest uncertainty. The nature of science is scepticism and science encourages argument and dissent. Scientific evidence is derived from repeatable and reproducible observation, measurement and experiment.

Evidence from my field (geology) is interdisciplinary, terrestrial and extra-terrestrial and shows the complex and fascinating intertwining of evolving natural processes on a dynamic planet. Some geological observations and measurements are validated with experiments. Scientists engage in healthy argument about the veracity of evidence.

In science, there is debate about whether the process of evidence collection was valid, what the errors were, what was the order of accuracy, what assumptions were made and whether such assumptions were valid. When I look at a scientific work, I like to know who collected the data, when it was collected, what equipment was used and where it was collected.

Much scientific data is collected by assistants and students and then given to the scientist to analyse. What happens if the primary data is incorrect? Would you trust an undergraduate to provide you with high-precision data? I wouldn't.

Much climate "science" research comprises a mathematical and computer model analysis of other people's data. In most cases, this data cannot be independently validated. Data for modelling has been "homogenised" hence it is no surprise that models and measurements don't agree. Much of the very small changes that have been predicted in climate "science" are within the order of accuracy of older measurements anyway and hence such predictions are knowingly fraudulent.

All scientific evidence and the ideas created from this evidence have a degree of uncertainty. To make strong statements or predictions without declaring the uncertainties is not science. True science is dispassionate and detached and there is no emotional attachment to the subject. This is why genuine scientists appear so boring.

Exploration geologists learn this very quickly. Testing of ideas based on the latest complex geophysics, geochemistry and geology hidden beneath the surface with a drill hole is a very humbling experience. There are always unexpected surprises. We often jokingly call a diamond drilling rig a rotary lie detector. Models based on geophysical measurements are often shown to be wrong by drilling. The new subsurface evidence from that drilling and measurements on the diamond drill core are then used to refine the geophysical model. The model may still be wrong but it is better than the first attempt and can be tested again by more drilling.

Scientific evidence must be reproducible. It matters not whether the evidence is from Chad, Colombia or China, the evidence must be able to be reproduced. Science transcends cultures, religion, gender and race because any validated scientific evidence can be independently validated by another person. This is why post-modernists, Marxists and nefarious others try to deconstruct science because it is underpinned by reproducible evidence and is one way of finding a truth.

When it is judged that there is a suitable body of evidence, this body of evidence is interpreted and explained. Primary or raw data is not adjusted, amended or "homogenised". This is fraud. Sometimes, data is judged unreliable and has to be recollected. There are no hard and fast rules for such judgments. Furthermore, some material just cannot be recollected (e.g. lunar, Martian, asteroid and space samples) and scientists have to be content with what limited data is available.

The interpretation and explanation of a body of data forms a scientific theory. In the law, some evidence may be deemed inadmissible which is why some legal decisions seem bizarre. Not so in science. A scientific theory is the best available explanation of all the evidence from a diversity of disciplines at the time. It may change with new evidence and must be in accord with the existing body of validated knowledge.

It is impossible in any scientific discipline for the science to be settled. If it is claimed that the science is settled, then we are dealing with propaganda and not science. If science is actually settled, we no longer have use for science and research grants and research institutions should not be funded.

Science is anarchistic, has no consensus, bows to no authority and it does not matter what a scientific society, government or culture might decide, only reproducible validated evidence is important. There have been aberrations such as Lysenkoism and climate "science". Like any other field, science suffers from fads, fashions, fools and frauds, has short-term leaders and can be cultish. Science advances after scientific leaders die.

The mainstream media views science as a popularity contest in which those with the greatest numbers win. Consensus is politics, not science. If there is a hypothesis that human emissions create global warming, then in science only one item of evidence is needed to show that this hypothesis is wrong. Dozens of items of evidence have been listed here and in other writings to show that the human-induced global warming hypothesis is wrong.

Climate "science", an infant scientific discipline, has been built from a synthesis of other well-established scientific disciplines. Mechanisms of climate science are complicated because they involve complex, non-linear, chaotically interacting systems which in turn involve poorly understood fluid dynamics.

It may take many decades before climate "science" can come to a better understanding of climate. In the interim, we are told that we have a "consensus" and "settled science". A large number of scientists, especially in non-Western countries, oppose the popular view that human emissions of carbon dioxide drive climate change.

This evolution of understanding of climate will be slowed down because of the financial interests of the IPCC, UN, renewable energy businesses, banks and by the misbehaviour the climate industry and scientific activists.

With climate "science", not only are there huge amounts of research funds sloshing around but the economic implications of a carbon trading scheme, new taxes, renewable energy subsidies and the transfer of wealth have brought together all sorts of disparate and desperate groups. You pay for all of this through taxes and unnecessarily excessive electricity costs.

Science funding opportunities, the rise of the Green Party and financial opportunities in the carbon market place have combined to produce an almost perfect storm of quasi-religious hysteria with all the hypocrisy that one associates with a fundamentalist religion. Funding, power and self-interest drive a consensus.

Fear of the environmental consequences of ever-expanding economies combined with the deterioration in science education encourages people to think of simple solutions to problems (or non-problems such as global warming). This is amplified by a lazy media that exploits fears to increase sales and can't really be bothered to get to the bottom of a story. As soon as an alternative view is aired in the media, there are howls of complaints from activists who scream that public discussion should be banned.[151] This could derail their gravy train.

There has been a great decrease in science ethics. We now have senior scientists seeking to suppress publication of research that undermines their beliefs. We have senior scientists making statements that the "science is settled" on human-induced climate change regardless of contrary findings of others. We now have previously well-regarded institutions discarding scientists whose work is against popular opinion. The age of open transparent science has arrived at a fork in the road. One points to Truth, the other to Oblivion.

The history of planet Earth has been ignored with the current popular catastrophist story of human-induced climate change. If large bodies of geological, solar and glaciological evidence and history are ignored, then this provides an unbalanced, misleading and deceptive view of global climate. If scientists ignore integrated interdisciplinary empirical evidence, then they have politicised science to gain government favours and they are operating fraudulently.

Scientists also argue about a scientific theory. Scientific theories are testable and once the scientific theory has

---

[151] http://www.radiotimes.com/news/2017-08-10/radio-4-labelled-ignorant-and-irresponsible-for-giving-airtime-to-climate-change-sceptic-lord-lawson

been tested over time, it becomes accepted into the body of knowledge. Just one contrary validated body of evidence can destroy a scientific theory. This has happened many times in all fields of science. Geology, astronomy, glaciology and history show that carbon dioxide (and human emissions of carbon dioxide) have not driven climate change in the past. Just because we live on Earth today does not mean that changes today on a dynamic planet are related to human activity.

Some scientific ideas such as cold fusion are very quickly abandoned when the phenomenon can't be reproduced and validated. The real peer review process actually occurs after publication of a scientific paper. Very often green left environmental activists have too much to say about the peer review system which only demonstrates that these folk have never had work peer reviewed, never been a reviewer and never been an editor.

At any scientific conference, one can see the various putsches prancing around narcissistically competing for attention, fame and research grant fortune while espousing their competing theories and bad-mouthing those with contrary views. This is normal human behaviour. Scientists do not hold the high moral ground and have the same weaknesses as their fellow man. They also cheat and lie, just like others.

Every active scientist is constantly involved in scientific disputes because science is never settled, new data and ideas are continually being aired and any new idea is generally critical of previous work. That is the nature of science. Many of us have written papers with our critics and rarely do scientific disputes become personal.

Scientific ideas can be tested. The idea examined in this book is that human emissions of carbon dioxide drive global warming and that this warming may be irreversible and catastrophic. This is easily tested by comparing with the reproducible validated evidence from past climate changes. This is the coherence criterion of science. Any new idea needs to be in accord with validated evidence from other areas of science.

Human-induced global warming is not in accord with geology. There are hundreds of examples that show that past events of global warming were not driven by carbon dioxide, that the planet has been far colder or warmer in past times and that the rates of temperature change have been far faster than any changes measured today. The past is a story of constant climate change and our present climate is just a repetition of previous ones and will change.

The idea that human emissions of carbon dioxide will lead to catastrophic global warming is therefore invalid and to continue to promote such an idea is ignorance, fraud or perhaps both. No wonder those who call themselves climate "scientists" don't want to debate geologists. They will not debate me in public because they are frauds. The public are not fools, they know when someone is telling the truth or telling porky pies. The climate "scientists" solution to this spot of bother is to ignore geology, astronomy and history, demonise dissent and yet claim that they are involved in science. All this shows is that they are involved in political and environmental activism.

There may be a body of evidence in support of a scientific idea. It only takes one piece of validated evidence to the contrary and that idea must be rejected. That is Karl Popper's concept of falsification or refutation.[152] Adolf Hitler sought to discredit the work of Albert Einstein by commissioning a paper[153] called "100 Authors Against Einstein".

Einstein responded: "... it doesn't take 100 scientists to prove me wrong, just one fact will do... ".

---
[152] K. Popper, *The Logic of Scientific Discovery*, Taylor and Francis, 2005.
[153] K. Israel *et al*, 1931: 100 Autoren gegen Einstein. *Naturwissenschaften* 19: 254-256.

This is rather like having a list of scientists and professional societies who support human-induced global warming. The Royal Society comes to mind, which is contrary to their motto[154] *"Nullius in verba."* The human-induced global warming farce has evolved in a similar fashion. Creationism has the same characteristics, long lists of allegedly eminent titled folk with impressive looking post-nominal sheep skins.

My criticisms of the climate industry are that much of the data (e.g. temperature and carbon dioxide measurements and corrections) is contentious, that computers have to be tortured to confess the pre-ordained result, that computer codes are not freely available, that neither the data nor the conclusions are in accord with what we know from the present and the past, that publication is within a closed system that challenges the veracity of the peer-review system, that financial interests are not declared, that the financial rewards for publishing a scary scenario are tempting and that the process of refutation seems to have been overlooked.

The climate industry needs to show it is dispassionate, has no pre-ordained conclusions, is unrelated to green left environmental politics, is driven by intellectual curiosity and can accept and critically analyse competing theories.

In the 1970s, the same people who were telling us we were going to freeze then started to tell us we will fry and die (e.g. Stephen Schneider). They exaggerated then and exaggerate now. Deceit is the one common thread that unites climate "science", as their own words show.[155] They left the sinking ship

---

[154] Take no one's word for it.
[155] *"We need to get some broad-based support to capture the public's imagination. That, of course, entails getting loads of media coverage. So we have to offer up scary scenarios, make simplified, dramatic statements and make little mention of any doubts we may have... Each of us has to decide the right balance between being effective and being honest."* Stephen Schneider, October 1989 interview for *Discover* magazine.

of global cooling and followed the money to the global warming ship.

In areas of science where much of the debate by experts is obscure to the layman and relies on truthfulness (including uncertainties), the climate "science" community has let down the side and operated as green left environmental activists for certain policies by singing the same tune as the funder of their "science" (i.e. governments). As Judith Curry[156] gently states:

> In the climate change problem, it seems that often one's sense of social justice trumps a realistic characterisation of the problems and uncertainties surrounding the science and the proposed solutions.

I am not so gentle. The global warming scare is the biggest scientific fraud in the 2,500-year history of science. And we are paying with taxes and ridiculously high electricity bills as a result.

For scientists to declare certainty, to make predictions and to ignore large bodies of contrary science shows that personality weaknesses overwhelm the scientific method. In my experience, academics have a huge chip on their shoulder and feel undervalued hence an opportunity to be in the spotlight and not be interrogated is tempting. Many feel that their work is so important that they are entitled to be funded by the taxpayer.

There is a view in society that a belief is evidence and a strongly held belief must be correct. This is emotion, not evidence. Most public discussions about climate change are underpinned by emotion and normally the intensity of emotion is inversely proportional to the amount and veracity of evidence.

---

[156] J. Curry, 2014 and 2015: *Ethics of climate expertise* and *Conflicts of interest in climate science*, http://www.judithcurry.com

The word belief is not used in science because belief is untestable. Belief is used in religion and politics. Testability has not taken place with the construction of what are popularly called the greenhouse effect, global warming, climate change or any other erroneous label. Furthermore, science is unable to make judgments about what is good or bad. These are judgments that vary with time and are based on contemporary politics, religion, aesthetics and culture.

That science should face crises in the late 20th and early 21st centuries is inevitable. Science is primarily funded by governments who are unable to appreciate that great scientific ideas and discoveries are made by individuals and not structured teams or committees following government policy.

These individuals are often dysfunctional loners, difficult to manage, fiercely independent, don't accept the popular paradigms and don't work well with others. Their discoveries are often serendipitous. These scientists are pilloried by peers, have difficulty in winning research grants, their work is often ignored for decades and they are grudgingly given accolades when recognition eventually comes.

The normal run of the mill scientist adds to the body of knowledge, does not solve significant problems and gives little return for a large taxpayer investment. Very little of today's scientific research is of use today yet there is the faint chance that some of it might be useful in future decades. This is why many scientific organisations have media units who loudly promote interim findings as a mechanism of winning future funding and of keeping the gravy train rolling.

Most scientists are paid to pursue their hobby and can end up in fields so narrow that only a few people in the world know their work. Maybe these scientists should fund their own hobby as the rest of the community does. In the 19th century, science was funded by individuals who had a burning curiosity. Corruption of science started when governments started to fund science after scientifically-driven war efforts.

Science is easily corrupted because governments want metrics such as publications, scientific citations, honours and awards – all of which operate in a closed cabal. Governments have difficulty with dispassionate science which may solve one problem, create another two and gives no definitive answer. Elected governments respond to publicity, media pressure and lobbying and hence fund science to avoid political embarrassment.

Power corrupts and science and much of the education system today is as the Catholic church was around the start of the 16th century. The church was used to having its own way by excommunicating heretics or even worse. Logical argument and knowledge was not used.[157] So too with science where individuals, heretics and non-conformists are frozen out of the research grant process and have difficulty in having work published in literature controlled by those in the club.

When some of us are senior in our field, we become editors of major scientific journals, sit on editorial boards, referee many scientific papers, sit as experts on research council panels, examine doctoral theses, are called as expert witnesses, advise industry and scientific institutions and continue as scientific authors. These activities involve the judgement of the veracity of scientific works of others. Such duties present a schadenfreudian opportunity to bury the careers of those not in the club.

Modern scientists have been polluted by the current education system which has been dumbing down for decades.

---

[157] David Gelernter, "The closing of the scientific mind", *Science Matters*, 12 June 2017

Kindergarten children are taught about the evils of human-induced global warming and that it is heresy to question such "facts". This process continues in high school and undergraduate studies. By the time a person has finished postgraduate studies and is poised for a research career, the ability to think critically, analytically and individually has been well and truly diminished. Pedestrian run of the mill science on a particular useless hobby subject becomes an easy option as it is non-competitive. This I saw with many of my staff in academic departments I led.

Many scientists promote that science is pure, truthful, free from bias and a noble quest for knowledge and that fraud is the preserve of business people, lawyers and crooks. How wrong can one be? Scientists are funded by the public to selfishly pursue their passion. They become narcissistic, incommunicative and out of touch with those who fund them.

Because science is a closed shop, it is far easier for fraud to flourish. Most scientific fraud involves omission and "homogenisation" of data, poor statistical treatment of data, stealing the work of others, ignoring contrary findings and producing pre-ordained conclusions. There certainly is a consensus in science and that is to keep the gravy train moving.

Governments may deem that a scientific area such as human-induced climate change is a national priority and should be preferentially funded. Immediately, all sorts of phenomena are claimed to be related to climate change. There is a very real danger that charlatans with self-interest persuade governments to follow a certain theme in science. We see this today with governments, self-interested businesses and government-funded scientists claiming that a conference in Paris can twiddle the planetary dials to stop a global temperature rise of 2°C.

This is why government funding for research into human-induced climate change gives the result that political policy requires. Everyone wins. Governments can continue a scare campaign, increase taxation revenue and restrict freedoms, scientists can be funded if they continue to produce the desired results and the public is duped into feeling that their taxpayer monies are creating a better society.

Government ideology and an uncritical media have led to the promotion of false scientific concepts such as human-induced global warming. This has led to billions of wasted dollars, poverty and now an electricity crisis. It's happened before. One example was Trofirm Denisovich Lysenko, a self-promoting Russian peasant who invented a process called vernalisation based on a trial of growing crops on half a hectare for just one season.[158] Seeds were moistened and chilled to enhance later growth of crops without fertiliser and mineral nutrients. It was claimed that the seeds passed on their characteristics to the next generation.

Stalin wanted to increase agricultural production, vernalisation never had scientific scrutiny, Lysenko became the darling of the Soviet media, was portrayed as a genius and any opposition to his theories was destroyed. Lysenko's theories dominated Soviet biology.[159] They did not work. Millions died in famines, hundreds of dissenting scientists were sent to the gulags or the firing squads and genetics was called the "bourgeois pseudo-science". History has an unfortunate habit of repeating itself.

Underpinning the global warming and climate change mantra is the imputation that humans live on a non-dynamic planet. On all scales of observation and measurement, sea level and climate are not constant. Change is normal and is driven by a large number of natural forces. Change can be slow or very fast.

We see political slogans such as *Stop Climate Change* or

---

[158] S. Ings, *Stalin and the Scientists: A History of Triumph and Tragedy 1905-1953*, Faber & Faber, 2016.
[159] The word: Vernalisation, *New Scientist*, 1 August 2007.

government publications such as *Living with Climate Change* demonstrating that both the community and government think that climate variability and change are not normal hence are the result of human activity.

If governments want to stop climate change, they need to stop continents moving, stop the shape of the sea floor changing, stop pulling apart the ocean floors, stop mountain building, stop volcanoes belching out greenhouse gases and dust, stop hot flushes of gas rising from the Earth's core and mantle, stop earthquakes, stop comets breaking up in the upper atmosphere, stop the changes in the Earth's orbit, stop the cycles of solar changes and stop radiation hitting Earth from deep space. Our generation did not discover climate change. The Earth's climate has always changed and this has been known for generations.

We have always lived with climate change and, whether we have governments or not, climate change will take place. By using the past as the key to the present, we know that we are facing the next inevitable glaciation yet the climate, economic, political and social models of today assess the impact of a very slight warming and do not evaluate the far higher risk of yet another glaciation. Geology, archaeology and history show that during glaciation, famine, war, depopulation and extinction are the norm.

In 1831, Admiral Sir James Robert George Graham had the Union Jack hoisted on a volcanic landmass that suddenly appeared near Sicily. It was called Graham Bank and was claimed by England. It was also claimed by the Kingdom of the Two Sicilies who called it Isola Ferdinandea, the French (L'Isle Julia) and other powers who variously named it Nerita, Hotham, Scicca and Corrao. In the subsequent dispute over ownership, France and the Kingdom of the Two Sicilies almost came to war and England and the Two Kingdoms of Sicily had a diplomatic row. During the intense diplomatic dispute, the island quietly slipped back underwater. In 1987, US warplanes thought the dark mass eight metres below sea level was a Libyan submarine and bombed it with depth charges.

In February 2000 when the volcano again stirred, Domenico Macalusa, a surgeon, diver and the Honorary Inspector of Sicilian Cultural Relics, took action. He persuaded Charles and Camilla,[160] the last two surviving relatives of the Bourbon Kings of the Two Sicilies to fund the bolting of a 150 kg marble plaque to the volcano at some 20 metres below sea level. The plaque pre-empted ownership if the volcano ever again rose above sea level. It was placed underwater in September 2001. By November 2002, person or persons unknown had smashed the plaque into 12 pieces.

This rock is worth nothing, is of no use as a territorial possession and is of no scientific interest and yet the French and Bourbons nearly came to war 170 years ago and the English and Italians are still in dispute.

Graham Banks serves to show that whatever political decisions we humans make, the land rises and falls, sea level rises and falls and climates change as it has since the dawn of time.

PEER REVIEW

The peer review process is meant to be the gold standard of science. Is it? The work of Newton, Darwin and some of that of Einstein was not peer-reviewed. It is wrong to impute that only good science is peer-reviewed.

Peer review depends on the integrity of the editor and of anonymous referees. If the editor has bias, it's easy to send a paper to peer reviewers who will give the desired expert opinions. A scientific paper or grant proposal can easily be

---
[160] Not the Charles and Camilla of the House of Windsor in the UK.

buried by sending the work to a referee in another camp. As an editor, member of a research grant committee or member of an appointment committee, it is useful to know who are the players in competitive tribalism because an opposing clique often buries good work. Peer-review is only an editorial aid. It involves rejecting or accepting a paper and making suggestions and changes to the submitted paper.

The number of reviewers that can be called on for a submitted paper is generally two or three people. Too often reviewers are not very diligent and only review a paper because, as a club member, they want a paper published later in the same journal.

Peer-review is conducted at the discretion of editors and the peer-review process is very susceptible to the influence and bias of small groups promoting and protecting their own interests. Most reviewers are anonymous. This only serves to strengthen the influences, biases and cowardice via anonymity. Pre-publication review is important but the most important process occurs after review.

This is when the scientific community at large can refute, check, confirm or expand on the ideas published. This is the real peer review process. Most scientific papers are read in their entirety by less than 20 people worldwide and most taxpayer-supported science is quickly forgotten. Questions asked after a conference presentation often show that attendees have not even read the conference abstract. Some science provides the building blocks for advancing new work.

Science has always self-corrected. It takes time for this long-term peer review. However, Western governments have uncritically and dogmatically embraced human-induced global warming as an excuse for increasing taxation, redistributing wealth, eroding freedoms, constraining liberal thinking processes and maintaining power by doing deals with groups allegedly concerned about the environment.

The climate clan is peer reviewing and publishing their own grant-earning dogma and some of the scientific community are showing how weak this science is post-publication. Peer-review, far from being the gold standard, is used by the climate industry to approve or censor science according to its own agenda. In the climate industry, it is not peer review, it is peer bullying and peer pressure to conform to the grant-earning orthodoxy.

There are many ways of showing that at times in the past, planet Earth experienced a warmer climate than today. This destroys the catastrophist narrative. The solution taken by climate "scientists" is to destroy the past. The spectacular finding in 1998 that rising temperature trends were much higher today than in the past by Michael Mann[161] underpinned political action on human-induced climate change.

Mann showed that rather than rather than having climate cycles changing between warm and cold periods, global warming was now out of control and his graph looked like a hockey stick. Mann's redrawn temperature graph was based on tree rings and expunged previous validated data from the temperature record. Editors and peer reviewers clearly did not question why Mann rejected previous knowledge which had actually been promoted by the IPCC. The evidence and the statistical methodology were re-evaluated.[162] The hockey stick was shown to be a fraudulent creation *ex nihilo*. Why did the peer-review system fail for a journal like *Nature*?

---

[161] M. Mann, R.S. Bradley and M.K. Hughes, 1998: Global-scale temperature patterns and climate forcing over the past six centuries. *Nature* 392: 779-787.

[162] S. McIntyre and R. McKitrick, 2003: Corrections to the Mann *et al.* (1998) proxy data base and Northern Hemisphere average temperature series. *Energy Envir.* 14: 751-771.

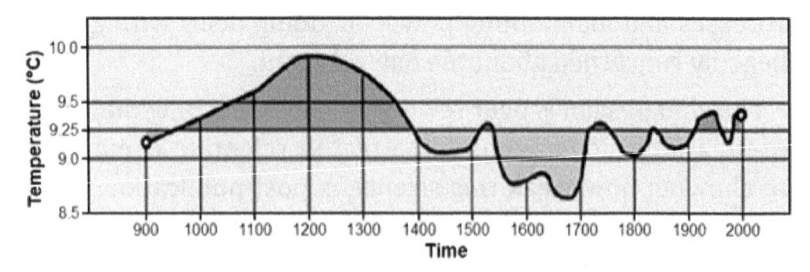

Figure 14: The IPCC plot of temperature of climate change in Europe over the last 1,000 years showing the Medieval Warming (900 to 1400 AD), the Little Ice Age (1400 to 1680 AD) and the 300-year period of warming since the coldest time of the Little Ice Age. The rate of warming and the temperature in the Modern Warming (1680 to the present) is lower than the Medieval Warming. If human emissions of carbon dioxide created the Modern Warming, then the Medieval Warming and Little Ice Age would have to be expunged from the record. This is exactly what Mann did.

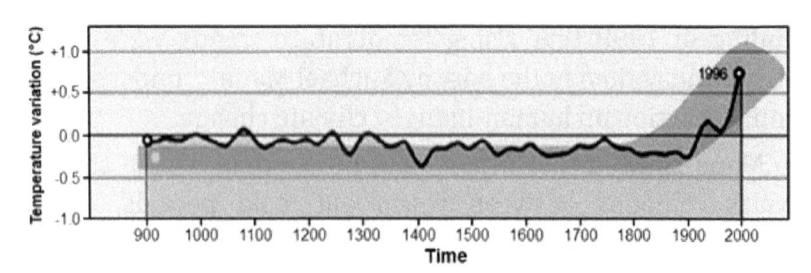

Figure 15: The fraudulent "hockey stick" of Mann showing the temperature variation from 900 to 1996 AD based on selective use of tree ring measurements and omission of volumes of data.

Scientific journals should take no position on any issue, they should be indifferent to politics, should publish the spectrum of competing hypotheses and leave it to the market (i.e. the scientific community) to distil, validate or replicate over time. Publication of Mann's paper and many later editorials shows that *Nature* has taken a partisan political view of human-induced global warming.

This controversy led to a US Senate Committee setting up an inquiry under Professor Edward Wegman. The inquiry showed that Mann's work could not be supported. In the Wegman Report,[163] a scathing review of the fraud of the Michael Mann hockey stick, the peer review process in the climate industry was questioned:

> One of the interesting questions associated with the "hockey stick controversy" are the relationships among authors and consequently how confident one can be in the peer review process. In particular, if there is a tight relationship among the authors and there are not a large number of individuals engaged in a particular topic area, then one may suspect that the peer review process does not fully vet papers before they are published.

and

> Indeed a common practice among associate editors for scholarly journals is to look in the list of references for a submitted paper to see who else is writing in a given area and thus who might legitimately be called on to provide knowledgeable peer review. Of course if a given discipline area is small and the authors in the area are tightly coupled, then this process is likely to turn up very sympathetic referees. These referees may have co-authored other papers with a given author. They may believe they know that author's other writings well enough that errors can continue to propagate and indeed be reinforced.

In the case of the "hockey stick", there were only 43 in the climate industry who could have been gatekeepers. With such a small club where each member knows everyone else, members review the work of their friends and everyone is dependent upon the others for favourable grant and paper reviews. It is not

---

[163] U.S Congress House Committee on Energy and Commerce; http://republicans.energycommerce.house.gov/108/home/07142006_Wegman_Report.pdf

surprising that shoddy "science" is published. Later work[164] has shown just how a small group of climate "scientists" managed to influence world thinking, especially amongst non-scientific green left environmental activists. This was until the "hockey stick" was proven to be fraudulent.

The IPCC gave Mann's "hockey stick" star billing until it was shown to be fraudulent and then quietly got rid of it. Mann has taken exception to scientific criticism and, rather than using data and reason to support his theories, has used the legal system to try to silence critics such as Tim Ball and Mark Steyn.

In 2007, Ross McKitrick spent two years trying to publish a paper to refute a claim by the IPCC that real-world data confirmed models predicting man-made global warming. The IPCC claim was based on fabricated evidence and was shown to be fabricated using statistical analysis. The climate cabal did not allow his work to be published in the mainstream literature. Ultimately, the paper was accepted for publication but not in the mainstream climate literature.

Over quite a period of time Ross McKitrick nibbled at Mann, his "hockey stick" and the IPCC's blind acceptance of Mann's fraud. McKitrick had great difficulty publishing papers in the mainstream climate "science" literature refuting claims by Mann regarding his "hockey stick" and the IPCC claims that real-world data confirmed models predicting man-made global warming. Ranks closed and the cosy catastrophic climate change club was protected.

Ultimately, McKitrick was able to show that this was not a minor academic spat in the hallowed halls of academe but was symptomatic of the whole climate industry closing ranks to protect their own, to inhibit free discussion, to hide or censor

---

[164] Mark Steyn, 2015: A disgrace to the profession. The world's scientists in their own words on Michael E. Mann, his hockey stick and their damage to science. Vol. 1 http://www.steynstore.com/product113.html

data and to stop data that leads to a different conclusion being published.[165,166,167,168]

Developments in climate "science" that challenge the findings of the IPCC continue. A 2011 paper by one of the world's most eminent meteorologists, Richard Lindzen, addressed concerns raised by critics of a 2009 paper. His research found that a doubling of carbon dioxide will increase temperatures by only 0.7°C, which is significantly below the lowest estimates of the models used by the IPCC. The climate industry closed ranks and it took about two years for Lindzen to find a journal willing to publish the paper.[169]

After publication, critics focused on the point that the paper was published in an Asian and not a Western scientific journal. This is the peer review process in operation. Even ignoring this racism, the scientific facts do not change. Fred Singer has documented how the journals *International Journal of Climatology, Geophysical Research Letters* and *EOS* buried his research work because it did not fit the popular paradigm.[170]

Climate "science" is a scare industry worth $1.5 trillion and has been built on the notion that human emissions of carbon dioxide are responsible for unprecedented and catastrophic changes to the world's climate. Scientists promoting this

---

[165] S. McKintyre and R. McKitrick, 2003: Corrections to the Mann *et al.* (1998) proxy data base and Northern Hemisphere average temperature series. *Energy and Environment* 14: 751-771.
[166] S. McIntyre and R. McKitrick, 2005: Hockey sticks, principal components, and spurious significance. *Geophys. Res. Lett.* 32: L03710, doi:10.1029/2004GL021750.
[167] Andrew Montford, 2010: The Hockey Stick Illusion. *Stacey International*.
[168] Ross McKitrick, 2015: The hockey stick: a retrospective. In: *Climate Change: The Facts* (Ed: Alan Moran). Institute of Public Affairs, 201-211.
[169] R.S. Lindzen and Y.S. Choi, 2011: On the observational determination of climate sensitivity and its implications, *Asian Pacific Joural of Atmospheric Science* 47: 377-390.
[170] http://americanthinker.com/articles/2015/08/peer_review_is_not_what_its_cracked_up_to_be.html

hypothesis are a small closed shop who validate one another's work in pal review rather than peer review. Response to criticism is to bully, smear and deny access to mainstream science journals. Peer review in climate "science" does not operate like peer review in other disciplines.

The skewing of the scientific literature is not healthy for a society that depends upon scientists and scientific literature for trustworthy advice for wise policy decisions. My advice to young climatologists: Do not write a paper questioning the IPCC, the popular paradigm of disastrous climate change or your host climate institute. You will destroy your career prospects.

The climate "science" industry has no parallels. There appears to have been a profession-wide decision that there can be nothing published that might threaten their fame- and fortune-giving ideology that global warming is driven by human emissions of carbon dioxide. There has also been an international "homogenisation" of raw data to give a warming trend to the degree that the historical temperature record is now probably useless.

This lack of scientific objectivity is such that one has to question whether any literature from the climate industry can even be treated seriously for another 30 years. The climate industry has openly and flagrantly violated every aspect of the core ethos of science. Climate "science" is technically-based advocacy in the service of politics, is not dispassionate and is no more valid than the tobacco industry publishing science about the harmlessness of smoking.

For the first 300 years after the Royal Society was founded, they adopted a position of aloofness from political debates refusing to become embroiled in the political controversies of the day. In fact, their journal *Philosophical Transactions of the Royal Society* carried a notice:

It is neither necessary nor desirable for the Society to give an official ruling on scientific issues, for these are settled far more conclusively in the laboratory than in the committee room.

This all changed in the 1960s and the Society gradually became increasingly involved in politics and policy. The last two presidents were especially active and it was only a revolt by Fellows that slightly changed matters. The Society now carries a damaged reputation because of its engagement in political controversies. The Royal Society is not alone. Almost all professional societies are involved in political lobbying, policy formulation and advocacy. The current President of the Royal Society told us that the science on human-induced global warming is settled.

A previous President[171] of the Royal Society also used his authority, this time to inform us that it is impossible for heavier than air machines to fly and that we know all that is to be known about physics. However, radioactivity was discovered a few years later and this has become a fundamental of physics. That was in the 1890s. So much for authority.

Many learned societies, including the Royal Society, are partially funded by government. Those societies to which I belong also have internal ructions when the executives and employees act against the wishes of members by taking a position on political controversies of the day.

The same has happened elsewhere such as the National Academy of Sciences in the US. The American Physical Society played the political game and in 2010 revised its statement on climate policy. In a scathing letter of resignation by one of its most eminent and senior members, Hal Lewis (University of California) commented on the popular paradigm of human-

---

[171] William Thomson, Lord Kelvin (1824-1907).

induced climate change: "It is the greatest and most successful pseudoscientific fraud I have seen in my long life as a physicist."

FRAUD, MISTAKES AND HOAXES

During my lifetime of science, I have seen lots of shoddy published science and failings in the peer-review system. Many papers I have read should never have been published, many papers contain obvious errors, reviewers often insist that their minor work be cited even if it is poor work or unrelated and editors often make decisions based on format, paper length and whether the subject is the hot topic of the day, rather than content. Some good work is published in obscure journals because the mainstream journals cannot accept alternative ideas.

A lot of what is published is incorrect. In medicine, maybe half of what is published is untrue.[172] Studies with small sample sizes, tiny effects, invalid analyses, flagrant conflicts of interest, or an obsession with pursuing fashionable trends of dubious importance seem to be the norm. Poor scientific methodology gets results and produces compelling stories. This is until they are checked or an attempt at replication is undertaken.

Incorrect publications

There is so much concern about published incorrect medical sciences that the Academy of Medical Sciences, Medical Research Council and the Biotechnology and Biological Sciences Research Council have now put their reputational weight behind such concerns. Statistical fairy tales abound, journals have poor peer review and editing processes and the processes of granting research monies all contribute to bad

---

[172] R. Horton, 2015: Offline: What is medicine's 5 sigma? *The Lancet* 385: 1380.

scientific practices. If this is the case for the medical sciences, what about other fields? I doubt it is any different.

There have been some high-profile errors in physics (e.g. cold fusion, particle physics) that have resulted in changing procedures such that intensive checking and rechecking of data takes place prior to publication.

Fraud

What about climate "science"? I suspect that with "homogenised" data, Climategate, failed climate models, failed predictions, exaggerations, omission of contrary data and unfounded statistical studies, this figure would be far higher because I suspect that the fields of physics and medicine are more rigorous than climate "science". Your guess is as good as mine but it's not a pretty sight anyway. Maybe with climate "science" it's a case of never let the science get in the way of a hefty research grant.

Jan-Hendrik Schön managed to publish a paper every eight days in major scientific journals between 2000 to 2002 on nanotechnology and single molecule behaviour. Every paper was a peer-reviewed paper, every paper was bogus and yet Schön's "great" works were published by *Nature, Science, Applied Physics Letters* and *Physical Review*. Schön won a number of prestigious prizes for his published work. This fraud was later detected by a student, not by the peer-reviewers. Schön showed that if you have influential mainstream mates, fraud can easily be published in the best peer-reviewed scientific journals.

This fraud was discovered and corrected by the post-publication evaluation of peer-reviewed papers. Peer review is not the gold standard that assures the quality of science publishing, which is why a number of journals and institutions have a science citation index that quantifies the scientific impact of a paper. Again, this is not the gold standard because in a small field the impact can be different from a big field and the various tribes are normally unwilling to cite the work of competitors.

Weak stomach publications

Sociology scholarly peer-reviewed journals have form. Here's the latest.[173] In 2017, two Australian academics published in a peer-reviewed journal called *Emotion, Space and Society*. The journal name gives it away. This work was supported by your taxes via the Australian Research Council and their wages from the Universities of Melbourne and Wollongong. It appears that climate "scientists" are snowflakes and suffer emotional stress because not everyone believes their fraud. If you have a weak stomach or pay too much tax, don't read their abstract below.

> This paper contributes to increased understanding of emotions and climate change through a study of the emotional management strategies employed by a sample of Australian climate scientists. We bring three broad areas of literature into conversation in order to think more productively about climate change and emotion: recent applications of the concept of emotional labour, studies of the role of emotion in science, and feminist perspectives on the performative role of emotions. In response to contextual drivers that include the social norms of science, a strong climate denialist influence and the preservation of self and family, these scientists mobilize a range of behaviours and strategies to manage their emotions around climate change and the future. These include emphasizing dispassion, suppressing painful emotions,

---

[173] L. Head and T. Harada, 2017: Keeping the heart a long way from the brain: The emotional labour of climate scientists. *Emotion, Space and Society* 24, 34-41.

using humour and switching off from work. Emotional denial or suppression of the consequences of climate change worked to enable the scientists to persevere in their work. This study suggests that painful emotions (anxiety, fear, loss) around climate change need to be acknowledged and discussed.

This is absolute trivial drivel. Is this scholarship? Is it any wonder that I claim that the education system has been dumbed down?

The Sokal hoax

How can we forget physicist Alan Sokal's sham article generously salted with nonsense, pseudo-babble and obvious howlers to the cultural studies journal *Social Text*? It was published in 1996.[174] The article was submitted to show the decline in standards, especially in the soft post-modernist bleeding heart hand-wringing disciplines. Sokal exposed his article himself, it was not discovered by readers, reviewers or editors.[175] The journal editors tried to justify publishing the article because it was:

> ... the earnest attempt of a professional scientist to seek some sort of affirmation from postmodern philosophy for developments in his field.

It was no such thing. It was a published paper showing the lack of scholarship in a major peer-reviewed journal of sociology. One of the editors even claimed that:

> Sokal's parody was nothing of the sort and that his admission represented a change in heart or a folding of his intellectual resolve.

---

[174] A.D. Sokal, 1996: Transgressing the boundaries – Toward a transformative hermeneutics of quantum gravity. *Social Text* 46/47: 217-252.

[175] A.D. Sokal, 1996: A physicist experiments with cultural studies. *Lingua Franca* May/June: 62-64.

It was no such thing, despite the editors of *Social Text* making valiant weak explanations.[176] It was not a parody. It was an exercise to show the lack of scholarship in sociology. Sokal was not admitting to being dishonest or changing his view. He was a rigorous physicist who showed that some sociology journals accept whatever nonsense is submitted to them as scholarship couched in post-modernist deconstructionist language. The title of the paper was absolute nonsense and any educated person would have seen that the text was nonsense.[177]

The penis and climate change

Two US academics, Peter Boghossian and James Lindsay had a peer-reviewed hoax paper published in 2017 in a social sciences journal. They argued that a penis is not in fact a male reproductive organ but merely a social construct and that penises are responsible for causing climate change. *"The conceptual penis as a social construct"* had two peer reviewers who were allegedly experts in gender studies, one of whom praised the way it captured "the issue of hypermasculinity through a multidimensional and nonlinear process" and the other who marked it as "outstanding" in every applicable category.[178]

---

[176] http://linguafranca.mirror.theinfo.org/9607/mst.html

[177] A quote by Sokal: *"The Einsteinian constant is not a constant, is not a center. It is the very concept of variability – it is, finally, the concept of the game. In other words, it is not the concept of something – of a center starting from which an observer could master the field – but the very concept of the game."* I have read this many times and, drunk or sober, it still means nothing to me. Another quote from the Sokal paper should have given the game away: *"the pi of Euclid and the G of Newton, formerly thought to be constant and universal, are now perceived in their ineluctable historicity."* I have been an editor of a major scientific journal for years and have seen some dreadful papers submitted. Why didn't the editor at least insist that such great revelations were intelligible? An editorial tip: If something has to be read more than once to be understood, then it is poorly written or nonsense.

[178] Peter Boyle and Jamie Lindsay, 2017: The conceptual penis as a social construct. *Cogent Social Sciences 2017*, 3: 1330439.

The conclusion should have set the alarm bells ringing: We conclude that penises are not best understood as a male sexual organ, or as a male reproductive organ, but instead as an enacted social construct that is both damaging and problematic for society and future generations. The conceptual penis presents significant problems for gender identity and reproductive identity within social and family dynamics, is exclusionary to disenfranchised communities based on gender or reproductive identity, is an enduring source of abuse for women and other gender-marginalized groups and individuals, is the universal performative source of rape and is the conceptual driver behind much of climate change.

Most of the cited references are quotations that make no sense in the context of the paper, others were obtained by searching plausible-sounding keywords and the authors stated that they did not read a single reference cited. Some references cite the Postmodern Generator, a website for hoaxing journals that produces a different fake post-modernist paper every time the page is reloaded. Fake journals such as *Deconstructions from Elsewhere* and *And/Or Press* are cited as is the fictitious researcher S.Q. Scameron.

This total nonsense published in a peer-reviewed journal was inspired by Sokal's (1996) hoax yet it was not seen as nonsense by the peer reviewers or editor. Sokal[179] argues that if ideas are fashionable, then critical faculties required for peer reviewing allow total nonsense to be published as long as it promotes certain values.

Another Sokal-inspired spoof was published in the Novem-

---

[179] Alan Sokal and Jean Bricmont, 1999: *Fashionable nonsense: Post modern intellectuals' abuse of science*. Picador.

ber 2009 *Scientific American* entitled "*A path to sustainable energy*". Authors Mark Jacobson and Mark Delucchi slipped in deliberate errors about the amount of concrete used to make wind- and nuclear-powered facilities, the amount of carbon dioxide released making a nuclear power station and the amount of down time of coal (12.5%), wind (2%) and solar (2%).[180] The *Scientific American* editors did not detect absurd statements such as the authors have a plan:

> ... to determine how 100% of the world's energy, for all purposes, could be supplied by wind, water and solar resources, by as early as 2030.

and

> How do you get all of the world's energy from wind, water and solar? Easy. 490,000 tidal turbines, 5,350 geothermal plants, 900 hydroelectric plants, 3,800,000 5-MW wind turbines, 720,000 wave converters, 1,700,000,000 3-kW rooftop solar PV systems, 49,000 concentrated solar power plants, and 40,000 300-MW solar PV power plants.

The spoof was not detected by *Scientific American* and was exposed in 2017, but not by the authors.[181] The authors published almost the identical paper in the *Proceedings of the National Academy of Sciences* which was robustly criticised by those without a sense of humour[182] and the authors won the Cozzarelli Prize of the Academy.

It's not hard to publish rubbish in peer-reviewed journals and even win prestigious prizes if the publication is in accord with the populist ideology.

---

[180] https://wattsupwiththat.com/2017/08/27/scientific-american-sokalized/
[181] Robert Bryce, "The appalling Delusion of 100 Percent Renewables, Exposed", *National Review*, 24 June 2017.
[182] C. T. M. Clack *et al*, "Evaluation of a proposal for reliable low-cost grid power with 100% wind, water and solar", *National Academies Press*, 27 June 2017.

Seinfeld public urination

A recent invited medical paper was published by a mythical author from a mythical institute based on a whole episode of the television sit-com Seinfeld (*The parking garage,* 1981).[183] References were hoax citations using the names of Seinfeld characters. The mythical medical affliction called uromysitisis poisoning was pulled out of thin air to describe urinating in public. The paper was peer reviewed and quickly published.[184] The conclusions gave it away:

> ... a national reciprocity program of public urination passes [...] so that people with uromycitisis can be free to urinate, if medically necessary, wherever and whenever they need to, not be burdened legally (or, indeed, psychologically) by existing local or state laws and regulations against public urination.

Now just tell me again about the gold standard peer review process.

Oh, and remind me again. Why is climate "science" not fashionable nonsense?

MODELS

Computer models do not constitute evidence. They are a method of trying to navigate through and understand a large body of data. Computer models are not the only way of trying to understand data. Although computer models have been used to try to make predictions, nature is very fickle, computer models suffer

---

[183] J.H. McNicol, "Opinion: Why I published in a predatory journal", *The Scientist*, 6 April 2017.

[184] M. van Nostrand, 2017: Uromycitisis poisoning results in lower urinary tract infection and acute renal failure: Case report, *Urology and Nephrology Open Access Journal* 4, 3, 00132.

from "unknown unknowns" and models suffer from garbage-in garbage-out.

Current models of climate are mere scenario predictions of the assumptions used to drive the model and are based on incomplete data. The fact that these are presented as accurate predictions of real climate underpinned by science rather than extensions of what seems to be unfounded assumptions is dishonest. A model is not scientific evidence.

Computer climate models cannot predict future climate and are not in accord with the past. Climate models tell more about the modellers than the climate as they produce pre-ordained conclusions and have been shown to be wrong by measurements.

Failure of models

In 1990, the IPCC predicted that the rate of global warming would be twice that which occurred. In 2007, the IPCC predicted that in the decade following 2005, there would be significant warming. There wasn't even any warming in that decade. In 2013, the IPCC were at it again and predicted short-term warming. We're still waiting. This was proven to be wrong as a result of measurement. The IPCC is embarrassed. It has now had to admit that the climate models built by their so-called climate scientists cannot predict the temperature that is measured. They meekly state:

> For the period from 1998 to 2012, 111 of the 114 available climate-model simulations show a surface warming trend larger than the observations.

This is the coward's way of saying that they were wrong.[185]

---

[185] http://www.economist.com/news/science-and-technology/21598610-slowdown-rising-temperatures-over-past-15-years-goes-being/

The pause

Why didn't the computer models predict (or even factor in) the hiatus in warming over the last two decades? When the standard climate models are run backwards, they do not record what has happened. Is this because the models are constructed to show that increasing human emissions of carbon dioxide gives increased warming and that other factors are ignored or minimised? Don't the modellers even think to consider that the great ball of fire and radiation in the sky that we call the Sun may be far more influential on surface temperature of the Earth, rather than traces of an invisible trace gas in the atmosphere?

When NASA satellite data showed a "huge discrepancy" between alarmist climate models and data, climate "scientists", the media and our elected officials would be wise to take notice. Whether or not they do so will tell us a great deal about the honesty of the purveyors of global warming alarmism.

When it was pointed out in the New Zealand press that the planet had not warmed for five years, David Wratt (an IPCC lead author) responded that climate models could not be expected to be accurate in anything less than a 10-year period. Now, after 20 years of no warming, Wratt's claim looks like a throw-away line unrelated to science. Either we can accept that the models are hopelessly wrong or we can continue to believe that there are fairies at the bottom of the garden where the unicorns graze.

Greens, climate "scientists", bureaucrats, politicians and many in the media still claim that we face catastrophic global warming. This is based on discredited models. To deny contrary data shows that global warming alarmism has nothing to do with science. A reasonable person would argue that after patiently waiting for more than 30 years for the planet to catastrophically warm, the most logical conclusion could be made: The prediction of human-induced global warming is demonstrably wrong.

Adaptation

Humans have adapted to live on ice, mountains, in the desert, in the tropics and at sea level and can adapt to future changes. History shows that during interglacials, humans create wealth which allows populations to grow whereas glaciation is associated with famine, starvation, disease and depopulation. The cycles of climate change suggest that the next inevitable glaciation will be little different from previous glaciations.

Imagine ice sheets covering the same area as in the last glaciation. Most of Europe, Canada and northern USA would be covered by ice. Other areas such as China, Mongolia and Australia would have howling cold winds and shifting desert sands. Alpine areas would be covered with ice and polar ice sheets would greatly expand. The food supplies for billions of people would be really under pressure. The best we humans can do is prepare for change (which we have not done in the past) and adapt (which we have done in the past). Because of modern technology, any adaptation to modern climate change would be far easier this time.

There are no models attempting to understand the effects of another glaciation.

Rainfall

The models also claimed that modest warming by carbon dioxide would be amplified by increased water vapour in the atmosphere. This can be tested by checking actual rainfall against the predicted rainfall. Out of eighteen regions tested in the USA, eight showed that rainfall would either increase or decrease depending upon the model used. One model predicted an 80% decrease in rainfall and another predicted an 80% increase in rainfall. Neither deserts nor swamps appeared.

The models were wrong.

Better models

Modellers would only have to spend a few hours looking at the scientific publications in astronomy to see that there is climate change on other planets and their moons. This extraterrestrial climate change could not possibly be related to human emissions of carbon dioxide and clearly the factors that drive extraterrestrial climate change are also important on Earth.

Only one model has shown the past temperature and climate over the past century. This is the model from Nicola Scafetta of Duke University and this is based on solar, lunar and planetary cycles and does not use carbon dioxide as the controlling variable.[186,187] The latest data from the CERN particle physics laboratory has also produced a model based on cycling and it foresees no runaway global warming. Instead, it sees an impending cold solar minimum.[188]

The hot spot

In testimony in 2015 to the US House Science Committee hearing on the Paris climate treaty, Professor John Christy (University of Alabama at Huntsville) showed a massive divergence between models and measurement for atmospheric temperature. Christy argued that he would not trust models and concluded that satellite and balloon measurements gave the most reliable measurement of the atmosphere's temperature.[189]

---

[186] N. Scafetta, 2016: High resolution coherence analysis between planetary and climate oscillations, *Advances in Space Research* 57, 2121-2135.
[187] N. Scafetta, 2017: Understanding climate change in terms of natural variability. In: *Climate change: The facts 2017*, (ed. J. Marohasy), IPA, 39-58.
[188] A. Orlowski, "CERN 'gags' physicists in cosmic ray experiment: What do these results mean? Not allowed to tell you", *The Register*, 18 July 2011.
[189] www.al.com/news/huntsville/index.ssf/2015/04/7_questions_with_john_christy.html

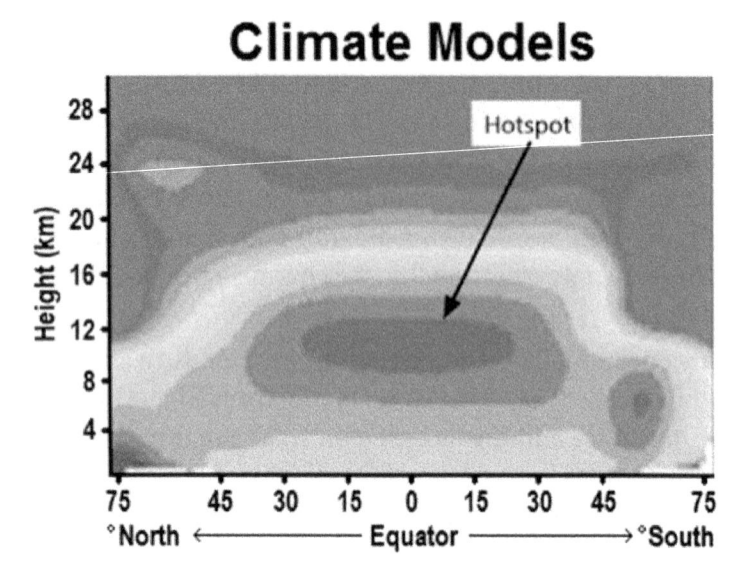

Figure 16: Predicted equatorial hot spot in the Earth's atmosphere from models.

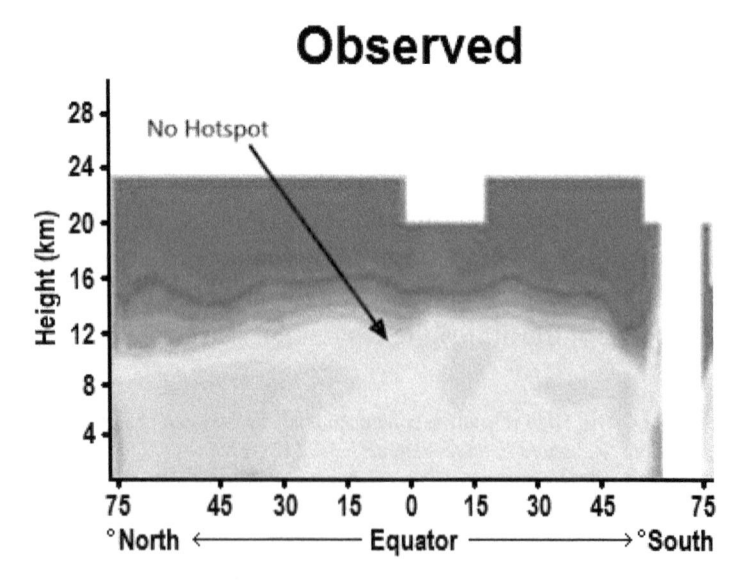

Figure 17: Atmosphere temperature from balloons and satellite measurements showing no equatorial hot spot in the atmosphere.

Christy also showed that there is a close correlation between satellite and balloon measurements which do not show global warming and show cycles of warming and cooling.[190]

The key pillar of the global warming theory is that carbon dioxide emissions should trap heat in the tropics at an altitude of 10 to 12 km in the Earth's atmosphere and prevent it from escaping into space.

Despite the release of about 30 million weather balloons since 1950, the modelled hot spot cunningly hid from every single one of these balloons. Balloon measurements of heat have been validated. NASA satellite data over the past decade has shown that our atmosphere releases much more heat into space than the computer models show. The missing heat is lost to space, not to the oceans and atmosphere as the models predicted. Each day aeroplanes fly at 10 to 12 km altitude in the tropics. No hot spot has been measured.

Climate, atmospheric temperature and ocean temperature models over the last few decades have all been checked with measurements. All models were wrong. The models all told us that we will fry and die and the measurements are telling us that we won't.

The models have been tried and tested. They have failed. They could not even confirm past climates by running backwards without substantial retuning. The real test is whether these models can predict future climate in say 100 years time. There are over 100 variants of climate models yet they did not even predict a period of no warming for 20 years. If they couldn't even get this right, then no matter how much tampering and adjustment, we can safely conclude that they can't predict what will happen in 100 years time.

---

[190] docs.house.gov/meetings/SY/.../HHRG-114-SY00-Wstate-ChristyJ-20160202.pdf

Violent climate activists

Shots in the dark

It appears that climate activists were not happy with Christy's House Science Committee testimony and seven high velocity bullets were fired at his office.[191] Dr Roy Spencer, another high-profile critic of climate "science" who has an office in the same building stated:[192]

> Given that this was Earth Day weekend, with a 'March for Science' passing right past our building on Saturday afternoon, I think this is more than coincidence. When some people cannot argue facts, they resort to violence to get their way. It doesn't matter that we don't 'deny global warming'; the fact we disagree with its seriousness and the level of human involvement in warming is enough to send some radicals into a tizzy. Our street is fairly quiet, so I doubt the shots were fired during Saturday's march here. It was probably late night Saturday or Sunday for the shooter to have a chance of being unnoticed. Maybe the 'March For Science' should have been called the 'March To Silence'.

The silence from climate "scientists" was deafening. This is commensurate with silence from climate "scientists" when environmental activists threaten contrarians with silencing or murder.

Assassination

The theory that human emissions of carbon dioxide drive global warming has been challenged by independent scientists.

---
[191] www.al.com/news/huntsville/index.ssf/2017/04/shots_fired_at_office_building.html
[192] Shots fired into the Christy/Spencer Building at UAH. drroyspencer.com

Some extremist environmentalists suggest that such scientists should be removed from employment, imprisoned and even assassinated. For example:[193]

> On September 20th the British newspaper, *The Guardian*, published an article on a letter from the Royal Society ... calling for the silencing of groups, organizations, and individuals who do not conform to their views on climate change and policy.

and[194]

> The British Greens have called for a purge of officialdom to get rid of anyone who doesn't accept scientific consensus on climate change.

Other scribes have had their two bob's worth on the same theme:[195,196,197]

> I wonder what sentences judges might hand down at future international criminal tribunals on those who will be partially but directly responsible for millions of deaths from starvation, famine, and disease in the decades ahead. I put [their climate change denial] in a similar moral category to Holocaust denial – except that this time the holocaust has yet to come and we still have time to avoid it. Those who try to ensure we don't will one day have to answer for their crimes. (Lynas, 2006)

and

> Stopping runaway climate change must take precedence over every other aim. Everyone in this movement

---

[193] Jeffrey Kueter, 29 September 2006. President of George C. Marshall Institute, letter to Congress.
[194] Brendan O'Neill, *The Weekend Australian*, 2-3 May 2015.
[195] Mark Lynas, *Dagelijksestandard*, 19 May 2006.
[196] George Monbiot, *The Guardian* 23 August 2008.
[197] Lovelock, James, *The Guardian* 29 May 2010.

knows that there is little time: the window of opportunity in which we can prevent two degrees of warming is closing fast. We have to use all the resources we can lay our hands on, and these must include both governments and corporations. (Monbiot, 2008)

and

I feel that climate change may be an issue as severe as war. It may be necessary to put democracy on hold for a while. (Lovelock, 2010)

Professor Richard Parncutt, a professor of systematic musicology, at Karl-Franzens-Universität Graz (Austria) has clearly critically evaluated the science of human-induced global warming and gave us his expert opinion:[198]

As a result of that process, some global warming deniers will never admit their mistake and as a result they will be executed. Perhaps it would be the only way to stop the rest of them. The death penalty would have to be justified in terms of the enormous numbers of saved future lives.

And if contrarian scientists are threatened with violence without a squeak of strong public objection coming from climate "scientists", then climate "science" is unrelated to science and is the preserve of thugs.

I wonder if these murderous anti-democratic scribes would be prepared to take the chop if they are wrong and the planet cooled. They are the deniers who try to tell us that there was no Medieval Warming, that there has been atmospheric temperature rise during the last 20 years and that there is a correlation between atmospheric carbon dioxide and temperature.

---

[198] Richard Parncutt, 25 October 2012, *Death penalty for global warming deniers? An objective argument...a conservative conclusion.* Internet text on website of Karl-Franzens-Universität Graz (Austria) until removed by order of university officials.

Green left environmentalist activism is the new Lysenkoism and such movements are a honey pot for all sorts of nutters. Many of them also proudly call themselves socialists.

When climate activists are confronted with contrary evidence, they resort to violence. This is the style of totalitarian religious fundamentalists and not the behavior of those in a civilised Western country. Why do we hear no loud public complaints from climate "scientists" about this behaviour?

It is safe to conclude that pretty well everything you hear, read or see in the popular media about climate change is fraudulent, wrong, exaggerated or made up. And if models are used, you can be certain that it is wrong.

PREDICTIONS

Models underpin scary predictions.

The UK Met Office wins the gold medal for failed predictions based on computer models. They undertook an investigation from 2004 to 2014 on the effects of future climate change. The Met Office's computer models assume that a rise in atmospheric carbon dioxide is driving climate change. Human emissions of carbon dioxide are rising, atmospheric carbon dioxide is rising (which may or may not be related to human emissions of carbon dioxide) and the average global temperature has not done what the computers told it to do. For nearly two decades, global temperature has not risen. Despite this, the Met Office has continued to tout scary scenarios for the future.

The Met Office receives £220 million *per annum* from the taxpayer and yet they want bigger and better computers in order to make more inaccurate predictions more quickly. According to the Met's chief scientist, their £33 million supercomputer is just not good enough for accurate predictions and they now want a £97 million supercomputer.

One prediction was that at least three of the years after 2009 would be hotter than the El Niño year of 1998 (when it is hotter anyway). And what happened? The Met Office claimed that 2010 and 2014 were hotter than 1998 but this was only after "homogenising" upwards the measured published sea and air surface temperatures. There was no scientific justification as to why this was valid. Furthermore, satellite measurements showed that 1998 was a hot year and 2010, 2014, 2015 and 2016 were nowhere near what were previous hot years. The Met Office conveniently ignored the more accurate satellite data.

Another Met Office claim was that there would be more heat waves like the 2003 heat wave when an extra 15,000 people checked out early. However, at the time, the same meteorologists stated that the 2003 heat wave was nothing to do with climate change and that it was due to an unusual influx of hot air from the Sahara. Extreme weather events, such as abnormal rainfall were another Met Office scare scenario. In 2014 there was heavy rain and flooding but the Met Office's own records show that far more rain fell between December 1929 and January 1930.

The global warming poster child, Greenland, is always good for a scare. The Met Office claimed that at some unspecified time in the future, there will be a melting of the Greenland ice cap, sea level would rise by six metres and coastal and estuarine cities will be engulfed. This is actually a correct prediction because, despite six major ice ages, ice has only been on Earth for less than 20% of time. Some time in the geological future when our current ice age stops, the polar ice will melt. I predict that this will be tens to hundreds of millions years away and will occur at five past on a Thursday. Maybe the Met Office was not aware of studies in Greenland that show since 1900 AD, the Greenland air temperature has not risen. Maybe the Met Office was not aware that the record low temperature for July was in Greenland in 2017 (-33°C).

During this period from 2004 to 2014 when the climate doomsday book was being compiled, the Met Office was making additional predictions by playing £33 million computer games. It was predicted that 2007 would be the "hottest year ever". This prediction was just before global temperatures decreased by 0.7°C. They also predicted that 2007 would be "drier than average". It was a very wet year with some of the worst floods in recorded history.

Between 2008 and 2011, the UK was to experience "warmer than average" and "hotter and drier" summers, or so the computer said. These were some of the coolest and wettest summers on record, even though there was a short time in 2009 that was called the "barbeque summer". In October 2010, the Met Office predicted that winter would be "two degrees warmer than average". It was the coldest and snowiest December since records began in 1659 AD.

Here is a credulity test for you, dear reader. A November 2011 Met Office computer forecast claimed that global temperature would rise by as much as 0.5°C by 2017. What do you think? Someone in the Met Office agrees with you and, after only a year, this prediction was removed from the Met Office web site. In 2017, temperature was falling.

In March 2012, we learned that spring would again be "drier than average". This was a prediction made just before the wettest April ever. Not to be outdone, in November 2013 it was predicted that the UK winter would be "drier than usual". Thousands of people breathing through snorkels disagreed as the UK had the wettest three months on record. The Met Office predicted a warmer, drier summer than average for 2015. The rain poured down in August and the summer was the 178th warmest since records began in 1659 AD. The English were left shivering in the rain. The Met Office has decided that scary predictions will enable them to gouge more money out of the poor suffering UK taxpayer.

The European summer of 2017 had some hot weather when the temperature was greater than 40°C. The media got into a lather, the world was going to end in what was clearly due to we sinful humans. In the Southern Hemisphere winter, it was bitterly cold in July. This was not predicted and was apparently due to global warming.

Climate "scientists" did not predict record crop yields. They did not predict a greening of the planet over the last 20 years. They did not predict record sea ice around Antarctica. They did not predict a rapid growth of Arctic sea ice. They did not predict fewer hurricanes and cyclones.[199,200] They did not predict colder European winters. They did not predict more snow in the Northern Hemisphere. They did not predict dam-filling rains in Australia. They did not predict a failure of the Great Barrier Reef to die. They did not predict that for more than two decades there would be no warming. Why should the taxpayer fund or even take any notice whatsoever of predictions by proven failures?

Green left environmental activists and their uncritical supporting media[201] are engaged in emotional posturing, selective use of information, omission of critical data, moralising, crying wolf and predicting doom and gloom. Global average sea surface temperatures have not increased in any significant way since 1998,[202] average global temperature has not increased

---

[199] R. Pielke, 2012: Hurricanes and human choice. http://www.wsj.com/articles/SB10001424052970204840504578089413659452702
[200] http://www.drroyspencer.com/2015/05/nearly-3500-days-since-major-hurricane-strike-despite-record-high-co2/
[201] Chris Kenny, "Media Watch's climate change obsession", *The Weekend Australian*, 1-2 August 2017.
[202] http://judithcurry.com

for more than two decades,[203] Arctic sea ice is increasing[204] and Antarctic sea ice is increasing.[205] The Arctic sea ice area initially decreased and now it is back to the 2006 levels.[206] Both the Greenland and Antarctic ice sheets are increasing in size. However, green left environmental activists see it differently and try to present heroic arguments that all is not well and we are doomed.

Their scary predictions come from models, not measurements. I think that there is a significant difference. Modellers actually don't collect primary data themselves and their claim to fame is to massage the data collected by others.

And why was 2°C important for Paris? This is just a number plucked out of the air. During previous times, the planet's temperature has changed up and down by 3.5°C, people didn't die like flies in warmer times and premature death is far more common in colder times. For example, in the warm up propaganda for Paris the French Foreign Minister Laurent Fabius stated to the US Secretary of State John Kerry on 13 May 2014 that "... we have 500 days to avoid climate chaos."

At the time of Fabius' comments, the UN had just scheduled a climate summit for Paris in late November-early December 2015, some 500 days from the Fabius' prediction. There has not climate chaos, just business as usual.

Climate conferences always attract predictions from those with a self-interest and all sorts of deluded people come out of the woodwork and make predictions. Not one previous prediction has been correct.[207] World leaders met in Copenhagen

---

[203] http://www.forbes.com/sites/jamesconca/2015/06/15/a-pause-in-global-warming-not-really/
[204] http://nsidc.org/arcticseaicenews/
[205] https://www.nasa.gov/content/goddard/antarctic-sea-ice-reaches-new-record-maximum/
[206] https://www.arctic.moaa.gov/detect/ice-seaice.shtml
[207] http://dailycaller.com/2015/05/04/25-years-of-predicting-the-global-warming-

in late 2009 to try to twiddle the dials to stop global warming when outside there was thick snow and it was bitterly cold. The leader of the greens in Canada wrote in 2009 that:

> ... we have hours to act to avert a slow-motion tsunami that could destroy civilization as we know it.

and

> Earth has a long time. Humanity does not. We need to act urgently. We no longer have decades; we have hours. We mark that in Earth Hour on Saturday.

This reads like the rantings of someone emotionally unbalanced. The only thing of interest with Earth Hour is the number of children conceived. April 22nd: Earth Day, March for Science, Lenin's birthday. Appropriate. The annual Earth Hour, when people in rich Western countries are supposed to turn off their lights for 60 minutes, is to repent for using fossil fuels to electrify our houses, schools, hospitals and businesses. Of course, refrigerators, heating and air conditioning are not turned off and toxin emitting candles are used by the deluded piously gathering in the semi-darkness. Some of us turn on every electrical gadget in the house during Earth Hour to celebrate humanity's incredible achievements since the Industrial Revolution that have given us modern living standards, increased health and longevity and the freedom for all people to improve their lives. The Western World may have had slight pollution problems, but these are nothing compared to the developing world and we have cleaned up our air, soils and water.

While the Earth Hour celebrants claim to be concerned about the poorest in the world, their policies produce soaring electricity prices, fewer jobs and lower living standards in the Western world and, in developing countries, perpetual poverty,

tipping-point/

disease, malnutrition and premature death. We are paying more and more for minimal improvements in the environment. This is combined with ever-expanding government and activist control of our lives and stonewall opposition to reliable, cheap energy for the Third World. Earth Hour should be renamed Green Energy Poverty Hour as recognition of how green schemes damage the poor, the elderly, the working class and developing world families.

In 1982 the UN was already telling the world that it only had a decade to solve global warming or face the consequences. Nature did the job and there has now been no global warming for more than the last two decades. We can thank nature for showing us how little we really understand about Earth's systems. In 1989, Noel Brown, a senior UN environmental official, claimed that:

> ... entire nations could be wiped off the face of the Earth by rising sea levels if global warming is not reversed by the year 2000.

We've had a few new nations appear rather than disappear since 1989 (e.g. South Sudan) and the prediction was wrong in 1989, it was wrong in 2000 and it is still wrong in 2017. The UN's IPPC former head, Rahendra Pachauri, stated in 2007 that:

> If there's no action before 2012, that's too late. What we do in the next two to three years will determine our future. This is the defining moment.

All that has happened since 2007 is that temperature has not increased. Pachauri's future was certainly determined in a defining moment. He resigned in 2015 as the IPCC head after allegations he sexually harassed many female colleagues. Again, in 2007, the UN reported that global emissions must peak by 2015 for the world to have any chance of limiting the expected temperature rise to 2°C. The temperature has not risen, the temperature has not fallen and carbon dioxide emissions continue to rise because more and more people in the Orient are enjoying a better standard of living. How long can the UN cry wolf?

Obama made a campaign promise to "slow the rise of the oceans" and Obama was advised in 2012 by the UN Foundations President (Tim Wirth) that his second term was the "last window of opportunity" to reduce the use of fossil fuels in the US. Wirth advised the President that it's the "last chance we have to get anything approaching 2 degrees Centigrade" and "if we don't do it now, we are committing the world to a drastically different place".

Obama attracted unsolicited advice and climate activist James Hansen warned in 2009 that Obama only "has four years to save the Earth". That was nearly a decade ago. Not only have I missed trains and planes, but it appears that I also missed the end of the Earth.

The UK has its share of doomsdayers. Prime Minister Gordon Brown in 2009 stated that there was only "50 days to save the world from global warming" and, following on from his hysterical prediction, claimed that there was "no Plan B". At that time there had been 12 years without warming. In July 2009, Prince Charles said that there would be "irretrievable climate and ecosystem collapse, and all that goes with it" and that we had 96 months to save the planet. I don't know how he arrived at that figure but those 96 months have been and gone with no change to the climate and ecosystem.

Green left environmental activist journalist George Monbiot, the master of failed predictions, wrote in 2002[208] that:

Within as little as 10 years, the world will be faced

---

[208] http://www.theguardian.com/uk/2002/dec/24/christmas.famine

with a choice: arable farming either continues to feed the world's animals or continues to feed the world's people. It cannot do both.

and

The impending crisis will be accelerated by the depletion of both phosphate fertiliser and the water used to grow crops.

Sorry George, the world's people 15 years later actually get a better feed than before and neither phosphate fertiliser nor water have run out. According to the UN, at the time of his 2002 prediction, about 930 million people were undernourished. By 2014 the number was 805 million despite a population increase from 6.28 to 7.08 billion people.[209] No wonder George is known affectionately in England as Moonbat. It's not hard to see why.

SCIENTIFIC CONSENSUS

A consensus on a scientific conclusion arrests the development of scientific advances.

Incorrect consensus

Consensus views in the past were that the Sun rotated around Earth, that burning material was due to a substance called phlogiston, that health was driven by humours, that leeches could cure most diseases, that malaria was caused by bad air, that earthquakes were an act of God, that heavier-than-air machines could never fly, that the continents were fixed and that duodenal ulcers were caused by stress. All these views strongly supported by the scientific consensus at the time have been shown to be wrong.

---

[209] http://www.faq.org/hunger/en/

Scientist Charles Darwin and his cousin Francis Galton invented eugenics to correct the fact that "poverty-stricken drunken masses bred in larger numbers than the nouveaux riches like the Darwins".[210] There was once a consensus about eugenics. The plan was to identify those who were feeble minded (and that variously included Jews, blacks and foreigners) and stop them from breeding by isolation in institutions or by sterilisation. George Bernard Shaw claimed that it was only eugenics that could save mankind. H.G. Wells spoke against "ill-trained swarms of inferior citizens" and Theodore Roosevelt stated that "Society has no business to permit degenerates to reproduce their kind."

Eugenics research was funded by the Carnegie Foundation and the Rockefeller Foundation. With some eminent citizens and respectable foundations supporting eugenics, how could it not be science? We now know it was a racist, anti-immigration social program masquerading as science.

What would have happened if, in 1900, we had assembled a few hundred geologists into a mythical International Panel on Continental Drift (IPCD) and asked the question: Do continents drift? The answer amongst geologists would have been a resounding no.

If we asked the same question again in 1920 after the great works by Alfred Wegener in 1913-1914 were translated from German to English after the 1914-1918 war, the consensus would have been negative. A few geologists might have said maybe, but we need more data and a more plausible mechanism. Even fewer would have stuck their necks out and said that the data supported continental drift. The majority of geologists would have said that continental drift was nonsense and used authority, professional societies and peer-reviewed science to

---

[210] A.N. Wilson, *Charles Darwin: Victorian Mythmaker*, Hodder and Stoughton, 2017.

support their position. Wegener and his supporters would have been ridiculed. And they were.[211]

In 1940, the consensus of IPCD would again have been no. By then UK's Harold Jeffreys had shown that the mechanism proposed by Wegener was impossible, Germany was again at war and there were strong anti-German feelings. The IPCD would have rejected the theory of continental drift and, by doing so, would have rejected all of the supporting data because the mechanism suggested by Wegener was not plausible.

In 1960, the IPCD consensus would still have been no. Maybe a few geologists, geophysicists and geochemists might have said possibly but better data and explanations are required. Again, the majority of geologists would have said that continental drift was nonsense and would have used authority, professional societies and peer-reviewed science to support their position. Arguments such as the "science is settled" would have been used. In the early 1960s as an undergraduate student, we were exposed to the theory of continental drift. Compared to today, those were enlightened times.

If the IPCD were asked the same question in 1980, as a result of the integration of many disciplines of geology and a huge volume of new data, the consensus would be that continents do indeed drift and that the mechanism is known. In the late 1970s, opponents of continental drift and plate tectonics were ridiculed in public by those who embraced the new paradigm.

In 2000, if the IPCD were asked the question again, the consensus would be yes. Some might give a cautious yes, until a better theory emerges. Others would argue about the details of plate tectonics and whether this all-embracing theory could explain all observations and, as a consequence, would enjoy the ridicule from their peers.

---

[211] www.age-of-the-sage.org

So too with "climate science", an infant scientific discipline, built from other disciplines. Mechanisms of climate change are probably far more complicated than the mechanisms of plate tectonics because they involve complex, non-linear, chaotically interacting systems which in turn involve poorly understood fluid dynamics. It may take many decades before "climate science" can come to a better understanding of climate. In the interim, we have a "consensus", "settled science" yet a large number of scientists, especially in non-Western countries, oppose the popular view that human emissions of carbon dioxide drive climate change.

Over 100 years, the hypothetical IPCD completely changed its position. The IPCC has only been alive for 25 years and it may take at least another 100 years before climate is better understood, as happened with continental drift (which evolved into plate tectonics). It was the integration of major disciplines of science that led to the theory of plate tectonics. There is no such coherent theory of climate.

The evolution of "climate science" needs the synthesis of other scientific areas such as geology. This evolution of understanding will be slowed down because the IPCC is an outreach arm of the UN which is trying to control all society by climate change treaties, governments and renewables. Businesses have billions to gain from the fraudulent self-serving behaviour of the climate industry, environmentalists and scientific activists.

With the idea of continental drift, there were not huge pots of money involved. Initially there were anti-German views. Only nationalism, egos, reputations and careers were at stake. With climate "science", there are attempts to control every country's energy use by the UN. There are attempts to transfer huge amounts of money from the Western world through sticky fingers to others and attempts by the UN to interfere with the internal politics in every country.

There are huge amounts of research funds sloshing around but the economic implications of a carbon trading scheme, renewable energy, batteries, electric cars, new taxes and the transfer of wealth have brought together all sorts of disparate and desperate groups. Science funding opportunities, the rise of the Green Party and financial opportunities in the renewable energy market place have combined to produce an almost perfect storm of quasi-religious hysteria with all the hypocrisy that one associates with a fundamentalist religion.

Fear of the environmental consequences of ever-expanding economies combined with the deterioration in science education encourages people to think of simple solutions to problems (or non-problems). Simple solutions to complex problems are invariably wrong. This is amplified by a lazy media that exploits fears to increase sales and can't really be bothered to get to the bottom of a story.

The 97% consensus fraud

We hear from the media that, in the scientific community, there is little controversy and 97% of climate scientists conclude that humans are causing global warming. Is that really true? It is a zombie statistic. However much you try to kill it off, it keeps coming back and it's simply not true.

In the scientific circles I mix in, there is an overwhelming scepticism about human-induced climate change and many of my scientific colleagues claim that the mantra of human-induced global warming is the biggest scientific fraud of all time. And future generations will pay.

If 97% of climate "scientists" agree that there is human-induced climate change, you'd think that the 97% of scientists who supposedly all agree about climate change would be busting a gut to vanquish climate "deniers" in debates. Apparently not. You'd think that the scientific establishment would embrace an opportunity to present their case to the public. After all, numbers and right is on their side. Instead, many scientists and activists are expressing outrage at this logical suggestion and are advising colleagues not to participate in debates.[212]

In my plus 40-year scientific career, there has never been a hypothesis where 97% of scientists agree. Just go to any scientific conference. Conferences are a collection of argumentative sceptical scientists who don't believe anything, argue about data and argue about the conclusions derived from data. Scientists, as well as lawyers, bankers, unionists and all other fields, can make no claim to being honest or honourable and various cliques of scientists have their leaders, followers and enemies. Scientists are no different from anyone else and they are human with all the human frailties. Scientists differ from many in the community because they are allegedly trained to think, analyse, criticise and be independent. Unless, of course, whacking big research grants for climate "science" are waved in front of them.

My own experience is that warmist scientists will not engage in open debates in front of those who pay them. Instead of dialogue, warmists want to preach and are offended at even the slightest suggestion[213] that there is something wrong with climate "science". To follow up on this, there was even more confected outrage by another climate scientist with his snout in the trough who stated,[214] "We can argue about what we should do or should not do ... but the argument is over."

The 97% derives from a survey sent to 10,257 people, of which the 3,146 respondents were further whittled down to 77 self-selected climate "scientists" of whom 75 were judged

---

[212] Julie Kelly, *The Federalist*, 6 July 2017.
[213] Naomi Orekes, Ben Santer and Kerry Emanuel, op. ed., *Washington Post*, 21 June 2017.
[214] Ken Caldeira, *Washington Post*, 1 July 2017.

to agree that human-induced warming was taking place. What were the criteria for rejecting 3,069 respondents? There was no mention that 75 out of 3,146 is 0.03%. We did not hear that 0.03% of climate scientists agreed that humans have played a significant role in changing climate despite the billions spent on climate research.

A recent paper on the scientific consensus of human-induced climate change was a howler.[215] Only such papers can be published in the sociology or environmental literature. The paper by Cook *et al.* (2013) claimed that published scientific papers showed that there was a 97.1% consensus that man had caused at least half of the 0.7°C global warming since 1950.

How was this 97.1% figure determined? By an inspection of 11,944 published papers. Inspection? Is this the way that rigorous scholarship is undertaken? This was not a critical reading and understanding derived from reading every one of the 11,944 papers. This was not physically possible as the study started in March 2012 and was published in mid 2013 hence only a cursory inspection was possible. What was inspected? By whom?

The methodology section of Cook's note tells it all:

> This letter was conceived as a 'citizen science' project by volunteers contributing to the Skeptical Science website (www.skepticalscience.com). In March 2012, we searched the ISI Web of Science for papers published from 1991-2011 using topic searches for 'global warming' or 'global climate change'.

This is translated as: This study was a biased compilation

---

[215] J. Cook, D. Nuccitelli, S.A. Green, M. Richardson, B. Winkler, R. Painting, R. Way, P. Jacobs and A. Skuse, 2013: Quantifying the consensus on anthropogenic global warming in the scientific literature. *Envir. Res. Lett.* 8: doi:10.1088/1748-9326/8/2/024024

of opinions from non-scientific politically-motivated volunteer activists who "inspected" 11,944 scientific papers, who were unable to understand the scientific context of the use of "global warming" and "global climate change", who rebadged themselves as "citizen scientists" to hide their activism and scientific ignorance, who did not read the complete paper and, if they did, were unable to critically evaluate the diversity of science published therein.

The conclusions were predictable because the methodology was not dispassionate and involved decisions by those who were not independent. If this was a financial study upon which investment decisions were made, people would have gone to gaol.

As part of a re-evaluation, the original 11,944 papers were read and the readers came to a diametrically opposite conclusion to Cook *et al.*[216] Of the 11,944 papers, only 41 explicitly stated that humans caused most of the warming since 1950 (i.e. 0.3%). Cook *et al.* had flagged that just 64 papers supported the consensus but only 23 of the 64 actually supported the consensus. Of the 11,944 climate "science" papers, 99.7% did not say that carbon dioxide caused most of the global warming since 1950. Just 1.0%, not 97.1%, had been found to endorse the claim that humans are to blame for a most of the current warming.[217]

Legates *et al.* stated:

> It is astonishing that any journal could have published a paper claiming a 97% climate consensus when on the authors' own analysis the true consensus was well

---

[216] D.R. Legates, W. Soon, W.B. Briggs and C. Monckton of Brenchley, 2013: Climate consensus and 'Misinformation': A rejoinder to Agnotology, Scientific Consensus, and the Teaching and Learning of Climate Change. *Sci. Educn* 24: 299-318.

[217] Joseph Bast and Roy Spencer, "The myth of the climate change '97%'," *The Wall Street Journal*, 26 May 2014.

below 1%. It is still more astonishing that the IPCC should claim certainty about the climate consensus when so small a fraction of published papers explicitly endorse the consensus as the IPCC defines it.

Not one paper endorsed a man-made global warming catastrophe. Not one! So what was the necessity to publish misleading and deceptive information?

Furthermore, Cook *et al.* used three different definitions of climate consensus interchangeably and arbitrarily excluded about 8,000 of the 11,944 papers that expressed no opinion on the climate consensus. It appears that Cook's lackeys (i.e. "citizen scientists") made scientific judgments yet they were not scientists. The Cook *et al.* paper supported predetermined conclusions which can be gleaned from the Skeptical Science website.[218] Readers can make up their own minds whether the 97.1% figure is misleading, deceptive, fraudulent or all three.

The Cook *et al.* (2013) paper is not in accord with some of Cook's other papers. In my lifetime of science, each new publication builds on previous work. In that way, there is no great Eureka moment in science but an increasing body of repeatable validated evidence that strengthens a theory. There are also publish or perish pressures and there needs to be a constant stream of publications to win new research grants and to stay employed. This pressure leads to decreasing the standard of science and contradictory publications. Cook is a prime example.

The Cook *et al.* paper certainly does not look very rigorous and looks like propaganda that an unknown journal accepted. This certainly is not the scholarship expected from taxpayer-funded university staff. Cook is far from a young man, was still undertaking PhD work in cognitive psychology when

---

[218] www.skepticalscience.com

the 2013 paper was published, he was at the University of Queensland's Global Climate Institute where he was a "Climate Communication Fellow" and ran an online course on *Making sense of climate science denial*. The course was developed "To further the work of educating the public and empowering people to communicate the realities of climate change." Cook's remaining career depends on the number of papers published, especially as he started his academic career in middle age.

This course looks like an imitation of Al Gore training presenters to promote the scares presented in his 2006 film *An Inconvenient Truth*. We taxpayers funded an under qualified person to undertake propaganda in support of a political ideology. As Cook's paper shows, evidence is in short supply. And now for a heretical thought: Maybe those who deny that human emissions of carbon dioxide drive climate change use evidence.

Claims that 97% of scientists agree are not backed up by any credible study or poll. The UN IPCC lead author Dr Richard Toll stated:

> The 97% is essentially pulled from thin air; it is not based on any credible research whatsoever.[219]

In the physical sciences, we would call Cook's work fraud. In the social sciences it is career advancement.

After the Cook *et al.* (2013) paper, he kept the research grant wheels spinning as coauthor of another paper on scientists' views about global warming.[220] This is not rigorous research as it is just a compilation of the opinions of 1,868 passengers on the gravy train who have a vested interest. However, the Verheggen

---

[219] UN IPCC lead author Dr Richard Toll. *Testimony to U.S. Congress: Full Committee Hearing – Examining the UN Intergovernmental Panel on Climate Change Process*, 29 May 2014.
[220] B. Verheggen, B. Strengers, J. Cook, R. van Dorland, K. Vringer, J. Peters, H. Visser and L. Meyer, 2014: Scientists' views about attribution of global warming. *Envir. Sci. Technol.* 48: 8963-8971.

et al. (2014) paper hints that the consensus among climate "scientists" on the gravy train might not be as strong as thought.

In 2013, Cook published that the consensus was 97.1% and in 2014 the consensus he published was now 43%. The authors also suggest that "as the level of expertise in climate science grew, so too did the level of agreement on anthropomorphic causation." Expertise was subjectively equated with the number of peer-reviewed papers that had been published in the climate "science" literature. Blimey, those in First Class on the gravy train are experts whereas those in Economy Class are not. You heard it here first. Papers published in sociology journals are always an amusing read, especially when one has a stiff drink at hand, and clearly demonstrate the dumbing down of our education system.

Such shoddy work raises many questions. How could the Cook *et al.* 97.1% suddenly drop to 43% in a year? This was not explained. The survey took place in March-April 2012 yet during March 2012 Cook was compiling data for his Cook *et al.* (2013) data. Cook would have known that his March-April 2012 data used in Verheggen *et al.* (2014) was not in accord with his March 2012 data used in Cook *et al.* (2013). Cook *et al.* (2013) did not explain the contrary results that he was acquiring in March-April 2012 for another publication.

I smell a rat. Did Cook get the idea from Verheggen, conduct his own hasty survey using "citizen scientists" and gain publication priority by beating Verheggen into print? If I ever shared the same funeral pyre with this mob, it still would be hard to warm to them. In the corporate world one can go to gaol for such behaviour and I see no reason why dodgy behaviour in the academic, political or union world should be any different from the corporate world.

So why was Verheggen *et al.* (2014) different from Cook

et al. (2013)? Neither paper had a disclaimer by Cook. Was there a sudden change in opinion by climate "scientists", was the methodology flawed, did the "citizen scientists" skew the data or do different data sets give different answers thereby rendering the whole process unreliable? The methodology of Verheggen *et al.* (2014) was very subjective. Some 6,550 people were invited to participate in the survey and only 1,868 participated. I often get asked to participate in such surveys and the invitation always ends up in the circular filing cabinet. A participation rate of 29% does not give confidence.

Respondents were picked because they had authored climate "science" papers between 1991 and 2011 that had key words such as "global warming" or "global climate change". This is a clumsy survey because Verheggen *et al.* (2014) were not required to read the papers, were not required to understand the science and were not able to understand whether the paper surveyed was good or bad science. A computer search of key words does not give confidence in either the methodology or the conclusions. It is a lazy person's methodology and certainly is not scholarship.

Fabius Maximus[221] analysed the Verheggen *et al.* (2014) study and showed that only 64% of passengers on the gravy train agreed that human emissions of carbon dioxide was the main or dominant driver of more than half the temperature rise. This was despite the fact that there has been no measured increase in the average global air temperature for more than 18 years. Of this 64% (1,222 participants), only 797 agreed that it was "virtually certain" or "extremely likely". That's only 43% of climate "scientists" that agree with the IPCC statement:

It is extremely likely (95% certainty) that more than

---

[221] http://fabiusmaximus.com/2015/07/29/new-study-undercuts-ipcc-keynote-finding-87796/

half of the observed increase in global average surface temperature from 1951 to 2010 was caused by the anthropogenic increase in greenhouse gas concentrations and other anthropogenic forcings together.

However, despite some pretty shoddy methodology on opinion compilation, if only 43% of climate scientists agree with the IPCC's 95% certainty, from my experience in science this is what would be expected.

Long gone are the days when experts can use their authority to state that "they know." Try that in court as an expert witness and you'll be chewed up and spat out in a few minutes. There is a huge body of literature that shows that expert opinions can be wrong. There is nothing wrong with saying "I don't know", as I have said to generations of students. This was done by 47% of the scientists surveyed by Verheggen *et al.* (2014).

As soon as there is a claim that 97% or 95% of scientists agree, then I disbelieve such a claim. And so should you. Many times I have had a group of geologists on an outcrop in the bush and heard the arguments. All the arguments are underpinned by the same evidence that is right in front of their very eyes. The only consensus achieved is that all participants agree to disagree.

Rather than argue using evidence, major trillion dollar policies are supported by stating that there is a 97% consensus. The policy debate starts with assuming there is a problem with climate. There isn't. Climate is complex, we don't fully understand it, models to try to understand it have failed and much climate "science" is fraudulent. All the calculations show that if one nation reduces its carbon dioxide emissions by 5% or even 50%, it will make no measurable difference to the modelled global average temperature. The developing world will continue to emit increasing amounts of carbon dioxide.

After all the policy fights, political battles and trauma and we eventually decide to have a carbon tax, emissions trading scheme, carbon capture and storage, renewable energy target, wind turbines, solar cells and everything else on Earth to stop using coal, will global temperatures be lowered? No.

So why do we do it? National pride at international climate conferences such as Copenhagen and Paris? Political prestige, ego, popularity and status? Greater government control of every aspect of our lives? Promotion of a global government? Shifting of more power to unelected bureaucrats, courts, academics, NGOs and green left environmental activists? Creating artificial carbon markets which both banks and socialists lust over? Destruction of capitalism? It's certainly not about the environment or the economy.

This will cost future generations dearly. Your electricity bills show that it is costing you now.

What is not mentioned is that in the US there were more than 30,000 scientists who signed a document claiming that human-induced global warming was nonsense. I am not playing a numbers game here, as I am only too well aware of Aristotle's logical fallacies. My point is that there is selective reporting of information and surveys.

By contrast, the American Physical Society (APS) which with more than 50,000 members is probably the largest society of physicists in the world, is formally examining competing views on the physical science basis of global warming. In the light of the failure of IPCC climate change models to predict reality, the APS has appointed a six-person panel including three independent scientists of a sceptical nature.

And what is a climate "scientist"? This is not a field of science. It is an invention of an exclusive club to exclude all those mathematicians, physicists, chemists, biologists, astronomers, geologists and meteorologists who don't follow the ideology. This club thrives off research grant cash grabs funded by taxpayers. And what does the taxpayer get in return? Climate "scientists" trying to frighten people witless so that there can be more taxpayer funded research. If one is to study climate, then almost every field of science needs to be studied and integrated. This is just not possible.

This is why the climate "science" clique, mainly comprising computer modellers and meteorologists, excludes those with the most to offer such as solar physicists, astronomers, geologists and carbon dioxide chemists. Climate "science" is geared to confirm the ideology of human-induced global warming and no other idea will be investigated. What about a study of the natural climate cycles? What about a study on the possibility of global cooling?

Following concerns that the so-called climate consensus was not reaching the public, a comprehensive opinion poll of 10,000 Europeans in 10 countries was conducted to establish levels of awareness, concern, and trust amongst different demographic groups and nationalities.[222] A majority (54%) of Europeans and two-thirds of Britons rejected the claim that climate change is mainly or entirely caused by humans. Europeans remain sceptical and unprejudiced, despite decades of climate alarmism and green indoctrination.[223] Maybe the public is not as stupid as climate "scientists" would like to believe.

The whole issue of human-induced global warming has become a gravy train. Even sociologists have swooped in to get funds to study why people may be sceptical about human-

---

[222] P.J. Buckley, J.K. Pinnegar, S.J. Painting, G. Terry, J. Chilvers, I. Lorenzoni, S. Gelcich and C.M. Duarte, 2017: Ten thousand voices on marine climate change in Europe: Different perceptions among demographic groups and nationalities. *Frontiers in Marine Science*, doi.org/10.3389/fmars.2017.0026

[223] Global Warming Policy Forum, 12 July 2017.

induced climate change. For example, from Tranter and Booth (2015):[224]

> Climate scepticism persists despite overwhelming scientific evidence that anthropogenic climate change is occurring (IPPC, 2013)[sic]. The reasons for this are varied and complex. Understanding why climate scepticism endures or is even on the rise, and why levels of scepticism vary across countries, requires accounts that recognise that biased information assimilation is mediated by cross-cultural and intra-national differences in values and worldviews. Fruitful explanations of scepticism must also account for the way in which partisans are influenced by their political leaders. Integrating such accounts may provide a way to both understand and address the social problem of climate scepticism.

This is post-modernist gobbledygook. Climate always changes hence there is no such thing as climate scepticism. There is not overwhelming evidence that anthropogenic climate change is occurring. The only reference used to support this statement is by an activist group with self-interest. There is a huge literature in the integrated interdisciplinary world of science (as I outlined in *Heaven and Earth. Global Warming: The Missing Science*)[225] that has well-founded but different conclusions. One doubts whether sociologists Tranter and Booth actually read the scientific literature, let alone understand it.

They certainly did not read the IPCC report, although they use an unknown IPPC 2013 report [sic] as a reference so presumably they have read this report because they claim it provides

---
[224] B. Tranter and K. Booth, 2015: Scepticism in a changing climate: A cross-national study. *Global Envir. Change* 33:154-164.
[225] Ian Plimer, *Heaven and Earth. Global Warming: The Missing Science*, Connor Court, 2009.

"overwhelming scientific evidence". It would be interesting to know how sociologists can evaluate science and conclude that there is overwhelming scientific evidence. In places, the IPCC report shows great reserve and shows that the magnitude of human influence on global climate is unknown.[226,227,228,229] The sociologists noted that when identifying "climate sceptics", their shoddy study did not mention that no scientist is sceptical about climate change but many of us are sceptical about human-induced climate change.

This world-shattering leading edge research was just a survey. The survey questions were trite, nonsensical and made unsubstantiated assumptions. The methodology of the survey from which conclusions are based is flawed. Natural climate changes and human-induced change (e.g. due to land clearing, the urban heat island effect) take place yet there is still no data to show that human-induced global warming from human emissions of carbon dioxide is significant or dangerous. A flippant concluding reference to an IPPC 2013 [sic] report shows the standard of scholarship, peer review and editorial rigour.

Science is underpinned by scepticism. The reason why public scepticism is on the rise is the average taxpayer has realised

---

[226] IPCC 2013, SPM, p. 3, Section B.1, bullet point 3, and in Synthesis Report p. SYR-6 ("... the rate of warming over the past 15 years ... is the rate calculated since 1951 ...").

[227] IPCC 2013, WGI, Ch 9, box 9.2, p. 769 and in Synthesis Report SYR-8 ("... an analysis of the full suite of CPIP5 historical simulations ... reveals that 111 out of 114 realisations show a GMST trend over 1998-2012 that is higher than the entire HadCRUT4 trend ensemble ...").

[228] IPCC 2013, SPM, D.1, p.13, bullet point 2 and Synthesis Report SYR-8 ("There may also be a contribution from forcing inadequacies and, in some models, an overestimation of the response to increasing greenhouse gas and other anthropomorphic forcing [dominated by the effect of aerosols]").

[229] IPCC 2013, WGI, ch 9, box 9.2, page 769 ("This difference between simulated [i.e. model output] and observed trends could be caused by some combination of (a) internal climate variability, (b) missing or incorrect radiative forcing and (c) model response error").

that climate "science" is exaggerated, embellished, wrong or fraudulent. Time and time again models and predictions have been wrong and the consequences of failed politicised science have started to become very expensive for the average person. They have become sick of being told that the world's going to end. Print, television and radio are now no longer the only source of information. The green left environmental alarmists have been crying wolf for a quarter of a century, the climate has not changed noticeably, people are struggling even more to pay their energy bills and the community now no longer listens to those who cry wolf.

People have not been influenced by their political leaders. Political leaders have responded to the community scepticism and have acted accordingly. In the UK, Australia and elsewhere some people have changed political allegiances because a particular party has uncritically embraced a policy on climate change. Maybe Tranter and Booth have a naïve understanding of politics and don't understand that many people cast their vote on a diversity of policies rather than down party lines.

Tranter and Booth label climate scepticism as a social problem hence they can rationalise opening the floodgates for further research funds. The research was not dispassionate otherwise it would not have been claimed that there is a social problem with climate scepticism. The authors forget to add their assumption in the paper: That climate change occurs and has been taking place for billions of years. Did the authors ever consider that scepticism underpins intellectual thought, that there is far more access to information today than ever before and the average person can now critically analyse claims made by green left environmental activists who claim that they are scientists? One does not have to be a titled person in a university to possess knowledge and the skills of critical analysis.

Independent thinkers such as sceptics do not accept being

told what to think, need evidence to underpin doomsday predictions and do not accept pronouncements made upon high by authorities. They actually think for themselves. The authors comment that sceptics tend to be older. Maybe older people had a more rigorous education where facts and thinking processes were paramount rather than feelings, beliefs and environmental ideology. Maybe older people have seen more fraud, fanatics, fools, fads and furied fundamentalism.

After nearly 45 years on the staff of various universities and a Chair for nigh on 30 years, all I can state is that many current university staff have been trained in a system that has been dumbed down. It shows. Some of them, although now middle-aged, tenaciously hang on to their undergraduate young and foolish Marxist view of the world to make their arrested development complete. A more fruitful line of sociology research would be to investigate why many people are so gullible and accept the unsubstantiated claims and models of the IPCC when there is such a large body of contrary evidence.

Tranter and Booth are clearly gobsmacked why the average taxpayer is not fooled. They need to get a life. They clearly treat anyone who is not a high and mighty academic sociologist and who does not follow the current fad with great disdain. A word of advice to Tranter and Booth: The public is not stupid. And what sort of journal publishes this tosh. The quote above shows that this article is certainly not scholarship, the facts were not checked, the refereeing was poor and the editing does not seem to exist.

If I was a peer reviewer of the Tranter and Booth paper, I would have suggested to the journal editors that "with a bit more hard work the paper would be of the standard of Sokal (1996)."

Some of us are unreconstructed sceptics. Some of us have a

deep-seated scepticism for anything produced by activists, career academics, governments, political parties, fundamentalist religious organisations, big business, self-interest groups, various environmental groups and anyone who rejects data, logic and rationality for emotional arguments. If in doubt, follow the money.

There is no substitute for common sense.

WHAT IS CLIMATE "SCIENCE"?

There is no such discipline as "climate science". Those trained in sociology, economics, politics, mathematics, physics, chemistry, oceanography and meteorology now call themselves climate "scientists". At times one is criticised because, as a geologist, it appears that I am not a climate "scientist". Well, have I news for you.

Geology and climate change

There have been hundreds of years of study of climate change by geologists. In studies of modern and fossilised plants, Georges-Louis Leclerc, Comte de Buffon (1707-1788) realised that climate had changed in Siberia and Europe. Buffon concluded that past Northern Hemisphere climates were once much warmer and that the Earth was cooling from its original molten state. As with all healthy science, this idea was criticised by Joseph Fourier (1768-1830). Studies of modern and extinct Siberian animals led Georges Cuvier (1769-1832) to conclude that there had been a past freezing event and argued that catastrophe was an important component of Earth history. This was the first time the concept of an ice age appeared in the earth sciences.

Catastrophism was criticised by James Hutton (1726-

1797) and Jean-Baptiste de Monet, Chevalier de la Marck, commonly known as Lemarck (1744-1829). In the Paris Basin, Lemarck recognised that many fossils were tropical species. Both Lemarck and Antoine-Laurent Lavoisier (1743-1794) recognised shallow and deep water sedimentary rocks in the Paris Basin and concluded that sea level rose and fell. The later mapping of the Paris Basin by Alexandre Brongniart (1770-1847) validated earlier work of climate and sea level changes. Baron Alexander von Humboldt (1769-1859) used integrated interdisciplinary science, as geologists do now, to show that the planet has climate zones.

The seminal works of Charles Lyell (1797-1875) showed that the planet is dynamic, that climate changes and that coral atolls do not die with sea level rise. Giovanni Battista Brocchi (1762-1826) also showed that there were tropical fossils in ancient rocks, this time in Italy suggesting that climate changes were global. The Danes (Henrick Beck, 1799-1863) and French (Gérard Deshayes, 1795-1875) independently validated that in the past Europe was tropical.

After discussions between Lyell and von Humboldt, Lyell realised that climate changes can be latitudinal and global and was getting close to thinking about continental drift in order to explain tropical fossils at high latitudes and polar fossils at low latitudes. Lyell realised that there had been ice sheets on Earth and the Swiss geologist Jean Louis Rodolphe Agassiz (1807-1873) concluded that a vast ice sheet had covered Europe in the recent past and left large blocks of rocks that had been transported from elsewhere (erratics). Agassiz showed that Swiss glaciers were retreating and concluded that Europe had been covered by an ice sheet from the North Pole to central Europe.

At that time, little was known about Greenland and its ice sheet and it was only when James Clark Ross (1800-1862) visited the Ross Sea in 1841 and 1842, that the frozen southern continent of Antarctica supported the view of giant ice sheets and zoned climate on Earth. There was great controversy about glacial erratics in Scotland. Lyell's former tutor, William Buckland (1784-1856), concluded that these erratics were relics from Noah's Flood. A visit to Scotland by Buckland, Agassiz and Roderick Murchison (1792-1871) resulted in decades of controversy about ice sheets and climate change.

It was the comprehensive work of Archibald Giekie (1835-1924) on Scottish glaciation that led to a recognition that sea level rises and falls, land level rises and falls and ice sheets retreat and advance as a result of climate change. Archibald Giekie's younger brother (James Geikie 1839-1915) published the definitive work on climate change (*The Great Ice Age*). Later work by James Croll (1821-1890) showed that calved icebergs carry rocks out to sea and these are later dropped on the sea floor when the ice melts. Even in the 19th century, geologists knew that ice sheets wax and wane and that the retreat of an ice sheet was a normal event unrelated to human activities.

Geology had a slow evolutionary advance. "Climate science", the new boy on the block, suddenly has all the answers. I think not.

Et moi

Why is it that after a lifetime of science I have little confidence in the data, the methods, conclusions and science communication of scientists in the climate industry and of those who promote the doom-and-gloom story? It is a long list. My criticism of the climate industry is:

(i) much of the temperature data has been "homogenised" to obtain a pre-ordained result,

(ii) Climategate has shown that data is "homogenised" internationally to achieve the required result, critics are prevented from publishing by corruption of the peer review process, dissent is demonised, information is withheld and fraud is committed,

(iii) the orders of accuracy, limitations and errors of measurements are unavailable,

(iv) the rates of past climate change and variability have been ignored,

(v) on all time scales the non-correlation of carbon dioxide and temperature has been ignored,

(vi) the Roman and Medieval Warmings have been dampened or deleted,

(vii) huge bodies of contrary validated data have been ignored (e.g. astronomy, geology, chemistry, history),

(viii) computers have to be tortured to obtain the preordained result,

(ix) computer codes are not freely available,

(x) computer models, even when contrary to measurements, are promoted as being a scientific conclusion,

(xi) neither the data nor the conclusions are in accord with what we know from the present and the past,

(xii) publication is within a closed system that abuses the peer-review system,

(xiii) the normal caution with scientific conclusions is not given,

(xiv) financial interests are not declared,

(xv) financial rewards for publishing a scary scenario are tempting and that the process of refutation seems to have been overlooked, and

(xvi) on a bad hair day, I might even describe the climate

industry as unscientific fraudulent environmental advocacy.

In the corporate world, the tampering, withholding or exaggeration of material information is illegal. If such actions took place in the corporate world, the gaols would be full. Not so in the world of academia where climate "science" is political activism dressed up as science and underpinned by fraud. If the same laws applied in the corporate and the climate "science" worlds, we would need more gaols.

Don't think that the media will keep you informed about the climate delusion and the real reason for your impossibly high electricity costs. It is a sad world when the only alternative information and opinions come from blog sites and not free-to-air radio and television (with a few notable exceptions).

The 2017 Associated Press annual style guide has been published. This is the Bible for many journalists. It again favours left wing bias over impartiality. It previously told journalists to refer to "global warming" as "climate change" and many journalists use these meaningless words.[230] Journalists' use of the words global warming infers that there is human-induced global warming. This has not yet been shown. Using climate change, the journalist infers that whatever change in climate or weather, it must be due to humans. Climate has changed for billions of years before humans were on Earth and there are cyclical warming and cooling periods. Is warming due to humans? Is cooling due to humans? Is any change whatsoever due to humans? The style promotes the use of propaganda rather than validated facts.

The guide now instructs lazy journalists to refer to climate sceptics who were previously labelled as "climate deniers" as "doubters". Associated Press has made itself the science overlord

---

[230] Thomas Richard, *PoliZette*, 13 July 2017.

and clearly does not want journalists to know that scepticism, doubt and criticism form the basis of science. The journalists who slavishly follow the style guide have not checked whether catastrophic global warming predictions have come true. Why bother? A sensational scare story with goodies and baddies keeps bread and wine on the table whereas impartial reporting could treaten employment.

Serious journalists don't need a style guide and can think for themselves.

The intellectual gains made since the Renaissance have been lost over the last two decades. Any system that allows questioning of beliefs is an enlightened system yet the climate industry is doing everything to stop the questioning of the basics of climate. Such behaviour is not that of scientists, but of paid political thugs. Much of the political and media pressure comes from full-time climate advocates paid to misinform.

The consumer and taxpayer, of course, will pay. That's you! And who will benefit? Banks, traders, brokers and insurance companies are lining up in the shadowy world of totally opaque carbon trading, renewable energy and electric cars and batteries to make astronomical profits.

The planet will also pay. Traders and banks have totally different objectives to governments, activists and global warming bureaucrats. No amount of profits made from trading will remove carbon dioxide from the atmosphere. It does not appear that we have learned very much from previous financial crises. The amount of funds transferred to bank profits means less money for genuine environmental programs. Government proposals for a green future are, in reality, proposals for opening the floodgates for the laundering of your tax money.

The tragedy of the climate industry, the politicisation of science and the grubby Climategate affair is the weakening of science. We can only feed ourselves because of science. It is science that gives us longer and better lives. When the next inevitable pandemic or global cooling hits humans, we are not going to solve problems by denigrating those with an opposing view and claiming consensus.

Governments may be so embittered by the human-induced global warming scam that there may be no funding for disciplines such as medical science. In a pandemic or war, we would need science, independent thinking, creativity and for scientists to tread where no one else has been. That is the only way to solve problems but these very problem-solving processes are being undermined and destroyed by the climate "science" industry.

What if I am wrong and a reduction of carbon dioxide emissions is absolutely necessary to save the planet? If Australia stopped all carbon dioxide emissions today, climate "science" models show that global temperature would decrease by 0.0154°C by 2050. Not only would Australia become bankrupt and could not feed itself, such voluntary acts of international environmental kindness would have absolutely no effect on the global climate. Maybe the Green Party would willingly commit economic suicide, no one else would least of all developing countries like China and India. They want our standard of living and nothing is going to stop them.

We cannot ignore social, political, economic and geologic history.

5
TEMPERATURE

If you don't want to read on, the nuts and bolts of this chapter are:

> (a) The official temperature record has been changed to show that past temperatures were colder and recent temperatures were warmer resulting in a warming trend. This is fraud.

As a result of this fraudulently concocted global warming:

> (b) Climate activists have argued that this warming trend is because humans emit carbon dioxide from coal-fired power stations.
>
> (c) It has been falsely promoted that human emissions of carbon dioxide are reduced by building wind and solar power complexes.
>
> (d) These complexes are subsidised and cheap reliable coal-fired electricity has been made uneconomic.
>
> (e) Energy prices have been jacked up and yet renewable energy is unreliable, horrendously expensive, puts people out of work and pushes up your electricity bills.

HOW IS GLOBAL TEMPERATURE MEASURED?
Who measures what and how?

Five organisations publish global temperature data. Two (Remote Sensing Systems [RSS] and the University of Alabama at Huntsville [UAH]) are satellite datasets and three are land-based sets (National Oceanic and Atmospheric Administration [NOAA], NASA's Goddard Institute of Space Studies [GISS] (both of which provide data to the National Climate Data Center [NCDC]) and the University of East Anglia's Climate Research Unit [CRU]). All land-based sets depend on data supplied by ground stations via NOAA. These data centres perform some final adjustments to data before the final analysis. Adjustments (called "homogenisation) are common and poorly documented.

Temperature measurements of the surface of the Earth are compiled into what is known as the HadCRUT data. The data derives from 5° latitude x 5° longitude grid cells covering the Earth's surface. This data gives a two-dimensional picture whereas balloons and satellite data give a three-dimensional temperature map of the atmosphere. There is a total of 2,592 cells.

Cells that entirely cover land comprise just 21.7% of all cells and 18.1% of the total surface area and their data comes from observation stations. Sea grid cells comprise 50.6% of the total cells and 53.8% of the total area, and their data comes from measurements of sea surface temperature. Cells that cover a mixture of land and sea account for 27.7% of cells and 28.0% of area, and their data might be from observation stations or sea surface temperatures, depending on what's available and what's regarded as most reliable.

The oceans cover 70% of the surface of the Earth. Both sea surface temperature and land temperature measurements have errors and the mixing data from two different measurement methods creates greater uncertainties. In any given month of measurements, coverage is around 80%, a slight fall from early 1980s levels. Data from the observation stations is commonly "homogenised" by the local meteorological authority. Data generally is not collected from more than 50% of the Earth's surface.

US climate research has received more than $US 73 billion in funding over the last two decades. And what do we have? Suspect data. In 1999, the NOAA administrator (Jane Lubchenko) paved the way for the politicisation of science by stating:[231]

> Urgent and unprecedented environmental and social changes challenge scientists to define a new social contract ... a commitment on the part of all scientists to devote their energies and talents to the most pressing problems of the day, in proportion to their importance, in exchange for public funding.

Those who promote a doom-and-gloom future in return for a job and research funds are the very same people caught fiddling the data in the Climategate fraud. Since the 1960s, some 62% of temperature-measuring stations have been removed and most of these were in the colder remote, alpine, polar and rural regions where temperatures were lower. Most measuring stations are now near the sea or are at airports and cities which bias measurements because of urban and industrial heat.

According to the calculations of Luboš Motl,[232] 31% of the stations used by the UK's Hadley Centre Climate Research Unit (HadCRUT), the standard for surface temperatures show that temperature has fallen since 1979. This agrees with the balloon and satellite temperature trends over the same period posted by John Christy and Roy Spencer (University of Alabama). The measurement of a global temperature is not simple. Historical instrumentally recorded temperatures exist only for 100 to 150 years in small areas of the world. From the 1950s to 1980s temperatures were measured in many more locations. Many measuring stations no longer exist. The main global surface temperature data set is managed by NOAA who state:

---

[231] www.noaanews.noaa.gov/stories2009/20090319_lubchenco.html

[232] http://motls.blogspot.com/2008/01/2007-warmest-year-on-record-coldest-in.html

The period of record varies from station to station, with several thousand extending back to 1950 and several hundred being updated monthly.

There now have been so many stations shut down, especially in cooler high elevation, high latitude, rural and remote stations, that a significant warming bias has entered the overall record from land stations. This is the main source of data for global studies, including the data reported by the IPCC.

Average surface air temperatures are calculated at a given station location. These are used to calculate the averages for the month which are then used to calculate the annual averages. This whole process is fraught with error and "homogenisation" can occur with any calculation.

The IPCC uses data processed and adjusted by the Climatic Research Unit of the University of East Anglia which states:

> Over land regions of the world over 3000 monthly station temperature time series are used. Coverage is denser over the more populated parts of the world, particularly, the United States, southern Canada, Europe and Japan. Coverage is sparsest over the interior of the South American and African continents and over the Antarctic. The number of available stations was small during the 1850s, but increases to over 3000 stations during the 1951-90 period. For marine regions sea surface temperature (SST) measurements taken on board merchant and some naval vessels are used. As the majority come from the voluntary observing fleet, coverage is reduced away from the main shipping lanes and is minimal over the Southern Oceans.

and

> Stations on land are at different elevations, and different countries estimate average monthly temperatures

using different methods and formulae. To avoid biases that could result from these problems, monthly average temperatures are reduced to anomalies from the period with best coverage (1961-90). For stations to be used, an estimate of the base period average must be calculated. Because many stations do not have complete records for the 1961-90 period several methods have been developed to estimate 1961-90 averages from neighbouring records or using other sources of data. Over the oceans, where observations are generally made from mobile platforms, it is impossible to assemble long series of actual temperatures for fixed points. However it is possible to interpolate historical data to create spatially complete reference climatologies (averages for 1961-90) so that individual observations can be compared with a local normal for the given day of the year.

The CRU station data used by the IPCC is not publicly available. Only the "homogenised" gridded data is available hence for global temperature data we must trust what the CRU at the University of East Anglia feeds us. Before Climategate I might have been nervous about their "homogenised" gridded data. But now that the keepers of the raw data have been shown to be frauds, we can only conclude that all the data used by the IPCC comes under a very big black cloud.

The NASA Goddard Institute for Space Studies (GISS) is a major provider of climatic data in the US (with NOAA as the source for GISS). Some 69% of the GISS data comes from latitudes between 30°N and 60° and almost half of those stations are located in the United States.[233] This certainly does not cover the globe and is totally unrepresentative for any global picture. If these stations are valid, the calculations for

---

[233] https://data.giss.nasa.gov/gistemp/stdata/

the US should be more reliable than for any other area or for the globe as a whole. However, many recording stations were closed down in the late 1980s and early 1990s.

In addition, the locations of many stations were moved to escape urbanisation. Some stopped collecting data during periods of conflict. With the number of temperature measuring stations changing over time, the so-called "global" record is not really global. Land surface thermometers increased from coverage of 10% of the land in the 1880s to about 40% in the 1960s. Since then, the coverage has been decreasing.

Coverage has been redefined as the "percent of hemispheric area located within 1,200 kilometres of a reporting station." This means that in remote areas where there may be no stations in a 5° x 5° grid box, the temperature is estimated from the nearest station within 1,200 km. I don't know about you but I think that there is a considerable variation in temperature over a distance of 1,200 km.

The closure of temperature measuring stations was not completely random. There has been a decrease in the number of Russian measuring stations and the Hadley Centre ignores many continuous long-term records. In Canada, the number of stations has decreased from more than 600 to 50. NOAA used only 35 of the Canadian stations. The percentage of those at lower elevation (less than 100 metres above sea level) tripled and those above 1,000 metres reduced by half. More southerly locations using population centres hugging the US border dominated over northerly areas.

In fact, only one thermometer remains in Canada for everything north of the 65th parallel. This site is called Eureka[234] because life is more abundant and the summers are warmer than elsewhere in the High Arctic thereby creating a

---
[234] https://wattsupwiththat.com/2010/04/24/inside-the-eureka-weather-station/

temperature bias for the whole region. Hourly readings from Russia and Canada can be found on the internet yet these are not included in the global data set.

China had a great increase in measuring stations from 1950 (100 stations) to 1960 (400) and then a great decrease to only 35 in 1990.[235] Recorded temperatures in China rose due to increased urbanisation during their current industrial revolution.

In Europe, high mountain stations were closed and there were more measurements from coastal cities, especially along the Mediterranean Sea. In northern Europe, the average results showed warming as the number of stations decreased. After 1990, Belgium showed warming yet the adjacent country (the Netherlands) showed no warming. A similar story is apparent for South America where alpine stations disappeared and the number of coastal stations increased. There was a 50% decrease in the number of measuring stations.

African raw data also shows a warming bias.[236] In North Africa, stations from the hotter Sahara were used in preference to those from the cooler Mediterranean and Atlantic coasts.

In Australia and New Zealand, some 84% of stations are at airports where there are hot fumes from planes and vehicles, concrete and buildings and stations that closed were generally in cooler areas.

In the US, some 90% of all measuring stations did not appear in the global historical climate network (GHCN) version 2 climate models. Most stations remaining were at airports and most of the high mountain measuring stations have closed. In California, only San Francisco, Santa Maria, Los Angeles and San Diego were used. The warming trend is still not significant despite this bias. Warmist James Hansen states:

---
[235] www.cma.gov.cn/en2014/
[236] tahmo.org/__african-climate-data/

The US has warmed during the past century, but the warming hardly exceeds year-to-year variability. Indeed, in the US the warmest decade was the 1930s and the warmest year was 1934.

Any computer projections of global climate using land temperature measurements are deeply flawed and we should treat all claims, models, trends and predictions based on these with a very large pinch of salt.

The oceans play the dominant role in perpetuation and mediation of natural climate change as they contain far more heat than the atmosphere. Density variations linking the Northern and Southern Hemispheres of the Pacific and Atlantic Oceans via the Southern Ocean drive the ocean circulation system that controls hemispheric and global climate. Differences in temperature and salt concentrations produce these density variations. The oceans both moderate and intensify weather and decadal climate trends due to their great capacity to store solar heat. The process involves global currents, slow mixing, salt concentration variations, wind interactions and oscillations in heat distribution over very large volumes.

Currents are global heat conveyor belts.[237] The Northern Pacific Decadal Oscillation, the El Niño-La Nina Southern Pacific Oscillation, the long period Southern Pacific Oscillation, the Gulf Stream-Northern Atlantic Oscillation, the Indonesian Through Flow, the Indian Ocean's Agulhas Current, the Southern Ocean Circumpolar Current and many other ocean currents and cycles have a large decadal-scale effect on weather, regional climate and global climate.

In some parts of the global ocean heat conveyor, natural variations in heating, evaporation, freshwater input, atmospheric convection, surface winds and cloud cover can greatly influ-

---

[237] https://oceanservice.noaa.gov/education/tutorial_currents/05conveyor2.html

ence ocean currents close to continents. In turn these modify carbon dioxide uptake or degassing, storms, tropical cyclone frequency, abundance of floating life, rainfall or droughts and sea level changes. The ocean transfer of heat is one of the major driving forces of climate. For climate "scientists", especially oceanographers, to suggest that it is only human emissions of carbon dioxide that drive global warming exposes their activist agenda.

The oceans occupy 70% of the Earth's surface. Sea surface temperature is measured. The Hadley Centre only trusts measurements from British merchant ships. These mainly use shipping routes in the Northern Hemisphere yet the Southern Hemisphere oceans occupy 80% of the ocean surface area. This produces a huge bias in measurements.

The change in methods over time from water collection in canvas buckets to engine water intakes at various water depths has introduced measurement uncertainties. Ocean temperatures from ships, buoys and satellite also present opportunities for "homogenisation", as the Climategate emails show.

Cherry picking the start and finish point of any series of measurements over time can give the required answer, especially if the maximum low or high points are chosen. These sort of statistical games have been played for decades[238] and the climate "scientists" use such techniques all the time. Karl *et al.* start their trend estimates in 1998 and 2000.[239] The year 1998 was an El Niño year. The Karl *et al.* 1998 to 2014 trend (0.106 ± 0.058°C per decade) is lower than the 2000 to 2014 trend (0.116 ± 0.067°C per decade). This is expected if the

---

[238] Darrell Huff, *How to Lie with Statistics*, Norton, 1954.
[239] T.R. Karl, A. Arguez, B. Huang, J.H. Lawrince, J.R. McMahon, M.J. Menne, T.C. Peterson, R.S. Vose and H-M. Zhang, 2015: Possible artifacts of data bias in the recent global surface warming hiatus. *Science* doi 10.1126/science.aaa5632

trend was started when sea surface temperatures were cooler, as they were in 2000. Again, we see that there is a thread of deceit through climate "science".

The more reliable ARGO buoy data was not used and only the longer-running and more unreliable historical records taken from bucket temperature measurements[240] and boat engine intake cooling water temperature were used. More reliable satellite data was not used.[241,242] Do you think that they were adjusted downwards or upwards? The end result was that the pause in warming over the last 20 years has now been "unfound". It disappeared and that means that there is no pause. Easy. Just change the temperatures upwards by 0.12°C. What does 0.12°C really mean? A temperature rise of 0.12°C is, with the order of accuracy, really 0.12 ± 1.7°C[243] hence any conclusions are totally meaningless.

Satellite and balloon measurements show that there has been no global warming for 20 years. This is called the pause. Many climate "scientists" argued very loudly that there was no pause. The uncritical usual suspects loved it because it satisfied their ideology and even claimed that the highly speculative Karl et al. paper should *"end the discussion of the so-called pause, which never existed in the first place"*[244] and *"US scientists: Global warming pause no longer valid"*.[245]

The so-called environmental writers of the global warming

---

[240] E.C. Kent, J.J. Kennedy, D.I. Berry and R.O. Smith, 2010: Effects of instrumentation changes on sea surface temperature measured *in situ*. *Climate Change* 1: 718-728.
[241] http://nsstc.uah.edu/climate/index.html
[242] http:///.remss.com/research/climate
[243] J.J. Kennedy *et al.* 2011: Reassessing biases and other uncertainties in sea-surface temperature observations measured in situ since 1850, part 2: biases and homogenisation. *Geophys. Res. Lett.* 116 (D4), doi 10.1029/2010JD015218
[244] http://www.theguardian.com/environment/climate-consensus-97-per-cent/2015/jun/04/new-research-suggests-global-warming-is-accelerating
[245] http://www.bbc.com/news/science-environment-33006179

media cheer squad showed that they know absolutely nothing about measurement in science, know nothing about statistical techniques used to weed out random "noise" and ignore the fact that variations are smaller than the order of accuracy.

Do the experiment yourself. Go to any temperature record. Pick an unusually cold year (say 2000) at the start of your plot and a warm year at the end (say 2016). For God's sake don't even think about order of accuracy, just plot your graph and claim that the world is warming. Oh ... and make sure you use the words "significant trend". It adds gravitas. It's easy. Climate "scientists" do it so why can't you. If you are silly enough to pick the last 2,000 years of temperature from proxies and measurements, then the story is different and is not worth writing about because it's cyclical and not sensational.

Karl *et al.* (2015) are not alone. The lack of warming for the last 20 years has stimulated some very creative explanations. For example, it now appears that the variability of climate in the 20th century, on inter-annual to multi-decadal scales, is due to changing ocean circulations.[246] However, no data exists to test this relationship on a longer time scale. It is interesting that Karl *et al.* who for decades had been promoting carbon dioxide-driven climate change now finds they are in a spot of bother and tells the world that the ocean now drives climate change. As long as the grants from taxpayers keep rolling in I suppose it doesn't matter to them.

For more than a decade, team climate fraud has been stating that there has been no hiatus in global warming. How could that be possibly so? It disagrees with our models was the argument. By using exactly the same data that climate sceptics used to show that there had been a pause in global warming, the normal suspects suddenly state that there really was a

---

[246] K.E. Trenberth, 2015: Has there been a hiatus? *Science* 349: 691-692.

hiatus and that climate models had failed to predict or replicate the hiatus. They were not gracious enough to acknowledge that many others had previously published that the hiatus was real. This is fraud.

Only a year earlier, Ben Santer and Carl Mears[247] had mocked Senator Ted Cruz for even mentioning a hiatus in during a 2015 congressional hearing and argued that "examining one, individual 18-year period is poor statistical practice and of limited usefulness" when evaluating global warming. It has been known for a long time that the models were wrong[248] yet they were supported in 2016 by Santer and Mears and Santer *et al. (*2017a).[249] These publications were used to criticise the EPA's Scott Pruit, the new and sceptical head honcho appointed by President Trump. At the same time Santer *et al.* (2017a) were writing that there had been no hiatus, Santer *et al.* (2017b)[250] published that there was a hiatus. As they say, go figure.

Either there are some very strange things happening between Santer's ears or he is just publishing for the sake of publish or perish to keep the research grants flowing in. Suddenly the consensus changed from there being no hiatus in the temperature record to a consensus that there really was a hiatus in the warming record. There was no new evidence to trigger this change. This is not the evolution of science but the continuing of fraud.

---

[247] B.J. Santer and C. Mears, "A response to 'data or dogma' hearing", *Skeptical Science*, 17 January 2016.
[248] P. J. Michaels and P. C. Knappenberger, 2015: Climate models and climate reality: A closer look at a lukewarming world. *Cato Institute Working Paper* 35.
[249] B.J. Santer, S. Solomon, F.J. Wentz, Q. Fu, S. Po-Chedley, C. Mears, J.F. Painter and C. Bonfils, 2017a: Trophospheric warming over the past two decades. *Scientific Reports* 7 doi: 10.1038/s41598-017-02520-
[250] B.J. Santer, J.C. Fyfe, G. Pallotta, G.M. Flato, G.A. Meehl, M.H. England, E. Hawkins, M.E. Mann, J.F. Painter, C. Bonfils, I. Cvijanovic, C. Mears, F.J. Wentz, S. Po-Chedley, Q. Fu and C-Z. Zou, 2017b: Causes of differences in model and trophospheric warming rates. *Nature Geoscience* 10, 478-485.

Matthew England, a coauthor with Ben Santer and Michael Mann in the Santer *et al.* (2017b) paper, stated:[251]

> What we're seeing in models is that the warming out of the hiatus is gonna be rapid, regardless of when the hiatus ends.

You'd better believe it folks. We're doomed, we'll fry and die and if you keep funding me at the Climate Change Research Centre (UNSW), I just might be able to solve your problems. Trust me with the data you've paid for, trust me while you continue to pay for me and trust my models. The hiatus has ended and global temperatures are now decreasing. Oh, and by the way, England and his mates don't get their precious hands dirty and actually do the measurements. They sit in an airconditioned office and computer massage the "homogenised" measurements of others. One can only read this sort of bilge when in a firm embrace with Bacchus.

Atmospheric temperature measurements are also recorded by satellites by the Remote Sensing Systems (RSS) and the Earth System Science Center at the University of Alabama in Huntsville (UAH). The RSS dataset does not resolve the discrepancy between observed and modelled temperature trends in the lower atmosphere. Greenhouse gas warming occurs in the lower troposphere and the key question is: Do emissions of carbon dioxide by humans cause dangerous global warming? The UN's IPCC and its followers claim that the warming trend is to be most pronounced over the tropics. The new RSS satellite dataset shows that the climate models are wrong.[252] The UAH data is superior to the RSS data because:[253]

---

[251] Chris Kenny, "Media Watch's climate change obsession", *The Australian*, 1-2 August 2015.
[252] http://www.sepp.org/the-week-that-was.cfm
[253] Jo Nova 2017: Satellite battle: Five reasons UAH is different (better) to the RSS global temperature estimates. joannenova.com.au

(a) UAH satellite data is validated by weather balloon measurements whereas the RSS data is not,

(b) UAH uses empirical data to adjust for diurnal drift whereas RSS uses model estimates,

(c) NOAA14 and NOAA15 satellite data is inconsistent, UAH has removed the data from the one they think is incorrect whereas RSS does not,

(d) RSS keeps the warming error before 2002 thereby giving a steeper warming trend, and

(e) UAH uses three satellite channels whereas RSS uses one.

Satellite measurements were removed from public scrutiny by NOAA in July 2009 after complaints of a cold bias in the Southern Hemisphere. Immediately after that, both ocean and global temperatures mysteriously increased. The 3,800 ARGO buoys have been measuring sea surface temperature since mid-2003[254]. There has been a slight cooling trend. ARGO buoy data is not used in monthly assessments of global temperature for some unstated reason.

In a comprehensive study[255] of the best documented and understood temperature data sets that are not contaminated from bad siting or urbanisation, it was concluded that the temperature data sets in the US and elsewhere are *"not a valid representation of reality."*

What is unprecedented is that in the history of science so much data has been tampered with by so few. The end result is that the present and historical climate record has been destroyed and any climate projections should be ignored.

All this has led to your electricity bill going through the roof.

---

[254] www.argo.ucsd.edu/
[255] J.P. Wallace, J.S. D'Aleo and C.D. Idso, 2017: On the validity of NOAA, NASA and Hadley CRU global average surface temperature data and the validity of the EPA's CO_2 endangerment finding. Abridged research report. www.theresearch.files.wordpress.com

Urban effects

Urban areas are warmer than surrounding rural areas. This is especially the case at night. Airports originally on the outskirts of urban areas have had cities grow around them and temperatures rise. A very large number of measurements for global temperature studies are from airports bathed in hot exhaust fumes. Oke created an equation for urban heat island warming.[256,257] A hamlet with a population of 10 has a warming bias of 0.73°C, a village with 100 people has a warming bias of 1.46°C, a town with a population of 1,000 has a warming bias of 2.2°C and a city of a million people has a warming bias of 4.4°C. This can only be an approximation as some cities grow upwards and others grow laterally into what were agricultural areas. Heated urban areas also increase plant growth.[258]

The World Meteorological Organisation (WMO) and NOAA supposedly have strict criteria for temperature measuring stations. Measurement stations should be located on flat ground surrounded by a clear surface and more than 100 metres away from local heat sources, tall trees, artificial heating or reflecting surfaces such as buildings, parking areas, roads and concrete surfaces. Temperature sensors should be shaded from direct sunlight, ventilated by the wind 1.5 metres above mown grass no higher than 10 centimetres.

Anthony Watts found that 89% of the US measuring stations (i.e. more than 1,000 stations) did not meet the official standards for temperature measurement regarding the distance between stations and adjacent heat sources.[259] They still don't, despite

---

[256] T. R. Oke, 1988: The urban energy balance. *Prog. Phys. Geogr.* 12, 471-508.
[257] Urban heat-island warming= 0.317lnP, where P=population.
[258] S.C. Zipper, J. Schatz, A. Singh, C.J. Kucharik, P.A. Townsend and S.P. Loheidell, 2016: Urban heat island impacts on plant phenology: Intra-urban variability and response to plant cover. *Envir. Research Letters* 11, 5.
[259] https://wattsupwiththat.com

the WMO and NOAA knowing that stations are poorly sited and will give anomalous warm measurements.[260] Many received back reflection from concrete and buildings, most received heat from hot car and aeroplane exhaust fumes and many were sited next to air conditioning units that emitted heat. Watts concluded:

> The raw data produced by the stations are not sufficiently accurate to use in scientific studies or as a basis for public policy decisions.

These biased measurements account for more than 50% of the measured warming since 1880. NOAA initially denied that it was an issue and then asked the government for $100 million to upgrade and correct the locations of 1,000 climate stations.

Many stations are so poorly sited that they do not have to undergo upward adjustments ("homogenisation").

ADJUSTED MEASUREMENTS

It is a cardinal sin of science to change raw data. It is universal in the climate "science" industry. Such fraud makes data useless.

After the raw temperature measurements are collected, warts and all, further adjustments are made. This process is called "homogenisation". Curiously, each adjustment produces more warming. MIT meteorologist Richard Lindzen commented:

> When data conflicts with models, a small coterie of scientists can be counted on to modify the data to agree with models' projections.

By extracting old data from papers of James Hansen and comparing them with data downloaded from NASA's GISS site, we can see a progressive global warming. This is certainly

---

[260] A. Watts, 2017: Creating a false warming signal in the US temperature record. In: *Climate change: The facts 2017* (ed. J. Marohasy). IPA, 75-91.

man-made warming, created by men making adjustments to the data, blending badly sited urban data with correctly-sited rural data and then, in 2007, removing the urban adjustment for US data sites. The frequency and direction of the NOAA "homogenisations" to US data were increased in 2007 at the same time as inarguable satellite data showing a pause in warming became public knowledge.

NOAA's "homogenisation" process has been shown to significantly alter the trends in many stations where the location suggests the data is now unreliable. In fact, adjustments account for virtually all the warming trend. Unadjusted data for the most reliable sites (i.e. rural sites) show cyclical multi-decadal variations and no long-term warming trend.

Former NASA scientist Ed Long showed that after "homogenisation", the rural data trend agreed with the urban data trend when an artificial warming trend was introduced[261]. Urban warming was allowed to remain in the urban data sets and a warming bias was artificially introduced to rural data sets that in their unadjusted state showed no warming.

Both NOAA and NASA resisted Freedom of Information requests for the release of all the unadjusted data and documentation for all the adjustments made. The US Data Quality Act requires that published data must be able to be replicated by independent audits. That is currently not possible given the resistance posed, despite promises of transparency.

Judith Curry (Georgia Tech University) comments:

> In my opinion, there needs to be a new independent effort to produce a global historical surface temperature dataset that is transparent and that includes expertise

---

[261] E.R. Long, 2010: Contiguous U.S. temperature trends using NCDC raw and adjusted data for one-per-state rural and urban station sets. *SPPI Original Paper*, 27 February 2010.

in statistics and computational science ... The public has lost confidence in the data sets ... Some efforts are underway in the blogosphere to examine the historical land surface data (e.g. such as GHCN), but even the GHCN data base has numerous inadequacies.

The term global warming is based on an increasing trend in global average temperature over time. This is based on measurements. Or is it? The IPCC reported in 2007 in Chapter 3 that:

Global mean surface temperatures have risen by 0.74°C ± 0.18°C when estimated by a linear trend over the last 100 years (1906–2005).

This is misleading because 100 years ago, thermometers did not have such a high order of accuracy.

The temperature record is not a good record and it is really only the long-term rural stations that provide meaningful raw data. Because of the closure of so many stations, by 2017 there were more measuring stations in the US than in the rest of the world. Since 2000, NASA has further "cleaned" the historical record by making adjustments for area, time of observation, equipment change, station history adjustment, by arbitrarily infilling missing data and urban heat island adjustment. The rationale for the temperature station adjustments was to make the data more realistic for identifying temperature trends.

The end result of all the "homogenisation" is that additional apparent warming has been created. For example in the US, the closest rural station to San Francisco (Davis) and the closest rural station to Seattle (Snoqualmie) have both had the older part of the record adjusted downwards.

The end result of this is to give a trend that looks as if there has been more than a century of warming in rural areas when the raw data shows a very different story.

Temperature

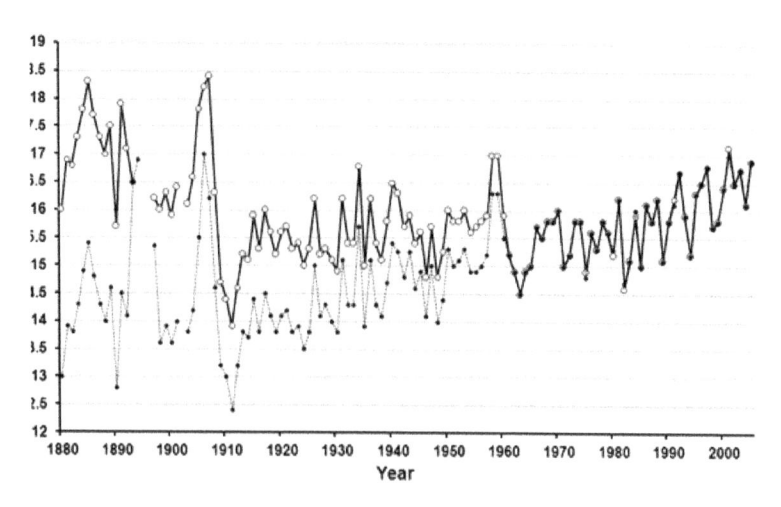

Figure 18: "Homogenised" (thin line, dots) and raw (thick line, circles) for urban measuring station, Davis, Ca., USA. Note that "homogenised" gives apparent warming.

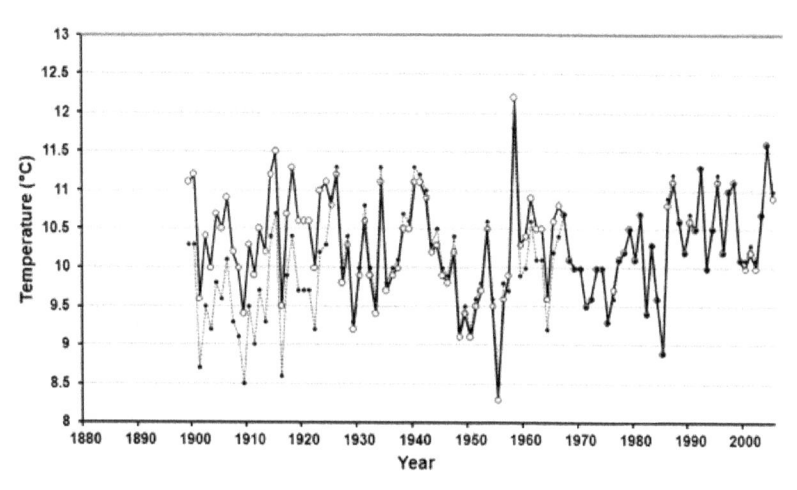

Figure 19: "Homogenised" (thin line, dots) and raw (thick line, circles) for rural measuring station, Snoqualme Falls, Washington State, USA. Note that "homogenisation" gives apparent warming.

Homewood[262] checked a swathe of South American weather

---
[262] https://notalotofpeopleknowthat.wordpress.com

stations. He found the "homogenisations" for each station gave a warming trend. These "homogenisations" were made by the US government's Global Historical Climate Network (GHCN). They were then amplified by two of the main official surface records, the Goddard Institute for Space Studies (GISS) and the National Climate Data Center (NCDC), which use the warming trends to estimate temperatures across the vast regions of the Earth where no measurements are taken. Yet these are the very records on which scientists and politicians rely for their belief that human emissions have produced global warming. As a result, you pay through the nose for your electricity.

Homewood also looked at weather stations across much of the Arctic, between Canada (51°W) and the heart of Siberia (87°E). In nearly every case, the same upward "homogenisations" were made, to show warming up to 1°C or higher than was indicated by the data that was actually recorded. This shocked Traust Jonsson, who was long in charge of climate research for the Iceland meteorological office. Jonsson was amazed to see how the new version completely "disappears" Iceland's "sea ice years" around 1970, when a period of extreme cooling almost devastated his country's economy.[263]

One of the first examples of these "homogenisations" was exposed in 2007 by the statistician Steve McIntyre, when he discovered a paper published in 1987 by James Hansen, the climate "scientist" who for many years ran GISS. Hansen's original graph showed temperatures in the Arctic as having been much higher around 1940 than at any time since. But as Homewood reveals in his blog post, *"Temperature adjustments transform Arctic history",* GISS has turned this upside down. Arctic temperatures from that time have now been lowered so much that that they are now dwarfed by those of the past 20 years.

---

[263] Paul Holmwood, 2012: Iceland's "Sea Ice Years" disappear in GHCN adjustments. https://notalotofpeopleknowthat.wordpress.com, 28 January 2012.

Temperature

Steve McIntyre has been asking for the CRU data set for years. He was refused public data collected by a taxpayer-funded organisation. However, in mid July 2017 it was accidentally left on the CRU server for a short time.[264] CRU has now plugged the leak lest raw data falls into the wrong hands.

Homewood's interest in the Arctic is partly because the allegedly vanishing ice (and polar bears) has become such a poster-child for telling us we will fry-and-die. Homewood chose that particular stretch of the Arctic because it is where ice is affected by warmer water brought in by cyclical shifts in a major Atlantic current. The ice-melt is cyclical[265] and is clearly unrelated to human emissions of carbon dioxide.

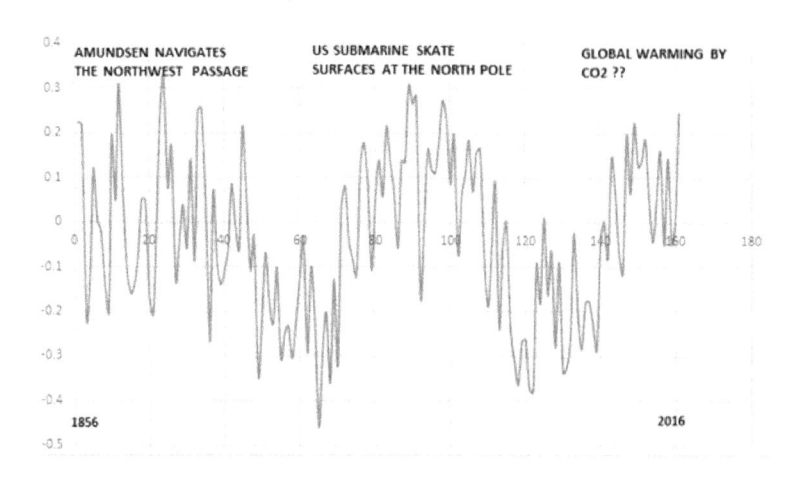

Figure 20: Cyclical warming and cooling of the Arctic depending upon oscillations in the Atlantic Ocean. Note that the recent warmer times in the Arctic were little different from the past.

Of much more serious significance, however, is the way this wholesale manipulation of the official temperature record, for reasons GHCN and GISS have never plausibly explained,

---

[264] https://climateaudit.org/
[265] nsidc.org/arcticseaicenews/2016/03/

has become the elephant in the room of the greatest and most costly scare the world has known.[266]

It appears that those in charge of the records and who promote human-induced global warming are climate "science" activists. This is a massive conflict of interest. Climate "scientists" have adjusted data and used poor measurements to change the past in order to control the present. This gives climate "scientists" the opportunity to bleat about a scary future in order to be funded more and more.

In my field, all exploration and mine data undergoes a quality assurance (QA) and quality control (QC) process. QA/QC is a normal due diligence process when dealing with data in case there is an error. Qualified scientists must sign off on the QA/QC in order to protect investors. If they get it wrong, they destroy their career. There is no QA/QC process in climate "science" and this allows the fraudulent "homogenisation" of data with no personal consequences. The taxpayer has invested orders of magnitude more funds in climate "science" yet the keepers of the data are not accountable.

Tampered Australian temperatures

In Australia, there are a number of city-based measuring stations. Measurements are adjusted. Australia certainly has manmade global warming. It has nothing to do with carbon dioxide and occurs at the stroke of a pen. If all 328 stations are "homogenised" from 1881 to 2016, the raw data shows a statistically insignificant warming of 0.07°C. Many historical measurements have an uncertainty greater than 0.5°C hence a warming of 0.07°C is well within the order or accuracy and is meaning-

---

[266] Christopher Booker, "The fiddling with temperature data is the biggest science scandal ever: New data shows that the "vanishing" of polar ice is not the result of runaway global warming", *Daily Telegraph*, 7 February 2015.

less. To promote that this figure has any significance is fraud. Even if it were significant, if I stood up there would be a temperature increase of greater than 0.07°C.

The analysis of the Rutherglen site in rural Victoria (Australia) is a well-documented example where measurements have been taken at the same site using the same equipment since 1912.[267] There were no equipment failures or breaks in the data. Rutherglen should provide the best temperature data available because of the lack of the urban heat island effect and consistency of measurement.

The data showed a cooling trend over the last century of 0.3°C, in accord with other measuring stations nearby.[268] The official temperature record had been "homogenised" to create a warming trend of 1.7°C over the same period. This was because the pre-1974 minimum temperatures had been reduced by the Bureau of Meteorology (BoM). This is fraud.

Surface air temperature measurements are administered by the taxpayer-funded BoM. There are almost 2,000 sites and up to 112 of the measurements are weighted and combined to create various temperature averages for a whole continent. This forms the Australian historical temperature record and is fed into international datasets that are used to measure and predict climate change.

When chased by Graham Lloyd, the environmental editor at *The Australian*, the BoM claimed that the change in recorded temperatures was because the site had moved. That was a lie. The BoM's own official Station Catalogue[269] claimed there were "no documented site moves during the site's history." At

---

[267] J. Marohasy, 2016: Temperature change at Rutherglen in South-East Australia. *New Climate* https://doi.org/10.2222/nc.2016.001

[268] Benalla (Vic), Echuca (Vic.), Deniliquin (NSW).

[269] www.bom.gov.au/climate/change/.../ACORN-SAT-Station-Catalogue-2012-WEB.pdf

Rutherglen, the BoM claimed that the Rutherglen trend needed to be consistent with trends at nearby stations. However, the trends from nearby station have been "homogenised" and the stations used at Wagga and Kerang are hundreds of kilometres away[270] and can hardly be claimed as nearby.

The closest station to Rutherglen (Vic.) is Deniliquin (NSW) and its cooling trend over the course of a century of 0.7°C was "homogenised" to a warming trend of 1.0°C.[271] In other areas of NSW, forty years of data at Bourke was just deleted, not even "homogenised". The 1.7°C cooling trend over a century at Bourke was "homogenised" to a slight warming trend and the site high of 51.7°C in 1909 was ignored, despite being recorded at nearby towns. Summer maximum temperatures at Bourke (NSW), Alice Springs (NT), Narrabri (NSW) and Hay (NSW) show a cooling trend.[272] This is ignored by the BoM.

The records from the western NSW town of Cobar are used to "homogenise" temperatures at Alice Springs (NT), 1,460 km away as the crow flies. There is absolutely no way data from a town in a semi-arid vegetated area can be used to "homogenise" data from a far larger town in the dead centre of a continent. Why were closer towns such as Broken Hill (NSW), Port Augusta (SA), Roxby Downs (SA) or Birdsville (Qld) not used? This is fraud.

Data from Amberley in south-eastern Queensland was "homogenised" using data from the Coral Sea. After all, the climate on a terrestrial landmass will be exactly the same as

---

[270] Brett Hogan, "Warming up in Rutherglen: A century-old weather station in Rutherglen has shown researchers how climate statistics are re-modelled to create a warming trend", *IPA Review*, July 2017, 38-42.
[271] J. Marohasy, 2016. Temperature change at Rutherglen in south-east Australia, *New Climate*, https://doi.org/10.22221/nc.2016.001
[272] E.L. Deacon, 1952: Climate change in Australia since 1880. *Australian Journal of Physics* 6, 209-218.

that on distant tropical islands.[273] Attempts to scrutinise the BoM met with delays then political obfuscation followed by inactivity. It is little wonder voters have lost confidence in politicians and institutions that were once proud of their independence.

Lloyd asked the BoM to justify its "homogenisation". The response was that its methodology had been published in the peer-reviewed literature and that "homogenisation" was the world's best practice. This is not justification. It is obfuscation. The BoM could not answer why all "homogenisations" gave a warming trend and why they changed cooling trends to warming trends in accord with "homogenised" warming trend of 1.0°C since 1910 elsewhere in the world. The climate science scare campaign is that temperature has increased by 1.0°C since 1910 and, if the data does not show such warming, then the primary data must be changed.

The BoM states that temperature adjustments are secret. [274] If they are secret, then the taxpayer should cease funding the BoM. If adjustments are made, then the tabulated data should show measured data, adjusted data, adjustment calculation and the reasons for an adjustment. This would happen if a QA/QC process was operative.

A recent study[275] has shown that "homogenisations" of global surface temperature readings by scientists in recent years "are totally inconsistent with published and credible U.S. and other temperature data." Climate "scientists" often "homogenise"

---

[273] J. Marohasy, J. Abbot, K. Stewart and D. Jensen, 2014: Modelling Australian and global temperatures. What's wrong? Bourke and Amberley as case studies. *The Sydney Papers Online*, 26.

274 J. Nova, 2015: If it can't be replicated it isn't science. BOM admits temperature adjustments are secret. http://joannenova.com.au/2015/06/if-it-cant-be-replicated-it-isnt-%20

[275] Michael Bastasch, "Study finds temperature adjustments account for "nearly all of recent warming" in climate data sets", *Daily Caller*, 7 July 2017.

primary data to account for what they perceive as biases in the data because, through their eyes, a measured cooling event must be "homogenised" to a warming trend because it is obviously wrong.

Each dataset pushed down the 1940s warming and pushed up the current warming, almost all "homogenisation" increases the warming trend and the paper authors[276] state that "nearly all the warming they are showing are in the adjustments".

When climate activists fill organisations entrusted with keeping official primary data and change the primary data, we can be very confident that political activism has replaced science and that we can't trust climate "science". Maybe, instead of global warming we have global fraud.

The weather station in Melbourne CBD[277] first started measurements when it was in open space in 1908. However, 109 years later it is surrounded by asphalt roads, concrete, tall buildings and vehicles emitting hot exhaust fumes. The data from this site are "homogenised", the older measurements are reduced and this gives the effect of a warming trend. Even without tampering with the primary data, there is a warming trend due to the urban heat island effect.

The raw data from Melbourne shows an increase in temperature from about 1950. This coincides with construction of large buildings thereby producing more back reflection, more vehicles and heating/cooling systems in office blocks. The raw Melbourne data shows the urban heat island whereas the adjusted data shows that Melbourne was warmer when there was little urban heat island effect and shows a warming from 1950.[278]

---

[276] Joe D'Aleo, *Daily Caller News Foundation*, 7 July 2017.
[277] Corner of La Trobe and Victoria Streets.
[278] T. Quirk, 2017: Taking Melbourne's temperature. In: *Climate Change: The Facts 2017* (ed. J. Marohasy). IPA, 101-116.

To make matters worse, the Melbourne CBD "homogenised" data was used to "homogenise" data from Cape Otway, 173 kilometres away on the isolated south-west coast well away from any civilisation. Why did primary data from Cape Otway need to be "homogenised"?

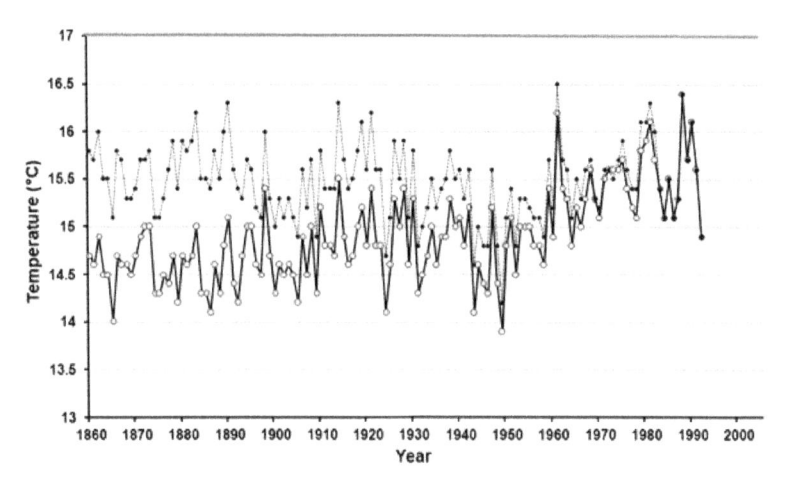

Figure 21 "Homogenised" (thin line, dots) and raw (thick line, circles) for urban measuring station in Melbourne CBD. The station location suggests influence by post-1960 increased traffic and buildings giving apparent warming.

The raw data from Darwin shows a slight cooling over the 137-year record. In the Second World War, the Darwin station was bombed and later moved from the city to the airport.[279] The pre-1950 Darwin records have been adjusted downwards. The Darwin data looks like there has been significant warming over the last 70 years.

Volunteer Australians in the outback provided daily meteorological data to the BoM for more than a century. They undertook what they considered was a public service in rain, hail or shine with the belief that they were adding to a permanent me-

---
[279] www.meteorology.com.au/local-climate-history/nt/darwin

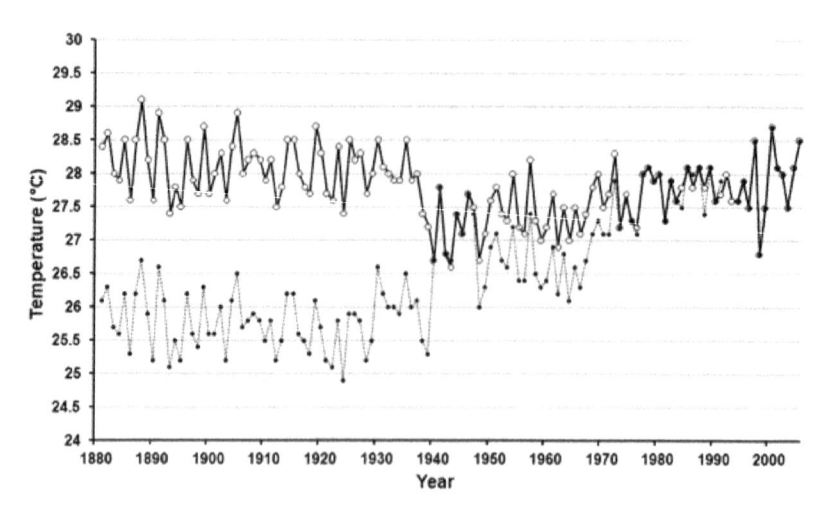

Figure 22 "Homogenised" (thin line, dots) and raw (thick line, circles) for urban measuring station, Darwin, Australia. Note that adjustment gives apparent warming.

teorological record of Australia. Their responsibilities were not taken lightly and measurements were done diligently and often under conditions of hardship. All this data and effort has been abandoned or "homogenised" by the BoM to fulfil their global warming ideology. This arrogance shows great disrespect for the taxpayer who funds the BoM and the volunteers who collected a century of daily measurements to create an accurate record.

In August 2017, the BoM was caught red-handed. During a cold spell in winter, temperatures lower than -10°C in Goulburn (NSW), alpine areas and Tasmania were made warmer or omitted from the record. The end result of this was in accord with the BoM "homogenised" warming trend. The BoM blamed equipment.[280] This does not explain why raw data first appeared on the BoM record, then disappeared and then appeared again as a warmer measurement. This could only have been the result

---

[280] Graham Lloyd, "Temperatures plunge after BoM orders fix", *The Australian*, 4 August 2017.

of humans changing the data. This is fraud. The BoM blamed equipment and stated that the solution to the problem was to buy new equipment. This is a lie.

The collectors and archivists of scientific data paid for by the taxpayer have changed the original raw data to show that there is global warming. Since 2015, the BoM have now only allowed access to the data by BoM staff[281] such that "homogenisations" cannot be checked by QA/QC or repeated thereby allowing the fraud to continue.

The BoM closed the station that records the lowest temperature in Australia (Charlotte Pass, NSW) and has established new stations (e.g. Borrona Downs, NSW). By removing low temperature measurements, a warming bias is created. On 19 August 2017, Borrona Downs recorded -62.6°C between 21:50:00 and 21:59:00, the temperature changed to +37.5°C at exactly 21:59:00 and then to -4.4°C at 22:00:00. The overnight minimum was recoded as 2.4°C.[282] The methodology of this adjustment is kept secret by the BoM.

Liquid in glass thermometers were installed in 1856. They have been replaced by with electronic measuring equipment which should be recording temperature according to the BoM's own 1997 guidelines.[283] New minimum temperatures are established every 30 minutes and refreshed every 10 minutes. The World Meteorological Organisation has best practice for integrating readings. A five-minute measuring time is not enough for an accurate realistic measurement and a seven-minute measuring time was more suitable for modern fast-response electronic measurement.[284]

---

[281] J. Nova, 2015: http://joannenova.com.au/2015/06/if-it-cant-be-replicated-it-isnt-%20
[282] http://jennifermarohasy.com/wp-content/uploads/2017/09/TMinIssues-JenMarohasy-20170831.pdf
[283] Instruments and Observing Methods Report No. 65, WMO/TD No 862.
[284] X. Lin, and K.G. Hubbard, 2008: What are daily maximum and minimum temperatures in observed climatology. *International Journal of Climatology* 28, 283-294.

The BoM records maxima and minima temperatures based on just one second of measurement and routinely destroys most of the other raw measurements. Spikes in temperature can be many degrees up or down and can occur for a diversity of reasons. Maximum and minimum temperatures resulting from measurement over one second produces a false result because there is no filtering of spurious results derived from instrumental error, no averaging and no order of accuracy. Furthermore, there is no measurement of temperature with electronic equipment side-by-side with thermometers thereby allowing a meaningful comparison with the historical record and allowing checking for spurious electronic measurements. Instrument error may account for 22 to 55% of the temperature trend and the last 20 years of instrumental measurement may be nigh on useless.[285]

The BoM data is used to compile global trends which are then used to scare us, create subsidised renewable energy and extract money from our pockets. When the BoM announces that we have the hottest day or year on record, I drop on my knees and pray for an independent audit of the BoM.

We audit banks, why not the BoM?

Kiwi "homogenisation"

Exactly the same adjustments were made for Auckland with a rural station suffering substantial adjustments in the pre-1950 record such that the adjusted record looks as if there has been a sustained period of warming. The National Institute of Water and Atmosphere (NIWA), a New Zealand government research organisation, recorded average NZ temperature for the past 100 years. They claim warming since 1900.

---

[285] http://joannenova.com.au/2017/09/bom-scandal-one-second-records-in-australia-how-noise-creates-history-and-a-warming-trend/replicated it isn't science. BOM admits temperature adjustments are secret.

Temperature

The NZ Science Coalition has obtained the raw data after long and painful political obfuscation by the Government and NIWA. The raw data, derived from seven long-term climate stations, have been "homogenised". A record of the process for making "homogenisations" just does not seem to exist. Every "homogenisation" showed an upward temperature trend.

A couple of peer-reviewed papers show that the historical weather data show a trend of 0.3 ± 0.3°C. This means that there was no change. This shows that there are no checks and balances operating for science funded by the public, that no QA/QC processes were in operation, that science has been politicised and that institutions, once regarded as the bastions of independent free thought and debate, are now promoting politicised and incorrect science.

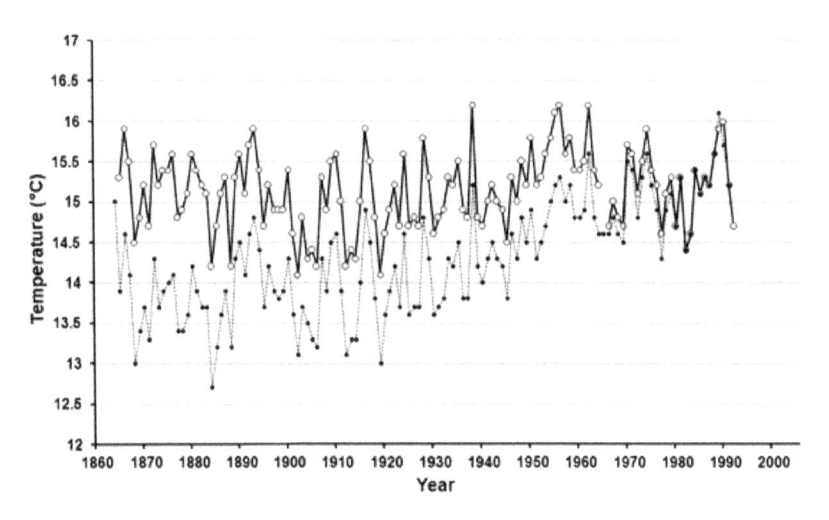

Figure 23 "Homogenised" (thin line, dots) and raw (thick line, circles) for urban measuring station, Auckland, New Zealand. Note that adjustment gives apparent warming.

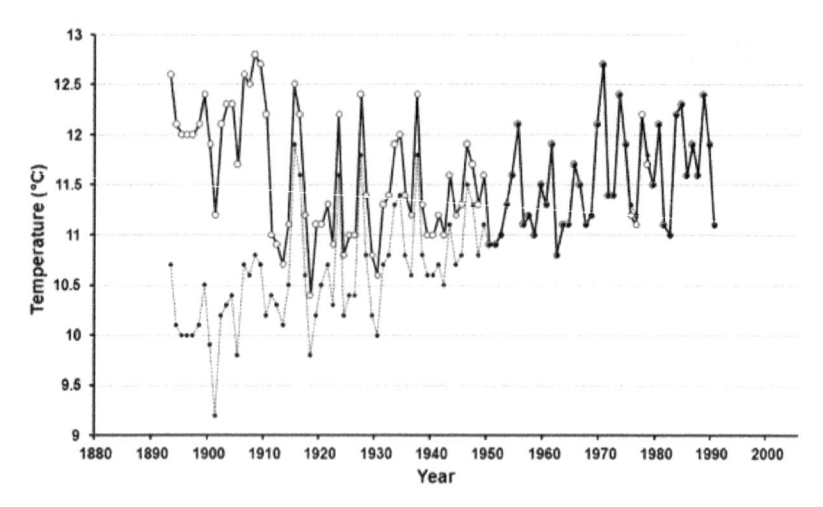

Figure 24 Adjusted (thin line, dots) and raw (thick line, circles) for rural measuring station, Hokitika, New Zealand. Note that the "homogenisation" gives a warming trend.

Temperature tampering in the US

In the US since 1985, temperature records over land have been steadily drifting higher than sea surface temperatures and are now about 1°C higher. This may be the result of closing rural and remote land stations, the urban heat island effect and the upward "homogenisation" of temperature trends. The US annual temperature from NOAA showed warming maxima in the 1930s while cooling minima peaked in the 1960s and 1970s. The NOAA web site had James Hansen stating:

> The US has warmed in the past century, but the warming hardly exceeds year-to-year variability. Indeed, in the US the warmest decade was in the 1930s and the warmest year was 1934.

There was constant friction as NOAA tried to infer warming from "homogenised" data. Indeed, by eliminating the urban heat island effect, warming since the 1930s suddenly appeared.

Emails derived from Freedom of Information requests have revealed that David Easterling admitted:

> One other fly in the ointment we have is a new adjustment scheme for USHCN(V2) that appears to adjust out some, if not most, of the 'local' trend that includes land use change and urban warming.

Clearly the new computer models are creating pre-ordained conclusions for use by scientific activists. This is contrary to the 2009 conclusions of Brian Stone (Georgia Tech University) who found in 2009:

> Across the US as a whole, approximately 50% of the warming that has occurred is due to land use changes (usually in the form of clearing forest for crops or cities) rather than the emissions of greenhouse gases.

and

> Most large US cities, including Atlanta, are warming more than twice the rate of the planet as a whole – a rate that is mostly attributable to land use change.

The claim that the Earth's atmosphere at the surface has warmed by 0.85°C over the last 100 years is unprecedented and is not in accord with evidence. Furthermore, with international "homogenisation" of data to upward trends, raw data is rendered unreliable.

The Central England Temperature record shows that between 1693 and 1733 AD temperature rose at a rate of 4°C per century. During the Roman and Medieval times it was warmer than now and the rate of warming at times was higher than in the 20th century.[286,287]

---

[286] B. Christiansen and F.C. Ljungqvist, 2012: The extra-tropical Northern Hemisphere temperature in the last two millennia: reconstructions of low frequency variability. *Clim. Past* 8: 765-786.

[287] F.C. Ljungqvist *et al.* 2012: Northern Hemisphere temperature patterns in the last 12 Centuries. *Clim. Past* 6: 227-249.

Global temperatures have increased some 3° to 8°C since the zenith of the last glaciation, global temperatures have a variability of about ± 3.5°C from the long term mean.[288] In the 21st century we are about 1°C above the long term mean but, before you go outside and slash your wrists, in the Medieval Warming the temperature was 2.5°C above the long term mean. There is nothing special about the 21st century temperature or temperature increase over the last 100 years. It is just the normal variability that planet Earth enjoys.

The only warming we see is the cooking of data.

WHAT WARMING?

Computer models show that the tell-tale signature of human-induced global warming would be a hot spot high in the atmosphere above the equator. Weather balloons have scanned the skies for decades and no hot spot exists. Ice core data showed that temperature rises were mirrored by rises in the atmospheric carbon dioxide content on a coarse scale. More detailed finer scale analyses showed that atmospheric temperature rises at least 800 years before the rise in atmospheric carbon dioxide.[289]

Warmists claim that even if carbon dioxide does not start the warming trend, it amplifies it. However, the past shows that for most of time the atmospheric carbon dioxide content has been far higher than now and we have not had runaway global warming. This means that such amplifications have not occurred in the past and yet we are expected to believe that they just happen now.

---

[288] J. Jouzel *et al.* 2007: Orbital and millennial Antarctic climate variability over the past 800,000 years. *Science* 317: 793-796.

[289] N. Caillon, J.P. Severinghaus, J. Jouzel, J-M. Barnola, J. Kang and V.Y. Lipenkov, 2003: Timing of atmospheric CO_2 and Antarctic temperature changes across Termination III. *Science* 299, 5613, 1728-1731.

If carbon dioxide amplifies the warming trend, why do the ice core records show that the warming trends reached the peak of the cycle then started to decline while the carbon dioxide content continued to increase? If carbon dioxide drives global warming, this could not occur.

Warming caused by carbon dioxide was predicted from computer models and how could they possibly be wrong? Something other than carbon dioxide caused the warming recorded in ice cores. Satellites circling the planet all the time have shown that there has been no increase in warming for two decades. However, atmospheric carbon dioxide continues to increase so clearly temperature is unrelated to atmospheric carbon dioxide content. Carbon dioxide has done its job. Any further increases in atmospheric carbon dioxide will produce an increasingly smaller temperature increase.

HOTTEST YEAR ON RECORD

If we keep getting told that each year is the hottest year on record, then surely we must believe that we have global warming? Have we really had the hottest years on record over the last few years?

The key scientific questions to be asked about the hottest year are:

(i) When did measurements start?
(ii) How were measurements made?
(iii) What is the order of accuracy of measurements?
(iv) Were measurements "homogenised"?

Attempts to validate[290] current surface temperature datasets managed by NASA, NOAA and the UK's Met Office failed. All these organisations make adjustments to raw temperature

---
[290] Joseph D'Aleo, *Daily Caller News Foundation*, 27 June 2017.

data ("homogenisation"). Adjustment of data is a cardinal sin of science and these adjustments dampen temperature cycles, increase the warming trend and give the pre-ordained catastrophist conclusion. Even with the doctored data, it is impossible to conclude from the datasets of NASA, NOAA and the UK's Met Office that recent years have been the warmest ever, despite claims of record setting warming. Global temperatures have dropped back to the pre-2016-2017 El Niño levels.

In the last 10,500 years of the current interglacial, 9,099 were warmer than now. Some 6,000 to 4,500 years ago in the Holocene Maximum, it was warmer than at present and sea level was about 1.2 metres higher than at present. This was the peak of the interglacial that we now enjoy. It was only 8,000 years ago that there was no summer ice in the Arctic. Many solar scientists are now predicting that we are on the downhill run towards the next cold cycle, glaciation or ice age. I hope not. We are hairless apes and some of us like it hot.

Normally the order of accuracy is greater than the suggested temperature rise hence the claim is scientifically invalid. The suggested annual temperature rise for Australia of 0.17°C is less than the order of accuracy of many measuring stations used to deduce this trend because older measurements are ± 0.5°C. This is meant to be an average temperature rise but how does one construct an average with few measuring stations and even fewer in remote areas? This is mathematical nonsense.

These speculations about the hottest year don't tell us about the natural thermostats such as precipitation and evaporation and ocean heat. They don't tell us about the past when there was far more carbon dioxide in the atmosphere than now yet there was no global warming. They don't tell us about the changes in the main greenhouse gas in the atmosphere (water vapour). They don't tell us that we had a solar maxima coincidental with warming. They don't tell us that we have had cooling with solar minima. They don't tell us that in recent years there were fewer typhoons, more Arctic ice, more Antarctic ice, a decrease in sea surface temperature and that sea surface temperature drives atmospheric temperature, not the inverse.

Australia has had some hot weather. When temperatures reach 45 or 50°C in Australia, history shows us that it is not the first time. Explorers Charles Sturt[291] recorded temperatures of 53°C, 54°C and 56°C, Mitchell[292] recorded 55°C as did Stuart.[293] Historical high temperatures have been adjusted downwards[294] but that didn't help people who died in the heat of the Federation Drought in the late 19th century.

In the warm times of 1910 to 1930, the raw data of the ACORN-SAT measuring sites had "homogenised" the minimum temperatures downwards[295] resulting in a temperature trend increase of 40%. The hottest ever temperature in Australia did not occur in the inland deserts at Marble Bar (WA), Marree (SA), Oodnadatta (SA) or Birdsville (Qld). Wait a minute – we have known for decades that these are the hottest places in Australia. According to the "official" records (ACORN-SAT), the hottest day ever in Australia was 50.7°C on 2 January 1960 at Oodnadatta (SA). Not so.

The highest temperature in the Australian temperature record was on the coast at Albany (WA). On 8 February 1933, the good folk enjoyed a summer day of 44°C. They were not to know that the BoM some 80 years later decided that

---

[291] C. Sturt, 1849: Narrative of an expedition into Central Australia. https://ebooks.adelaide.edu.au/s/sturt/charles/s93n/
[292] T.L. Mitchell, 1846: Journal of an expedition into the interior of tropical Australia. gutenberg.net.au/ebooks/e00034.html
[293] J. N. Ewart, "Mr Stuart's party", *The Cornwall Chronicle*, 23 January 1861.
[294] J. Nova, 2017: Mysterious revisions of Australia's long hot history. In: *Climate Change: The Facts 2017* (ed. J. Marohasy), IPA 117-131.
[295] K. Stewart, 2010: *The Australian temperature record – Part 9: An urban myth.* https://kenskingdom.wordpress.com

the temperature was really 51°C and the original record of temperature measurements was changed. The ACORN revision of the BoM record clearly introduced errors or deliberate changing of data that have not been corrected. There is clearly no quality control at the BoM and that institution needs a QA/QC audit.

An independent audit[296] of the BoM temperature records stated that:

> The global temperature trends provided by the Australian Bureau of Meteorology are artificially exaggerated due to subjective and unidirectional adjustments of recorded values.

and

> Over the full length of the long Australian records since the end of the 1800s, there is no sign of warming or increased occurrence of extreme events.

In the interim, the ABC is telling their decreasing coterie of followers that the BoM needs more climate scientists to avoid expensive mistakes.[297] A review by the Australian Academy of Sciences, lobbied for 77 extra positions in climate research at the BoM because it was claimed that accurate climate modelling can potentially avoid costly, unnecessary investments. What planet do these people inhabit? The investment in the BoM is already large, mistakes are numerous and the historical results are tainted with fraud. The climate modelling has already been shown to be inaccurate and the data is cooked. Any extra budget should be spent on severance payments.

Investment in wind and solar energy resulting in high

---

[296] A. Parker and C.D. Ollier, 2017: Discussion of the "hottest year on record" in Australia. *Quaestiones Geographicae* 36(1), 79-91.

[297] http://mobile.abc.net.au/news/2017-08-03/australia-needs-more-climate-scientists-review-urges/8767004?pfmredir=sm&WT.mc_id=-newsmail&WT.tsrc+Newsmail

electricity prices is because data is "homogenised" to show alleged global warming is driven by human-induced carbon dioxide emissions. The costs to Australia are billions of dollars yet the Academy lobbies to have more of their mates employed to provide more "homogenised" data and models. Maybe the original historical climate data should be preserved and not destroyed by the BoM rather than using "homogenised" data to continue models that long ago were shown to be wrong.

While Australia is huffing and puffing about the warmest year, warmest summer, the angry summer or warmest month, the US National Climate Data Center had more record low temperatures than measured before. Bitterly cold winters have occurred in the Northern Hemisphere over the last few years. Are these a result of human emissions of carbon dioxide that seem to produce both hot summers in Australia and perishingly cold winters in the Northern Hemisphere? Predicted winter droughts in the Northern Hemisphere were heralded by record rains and flooding.

In May 2017, the Australian Bureau of Meteorology released their winter outlook and predicted a warmer winter in southern Australia. In fact, Southern Australia had one of the coldest winters on record and northern Australia had a warmer winter.[298] So much for models and long-term predictions.

Every month, season and year, the world data centres release their assessment of the historic ranking for that period. NOAA announced that December 2009 was the eighth warmest December and the winter was the fifth warmest for the planet. No one believed NOAA. The Northern Hemisphere was then suffering its third consecutive brutal winter. Many places were the coldest for decades and some had the highest snowfall in history.

---

[298] http://thenewdaily.com.au/weather/2017/09/01/australia-hottest-winter-record/?

What NOAA did not tell us is that if "unhomogenised" temperature data is used, the most notable change that has occurred in the US over the last 80 years is that summer heat waves have decreased in frequency, intensity and extent.[299] American folklore has horrific stories of dust bowls and heat for the 1930s. Since the 1930s the average percentage of days over 35°C has halved. Up until GISS data started to be "homogenised", the 1930s were relatively warm and then there was cooling until the 1970s.[300] It's marvellous how just a little bit of fraud can rewrite the past, a process used for 70 years in the USSR and communist eastern Europe.

These shock-horror revelations that a particular year is the warmest on record implies that it is due to us humans emitting carbon dioxide and, as a result, the planet is getting warmer and warmer. But what does this really mean? We measure temperature by thermometer but sediments, fossil life, ice, caves and soils also provide us with a record of past temperatures. It was far warmer for most of the history of planet Earth than now. In the last interglacial some 120,000 years ago, it was warmer than now. Most of the past 10,500 years have been warmer than the present. With the exception of a few short sharp cold periods, Greenland temperatures were considerably warmer than now for at least the last 9,100 years. Around 6,000 years ago, it was far warmer than now.

It was announced on 28 December 2010 that 2010 was the hottest year since records had been kept. A decade before it was the super El Niño year of 1998 but even this year trailed 1934 by 0.54°C and a decade after it was the 2016-2017 El Niño. NASA has "homogenised" the US data for 1934 downwards and

---

[299] T. Heller and J. Marohasy, 2017: It was hot in the US – in the 1930s. In: *Climate Change: The Facts 2017* (ed. J. Marohasy), IPA, 92-100.
[300] J. Hansen, R. Ruedy, J. Glascoe and M. Sato, 1999: www.gis.nasa.gov/research/briefs/hansen_07/

the 1998 and 2016-2017 data upwards and just forgot to tell us about the strong El Niños in 1998, 2010 and 2016-2017.

"Homogenisation" now means that 2010 was not as hot as 1998, 2014, 2015, 2016, all touted as the hottest years on record, and all El Niño events are now hotter than 1934. This makes it look as if temperature is increasing. Regardless of which year wins the temperature "homogenisation" battle, how significant is it that we have warmer temperatures after a 300-year period of gradual warming since the Maunder Minimum? Are you surprised that temperature has risen since the zenith of the Little Ice Age? Despite the cooking of the books, 2010 was only the 9,099th hottest year in our current 10,500-year long interglacial.

To support the story that a particular year was the hottest year on record, we get told that there was less Arctic ice than ever before. What we were not told was that the loss of Arctic sea ice a couple of years ago was within the range of variability measured from over 30 years of satellite measurements, that there was an equally rapid gain of Antarctic sea ice and that now the volume of Arctic ice is increasing. Even the Minoan, Roman and Medieval Warmings were warmer than now and people did not fry-and-die.

The year 2010 was billed as the poster boy of global warming. Was 2010 a hot year? Definitely not. The Australian media was in a frenzy reporting Hansen's declaration that 2010 was the hottest year on record (contrary to his earlier statements). They did not know that Hansen's year was the meteorological year that ended on 30 November and not a calendar year and hence he managed to avoid the cool Southern Hemisphere summer and the bitterly cold Northern Hemisphere December. Furthermore, Hansen compared 2010 to 2005 by using GISS records but did not state that these records have been "homogenised" so much that they differ greatly from the available raw data. Even forgetting the quibbles about "homogenised" data, the warmest year since 2000 has only exceeded the coldest year by 0.12°C. I am sure that 0.12°C is not life threatening and that we might just be able to adapt to such a change.

NASA's climate group is run by James Hansen, an outspoken environmental activist. He exaggerates and has been caught out in public many times. It is strange that the areas of the world where NASA detected the most significant warming in 2010 were also the areas where there were no weather stations. Yes, you read it correctly.

Many areas of the world have no historical temperature measurements (e.g. Third World countries) yet it was claimed that these areas showed great warming. It appeared that Greenland was much warmer but most of the 5° x 5° grid boxes in Greenland have no measuring stations and most of the other grid boxes have only one station. The two hottest areas of Greenland shown by NOAA showed 5°C warming yet had no measuring station. Would it be wrong to state that these numbers were just plucked out of the air and, to our great surprise, showed significant warming that could not be validated by measurement?

It is the same for Siberia. In the 1940s, Siberia had areas that showed differences in warming and cooling by a few degrees. The measuring stations have now been closed down. Siberia appears now to have warmed in 5° x 5° grids where there are now no measuring stations. As the Siberian historical data only goes back to the 1940s, it is hard to calculate a trend of warming or cooling. It appears that the Arctic of Canada is also sweltering. Almost all 5° x 5° grids have at least one measuring station. The only grid without a measuring station just happened to have warming of 5°C. Many stations are no longer maintained in the

GHCN or CRUTEM3 databases yet NOAA claims that there has been a 4°C warming over the last 40 years. The record shows that similar warming occurred in the 1930s.

Temperature changes in the Canadian Arctic correlate with the El Niño-Southern Oscillation. The HadCRU data shows 60-year cycles of temperature. As temperature has been increasing for 300 years since the Maunder Minimum at the end of the Little Ice Age, the cycles are secondary to the main trend. These 60-year cycles have been measured as far back as 2,000 years.

Just in case people did not understand that we would fry-and-die, it was announced that 2014 was the hottest year and again 2015 was the hottest year. The media went into overdrive. And then it was announced that 2016 was the hottest year[301] and was 0.99°C above the 20th century average. The media again went into scary campaign mode. The UK Met office reported[302] that 2016 was one of the two hottest years on record but did qualify their statement by saying it depends on which baseline is used. You bet. It also depends upon how much opaque "homogenisation" took place and the fact that 2016-2017 was an El Niño year. It was not explained why surface temperature measurements were not in accord with balloon and satellite measurements.

The surface temperature records in the pre-satellite era (1850 to 1980) have been so widely, systematically and "homogenised" to a warming trend, that it cannot be claimed that there is human-induced global warming in the 20th century. These changes to the historical records mask cyclical changes that can be explained by oceanic and solar cycles. To increase the apparent trend of warming, earlier records of warming have been adjusted downwards. It is probable that

---
[301] *NASA, NOAA data show 2016 warmest year on record globally.* nasa.gov release 17-006, 18 January 2017.
[302] 2016: one of the warmest two years on record. MetOffice.gov.uk

these temperature databases are now useless for determining long-term trends.

More than three-quarters of the 6,000 stations that once reported are no longer being used in data trend analysis and some 40% of stations now report months when no data is available. These months where no data is reported required "infilling" which only adds to the uncertainty. The exaggeration of long-term warming by 30-50% is probably because of urbanisation, changes in land use, incorrect positioning and inadequately calibrated instrument upgrades. When changes in data sets were introduced in 2009, there was a sudden warming. However, satellite temperature monitoring has given an alternative to land-based stations and this record is increasingly diverging from the land-based measurements consistent with adding a warming bias to the land records.

When interviewed by the BBC, Phil Jones from the Climate Research Unit of the University of East Anglia and of Climategate infamy stated:

> Surface temperature data are in such disarray they probably cannot be verified or replicated.

If you cannot replicate the data, no matter how much it has been tampered with, then it is not science. The data is useless. No conclusions, trends or predictions can be made.

You heard it from the horse's mouth.

6
SCARY SCENARIOS

A few summary pointers:

(a) Climate always changes and has been for 4,567 million years.

(b) Mention of carbon pollution is a display of ignorance.

(c) Sea and land levels rise and fall, there can be no conclusion about sea level change unless the local land level change is known; sea level has been rising since the last glaciation.

(d) Mention of ocean acidification is fraud, the oceans have been alkaline since the beginning of time and during times when the atmospheric carbon dioxide content was hundreds of times higher than now.

(e) We are in a period of normal species turnover and not in a period of mass extinction.

(f) Reefs have been on Earth for 3,500 million years, coral reefs have been around for 500 million years, reefs come and go, reefs are constantly under attack from other critters, sedimentation and storms; reefs are killed by sea level fall and land level rise.

(g) Polar ice is increasing, polar temperatures are decreasing and some Antarctic ice is underpinned by water, hot rock and volcanoes.

(h) Extreme weather conditions have been decreasing over the past few decades.

(i) Droughts come and go and long intense global droughts occur during periods of glaciation.

WHAT IS NORMAL CLIMATE?

The question often asked is: *"Do you believe in climate change?"* This is known as the moron's question. Believe is a word of religion and politics, not a word of science. Climate change has been taking place for thousands of millions of years. The questioner has demonstrated that they don't know how the planet works and that climate "science" is the new fundamentalist political environmental religion with no knowledge base.

The dim distant past

For 80% of time, planet Earth has been a warm wet greenhouse planet. Polar icecaps are rare, plants have only been on Earth for 10% of time and 99.99% of all life that has ever existed is extinct. Global atmospheric carbon dioxide and methane have been variable over time and have decreased over time whereas oxygen has been in the atmosphere for 50% of time, has greatly fluctuated and has increased over time. There have been five major and many minor mass extinctions of complex life, extinction opens new environments for colonisation and, because former terrestrial animals have become extinct, we humans now have a habitat.

Sea levels have risen and fallen thousands of times by up to 600 metres, land levels constantly rise and fall and massive rapid climate changes derived from supernovae, solar flaring, sunspots, asteroids, comets, uplift of mountain ranges, pulling apart of oceans, stitching together of land masses, drifting continents, orbital changes, changes in the shape of Earth, ice armadas, changes in ocean currents and volcanoes.

The major components of the atmosphere have been added by volcanicity and oxygen has been added by life, principally from the organisms that have ruled and continue to rule the world (bacteria). The messages written in stone show that the lithosphere, hydrosphere, cryosphere, atmosphere and biosphere are constantly interacting on our dynamic evolving planet. There is no evidence to suggest that the future of planet Earth will be significantly different from its past. However, planet Earth is not a spaceship and some great environmental changes in the past have been related to rocky and icy visitors from space. Climate cycles will continue.

Planet Earth condensed 4,567 million years ago from recycled stardust. Since that time, the continents have been enlarging, Earth materials have been constantly recycled and the Earth and all associated systems have been dynamically evolving. The Earth has not stopped being an evolving dynamic system just because humans are now on the planet.

During the first 800 million years of planetary history, Earth was bombarded by large asteroids, one of which broke off a mass that condensed to form the Moon. Volcanic activity was degassing Earth. The main gases emitted from volcanoes were and still are the greenhouse gases water vapour, carbon dioxide and methane together with minor helium, nitrogen compounds, sulphur compounds, acids and rare gases. The earliest atmosphere probably had a high ammonia and sulphur dioxide composition and these gases were quickly scrubbed out of the atmosphere by rainfall.

The main greenhouse gas on Earth is water vapour. Volcanic gases accumulated to form a primitive oxygen-deficient greenhouse gas atmosphere. Early accumulations of water in oceans were vapourised by large asteroid impacts. The intensity and frequency of impacting has decreased over time and now only 40,000 tonnes of extraterrestrial material is added to Earth each year.

As soon as there was liquid water on Earth, there was life.

This early bacterial life thrived, the oxygen-deficient atmosphere was hot and carbon dioxide- and methane-rich. Rain was extremely acid. Early Earth was very warm and wet however there is some evidence to suggest that there was an ice age about 2,900 million years ago.

Bacteria slowly diversified and, by the time the Earth was middle aged, one group of bacteria had emitted large quantities of oxygen and the atmosphere became slightly oxygenated. Some of this excess oxygen was trapped in rocks by weathering, most dissolved in the oceans resulting in the precipitation of iron oxides. It is these iron oxides that form the great iron ore fields of planet Earth (e.g. Pilbara, WA). Life, the atmosphere, the oceans and the rocks interacted, a process that has been occurring for at least 2,500 million years on our dynamic evolving planet. There was a long ice age from about 2,400 to 2,100 million years ago. Ice sheets were at sea level and at the equator.

For at least the last 2,500 million years, the continents have been pulled apart and stitched back together. Every time the continents are pulled apart, huge quantities of volcanic water vapour, carbon dioxide and methane are released into the atmosphere. When continents stitch together, mountain ranges form. Mountains are stripped of soils, new soils form and remove carbon dioxide from the atmosphere, these soils are stripped from the land and the carbon dioxide becomes locked in sediments on the ocean floor. Because of the inverse solubility of carbon dioxide, events of glaciation remove carbon dioxide from the atmosphere whereas carbon dioxide is added back to the atmosphere during the warmer interglacials.

The origin of the greatest climate change on Earth is an enigma. A remarkable event took place. Ice sheets many kilometres thick formed at the equator at sea level some 700 million years ago. This was the greatest climate change that our planet had ever enjoyed and yet we don't know exactly why this ice age started and why it ended after tens of millions of years. This was the time of snowball Earth. A huge amount of glacial debris was left behind by the retreating ice. Soon after this ice age ended, there was an experiment with multicellular life during the interglacial (Arkaroola Reef) when the global temperature had risen just slightly from about -40°C to +40°C and the oceans had been filled with nutrients from melt waters that washed very fine particles of rock flour into the oceans. Environmental activists are worried about a 0.7°C temperature rise over a century (which may be natural anyway) and don't seem to know that a temperature rise of 0.7°C takes place by standing up or walking into another room.

The Arkaroola Reef formed at about 650 million years ago and fossils of multicellular sponge-like organisms unlike any later life on Earth are preserved. The reef was killed off by another ice age about 600 million years ago. Again, we do not know exactly why we had an ice age and why the ice age conditions changed to warmer conditions. The Arkaroola Reef is one of the few known well-preserved failed experiments with multicellular life on Earth.

After the second major glaciation, the Earth became warmer and again, there was another experiment with multicellular life. This was the Ediacaran fauna, which existed from about 583 to 542 million years ago. The soft-bodied jellyfish-like Ediacaran fauna are they key to understanding how animal life evolved on planet Earth. Predation led to the end of the Ediacaran fauna that ate algal mats, the new predators grew protective scales, shells and skeletons and the soft-bodied Ediacarans were no match. Sea floor sediments older than 542 million years in age are finely banded whereas anything younger has been turned over by sea floor life showing that since the Cambrian explosion of life there was a meal in eating the bacteria in mud as well as soft-bodied critters. Today the greatest biomass on Earth is still bacteria with most life living in rocks down to 4 km depth.

This sequence of glacial rocks is very well displayed in the Flinders Ranges. Sea level changed by up to 600 metres and interglacial sea temperatures were +40°C. Furthermore, glaciation was at sea level and at the equator. Bacterial life survived, which is no surprise, because we now know that bacteria live in oxygen-poor and oxygen-rich conditions in acid and alkaline hot springs, in highly radioactive water, deep in the Earth in cracks in rocks, in volcanic rocks in mid ocean ridges, in clouds, in ice and in every environment on Earth. Bacteria always have been the dominant biomass of Earth and yet their role in interaction with the atmosphere is unknown. Bacteria some 400,000 years old have been resurrected from the Greenland ice sheet and from salty liquid inclusions of water in rock salt in Texas.

After this ice age, the atmosphere had perhaps 20% carbon dioxide and bacteria thrived and diversified in the warm oceans. An explosion of life from 543 to 520 million years ago in the Cambrian gave us all of the major life forms currently present on Earth. It was also an explosion of predation.

The record written in stone by fossils in the period 520 million years ago to the present shows that the planet is a warm, wet, greenhouse, volcanic planet with the normal cycles of rising and falling sea levels, rising and falling land levels and changing climates.

Plants appeared at 470 million years ago, there was another ice age, there were numerous minor mass extinctions and there was a major mass extinction of multicellular life at 450 to 430 million years ago. Vacated ecologies were quickly filled and life continued diversifying. More minor mass extinctions followed and there was another major mass extinction at 375 to 368 million years ago. Between 368 and 251 million years ago, massive coal deposits formed, there was a major 50 million year long ice age and the atmosphere was blessed with a very high carbon dioxide and oxygen content. Life continued to diversify.

Minor mass extinctions continued and, at 251 million years ago, the biggest major mass extinction on Earth took place. Some 96% of species became extinct, evidence for impacting is weak and the smoking gun is volcanicity in Siberia. Massive volcanic activity over a very short period of time exhaled huge volumes of carbon dioxide and this was counterbalanced by the release of sulphur compounds that reflected heat and light. The resultant climate was cooler, acid rain may have destroyed large tracts of vegetation thereby creating a collapse of terrestrial environments. Life diversified quickly to fill the vacated ecologies.

Another major mass extinction took place at 217 million years ago. A swarm of asteroids hit the Northern Hemisphere, a continental mass was fragmented, large volumes of lava were released and the Atlantic Ocean formed. The planet was still a warm wet greenhouse planet with the normal cycles of rising and falling sea levels, rising and falling land levels and changing climates.

From 251 to 120 million years ago, the global carbon dioxide content varied greatly and increased to a peak 6% carbon dioxide 120 million years ago. This derived from intense volcanic activity associated with continental fragmentation. Thick vegetation covered the land masses. The atmospheric oxygen content greatly increased to 35% at 300 million years ago, decreased and then increased to 27% at 150 million years ago. It is currently 21%.

During times of high atmospheric oxygen content, there was spontaneous combustion of the atmosphere, global bushfires and increased erosion. Some 120 million years ago, Australia was at the South Pole enjoying a temperate climate. There was a short ice age with minor glaciers in the highlands. Volcanoes were active and dinosaurs adapted to the long periods of darkness by evolving enlarged optic nerves. Global sea level was more than 100 metres higher than at present, the sea surface temperature was 10 to 15°C higher than now and many continents were covered by shallow tropical seas. Planet Earth was a warm wet greenhouse paradise and thick vegetation covered the land masses. Australia started to pull away from Antarctica at about 100 million years ago. It drifted northwards at 7 cm/year, the Tasman Sea opened and the Indian Ocean opened with India starting to drift away from Western Australia.

The opening of the Tasman Sea produced the rise of the Great Dividing Range, the diversion of the major river systems and changes to the climate of eastern Australia. A minor mass extinction of life 90 million years ago was the result of volcanoes in the Indian and Pacific Oceans belching out gases into the oceans and atmosphere. Some 26% of advanced life became extinct and there was a runaway greenhouse until volcanism waned. Volcanic emissions of carbon dioxide are common. In 1984 and 1986, burps of carbon dioxide from the volcanic crater lakes of Monoun and Nyos respectively killed thousands and added carbon dioxide to the atmosphere. Near Mt Gambier, carbon dioxide is commercially extracted from rocks, one small hot spring on Milos (Greece) contributes to 1% of the planet's volcanic carbon dioxide and huge quantities carbon dioxide, the planet's second most common volcanic gas, constantly leak from at least 3.4 million unseen submarine volcanoes.

An extraterrestrial visitor at 65 million years ago impacted Earth but the major mass extinction of life at that time probably derived from a huge volcanic event in India with the associated release of carbon dioxide, methane and sulphur gases into the atmosphere. Ancient soils, vegetation and rock chemistry show that conditions were tropical. Another minor mass extinction at 55 million years ago was caused by a Caribbean volcano. There was a rise in sea surface temperatures by up to 8°C for 100,000 years, atmospheric carbon dioxide was 10 times that of today, the oceans became acid for a geological instant, the oceans lost dissolved oxygen and the ocean floors released methane into the atmosphere. During these warmer times, plankton sucked up the atmospheric carbon dioxide, mammals thrived and life filled the vacated ecologies. Atmospheric carbon dioxide decreased from 3,500 to 700 parts per million (ppm) within a million years, stayed low until 47 million years ago and went up and down to about the present level (400 ppm) about 40 million years ago.

India collided with Asia at 50 million years ago. Uplift produced the Tibetan Plateau that started to scrub carbon dioxide out of the atmosphere. The Tibetan plateau is still rising and carbon dioxide is still being scrubbed out of the atmosphere. The Drake Passage opened as South America drifted from Antarctica, a circum-polar current developed and Antarctica refrigerated. This was the start of the current ice age 34 million years ago.

Changes in submarine topography along the Tasman Rise, the closing of the Mediterranean Sea, the onset of polar glaciation and the flow of polar bottom waters formed climate zones, a feature that had not previously existed during the previous long periods of warm wet tropical climates. There were numerous minor mass extinctions, comet and meteorite impacts and sea level changes. For example, the Murray Basin became a large sea, retreated and then advanced again only to start its final retreat at five million years ago.

Warm currents in the Indian Ocean were deflected and drifted through the Great Australian Bight and up the Pacific coast of Australia. Southern Australia from 17 to 14.5 million years ago was again tropical with mid-latitude temperatures 6°C warmer than today. Atmospheric carbon dioxide was 180 to 290 ppm. This greenhouse occurred when atmospheric carbon dioxide was 30 to 50% lower than today. Land changes closed the Straits of Gibraltar at seven million years ago, the Mediterranean Sea again dried and because there was less salt in the oceans, parts of the oceans froze. Both the ice and salt reflected sunlight and the planet cooled further.

By five million years ago, Earth was so cool that very slight orbital wobbles every 41,000 years now had a bearing on climate. It suddenly changed about 1 million years ago and from then on, every 100,000 years due to the Earth's elliptical orbit. Climate cycles were now 90,000 years of glaciation and 10,000 years of interglacial. We are currently in one of those interglacial periods. Cooling changed forests to grasslands and primate extinction and diversification to upright bipeds took place.

By 2.67 million years ago, central American volcanoes had closed the seaway between the Pacific and Atlantic Oceans, explosive volcanoes in Kamchatka had added dust to the atmosphere, dust reflected sunlight and the planet cooled further. The Northern Hemisphere polar ice cap formed. The 100,000-year cycles of interglacials and glaciation continued.

The penultimate interglacial was 120,000 years ago. *Homo erectus, Homo neanderthalensis. Homo floresiensis* and *Homo sapiens* coexisted, sea level was at least six metres higher than at present and the planet was far warmer and wetter than now. We live in unusual times when only one species of *Homo* is on Earth. During this interglacial, there was more vegetation than today and atmospheric carbon dioxide was 78% of today's concentration. After warming, the atmospheric carbon dioxide and carbon dioxide content increased suggesting that atmospheric temperature rise drives an increase in atmospheric carbon dioxide and methane contents.

Orbital-driven cooling commenced, the eruption of Toba (Indonesia) produced dust that reflected sunlight, sea level dropped and glaciation continued. Only *Homo sapiens* survived the last glaciation. We very nearly became extinct when the atmosphere was filled with 3,000 cubic kilometers of dust after the Toba eruption 74,000 years ago. Geneticists have estimated that there were only 8,000 humans left on Earth after this eruption.

During the history of the latest glaciation, armadas of ice were released into the sea every 7,000 years resulting from the physical failure of thick ice sheets. These had a profound effect on climate. Small cool periods occurred every 1,100 to 1,300 years.

The day before yesterday

Some 24,000 years ago, during coldest part of the last glaciation, there was a sudden warming of about 15°C. Now that is global warming! Was this warming due to industrial smoke stacks emitting carbon dioxide? Shortly after, temperatures dropped suddenly by about 7°C. Temperatures remained cold for several thousand years and fluctuated between 3°C warmer and 3°C colder than now.

The zenith of the last glaciation was 20,000 years ago. Sea level was 130 metres lower than today, temperature was 10 to 15°C lower than today and there were very strong cold winds. The Northern Hemisphere was covered by ice to 38°N with more northern areas such as Scandinavia covered by more than three kilometres of ice. The loading of the polar areas with ice changed the shape of the planet, the planet's rotation changed and as a result ocean currents distributing heat across the

Earth were changed. Humans lived very short lives around the edge of ice sheets.

Australia was buffeted by anti-cyclonic winds that deposited sand dunes and carried sea salt spray that was trapped in the inland basins such as Lake Frome and Lake Eyre. Tasmania and parts of the south eastern highlands of Australia were covered in ice and sea level was so low that Aboriginals walked to Tasmania from mainland Australia. Rainforests disappeared and the Amazon Basin consisted of grasslands and copses of trees. Coral reefs were stressed and disappeared in colder conditions when sea level was lower. Coral still existed in equatorial waters.

About 15,000 years ago, there was another sudden rapid warming of about 12°C. This is far greater than anything measured today or predicted by the most scary catastrophist for the future. Was this warming due to heavy industry emitting carbon dioxide? The large ice sheets that covered Canada and the northern USA, all of Scandinavia and much of northern Europe and Russia started to melt quickly and glaciers retreated.

The northern polar ice sheet started to melt 14,700 years ago. A few centuries later, temperature suddenly fell about 11°C and glaciers re-advanced. About 14,000 years ago, global temperatures again rose rapidly by about 4.5°C and glaciers receded. Was this due to human emissions of carbon dioxide?

There were very rapid and major temperature fluctuations, sea level rose and fell and the total sea level rise over the last 14,700 years has been at least 130 metres. Land masses previously covered with ice started to rise. For example, Scandinavia is still rising and has risen more than 340 metres over the last 14,700 years. As a counterbalance, the Netherlands, southeastern England, Schleswig-Holstein and Denmark are sinking. One of the consequences of a massive sea level rise over the last

14,700 years is that the West Antarctic Ice Sheet was no longer unpinned by land. Two-thirds of the West Antarctic Ice Sheet collapsed into the oceans and sea level rose 12 metres. The final third of the West Antarctic Ice Sheet has yet to collapse to produce a six metre sea level rise as part of the dynamic post-glacial climate on Earth. If the last third of the West Antarctic Ice Sheet collapsed, sea level would be back to where it was 120,000 years ago in the last interglacial.

About 13,400 years ago, global temperatures dropped, this time by 8°C and glaciers re-advanced. At 13,200 years ago, global temperature again increased rapidly by 5°C and glaciers re-advanced. About 12,700 years ago, temperatures fell quickly by 8°C and a 1,300-year cold period, the Younger Dryas, started. This was probably due to the breaching of a meltwater dam that produced a sudden sea level rise. Climate changes induced by changes in ocean currents cooled North Africa, grasslands changed to a desert, humans migrated and the great Mesopotamian cities were established.

After this intense cold period, global temperatures rose very quickly by about 12°C with a 500-year cold period from 8,500 to 8,000 years ago, again from the breaching of a meltwater dam producing another sudden sea level rise. Alpine glaciers advanced. The interglacial sea level rise resulted in the earthquake-triggered breaching of the Mediterranean along the Bosphorus into the Black Sea Basin some 7,600 years ago and is probably the origin of the Sumarian, Babylonian and biblical stories of a great flood. The filling of the Black Sea Basin with water resulted in a sea level fall.

There was then prolonged warming until about 6,000 years ago. This was the Holocene Optimum. It was a few degrees warmer, there was 20% more rainfall and sea level was 1 to 3 metres higher than now. As there was no industry pumping traces of a trace gas into the atmosphere, it can only be concluded that the cyclical warmings and coolings in postglacial times were natural.

Cold dry periods, glacier expansion and crop failures between 5,800 and 4,900 years ago resulted in deforestation, flooding, silting of irrigation channels, salinisation and the collapse of the Sumerian city states. Long periods of El Niño-induced drought resulted in the abandonment of Middle Eastern, Indian and North American towns. About 1,470 BC, Thira exploded and threw 30 cubic kilometres of dust into the atmosphere. The tsunami, ash blanket and destruction of Thira greatly weakened the dominant Minoans. This led to the rise of the Mycaeneans and Greeks. One volcano changed the course of Western history.

Global cooling from 1,300 to 500 BC gave rise to the advance of glaciers, migration, invasion and famine. Global warming commenced again at 500 BC, there was an excess of food and great empires such as the Ashoka, Ch'hin and the Romans grew. Contemporary records of crops and Roman clothing shows that conditions were some 5°C warmer than today.

There were again periods of warming and cooling in Egyptian and Minoan times. Warmings and coolings were natural and had nothing to do with human emissions of carbon dioxide. Prior to the founding of the Roman Empire, Egyptian records show a cool climatic period from about 750 to 450 BC. If science is not your thing, then history has some good examples of how climate has changed quickly and naturally. Hannibal was able to take his army and elephants across the Alps in the winter of 200 BC. This would not be possible now.

Julius Caesar conquered Gaul in about 50 BC. He had to build a bridge across the Rhine River (which now separates France from Germany) in order to wage war. The Rhine acted as a natural protective barrier for nearly 500 years but, as the

Roman Empire was disintegrating in Gaul in the fifth century AD, the Vandals were able to walk across the frozen Rhine River to wage war. The Rhine has not been frozen in modern times.

The Romans wrote that the Tiber River froze and snow remained on the ground for longer periods. This does not now happen in Rome. During the Roman Warming after 100 BC, the Romans wrote of grapes and olives growing much farther north in Italy than had been previously possible because there was little snow and ice. Wine grapes were grown in northern England where it is now too cold to grow wine grapes. In 29 AD, the Nile River froze. It has not frozen since. Sea levels did not rise during the 600-year long Roman Warming. Although the Romans did emit some carbon dioxide from smelting and grape fermentation, the amounts emitted were tiny compared to now. In Roman times it was warmer than now but this could not be due to human emissions of carbon dioxide.

In 535 AD Krakatoa exploded, as did Rabaul in 536 AD. The Earth passed through cometary dust in 536 AD and the Sun was in a solar minimum. The dusty atmosphere reflected heat and darkness prevailed and, as a result, the climate cooled and there was famine and warfare. During the Dark Ages, it was very cold. A strange event occurred in 540 AD when it became even colder. Trees grew very slowly, the Sun appeared dimmed for more than a year and temperatures dropped in Ireland, Great Britain, Siberia, and North and South America. In 800 AD, the Black Sea froze. It has not frozen since.

Changes in solar output resulted in the Medieval Warm Period from 900 to 1300 AD. The first to feel the change were the Vikings who were able to navigate the northern waters, colonised Newfoundland and Greenland and established extensive trade routes as far south as the modern Gulf States. The warmer wetter climate of Europe produced excess crops and wealth that resulted in the building of castles, cathedrals and monasteries. As with previous warming events, there was great prosperity.

In the Medieval Warm Period in Europe, grain crops flourished, the tree line in the Alps rose, the population more than doubled and many new cities arose. Great wealth was generated, as always happens in warmer times. The warmer climate allowed the Vikings to colonise Greenland in 985 AD. Ice-free oceans enabled Viking sea travel as far as Canada and the fishing grounds were enlarged. The Vikings grew barley and wheat and grazed sheep and cattle in places that are now ice-covered. The depths of the Viking graves show that there was no permafrost. In France and Germany, grapes were grown some 500 kilometres north of the present vineyards, again showing that it was far warmer than now.

In Germany, where grapes are now grown at a maximum altitude of 560 metres, the Medieval vineyards were up to 780 metres altitude showing that temperature was warmer by about 1.0 to 1.4°C. Wheat and oats were grown at Trondheim (northern Norway), again suggesting climates 1°C warmer than now. Prolonged droughts affected southwestern USA and Alaska warmed. Lake sediments show that the Medieval Warming had a considerable warming effect on South America. Lake sediments in central Japan record warmer temperatures, sea surface temperatures in the Sargasso Sea were approximately 1°C warmer than today and the climate in equatorial east Africa was drier from 1000 to 1270 AD. Ice core from the eastern Antarctic Peninsula shows warmer temperatures during this period. The Medieval Warming was just not some local warming event in Europe. It was global.

In modern times, the Modern Warming has been no greater than 0.8°C. This we are told is due to human activities yet the Medieval Warming of almost twice this amount was natural. The obvious question arises: Which part of the Modern Warming is natural and which part is due to humans? I suspect that all the modern warming is natural. The Medieval Warm Period was warmer than now, was global and clearly was not driven by power stations and factories emitting carbon dioxide. Sea level did not rise during this 400-year period of warming.

It is absolutely vital for warmists to expunge the Medieval Warming from the record, as shown by Mann's fraudulent hockey stick. If there is no Medieval Warming, then it can be argued that the Modern Warming is unprecedented and related to modern industrialisation. Time and time again, scientific papers appear showing that the Medieval Warming was global and that temperatures exceeded those of today.[303]

The Little Ice Age (1300 to 1850 AD) started with a rapid decrease in temperature of 4°C. Glaciers expanded worldwide. This was not due to a sudden decrease in atmospheric carbon dioxide. Greenland glaciers advanced and pack ice expanded. In 1280 AD, volcanic eruptions on Iceland, reduced sunspot activity and a change in ocean currents started the Little Ice Age which finished in 1850. The Gulf of Bothnia between Finland and Sweden froze in 1303 and 1306 to 1307. Three years of torrential rains led to the Great Famine of 1315 to 1317 and the population was further reduced by the spread of the Black Death plague that attacked the weakened population in 1347 to 1349. The grain-dependent population of Europe suddenly faced a cold and variable climate with early snows, violent storms, catastrophic flooding and massive soil erosion. There were constant crop failures, livestock died and famine became a great killer.

Wine production decreased, the growing season for all crops was shortened, rivers such as the Thames in London and

---

[303] J. Abbot and J. Marohasy, 2017: The application of machine learning for evaluating anthropogenic versus natural climate change. *Geophys. Res. Jour.* doi:10.1016/j.grj.2017.08.001

canals in the Netherlands froze. All travel became hazardous. There was massive depopulation and it took Europe 250 years to reach the population of 1280 AD. During the Little Ice Age, there were warmer periods associated with sunspot activity. During minimum sunspot activity (1440-1460, 1687-1703 and 1808-1821), the intensely cold conditions were recorded by the Dutch masters and King Henry VIII was able to roast oxen on the frozen Thames. There were food shortages.

Glaciers advanced in the Alps of Europe, Papua New Guinea, the Andes and New Zealand and they buried villages in the Swiss and Austrian Alps. When New York Harbour froze in the winter of 1780 AD, people could walk from Manhattan to Staten Island. In June 1644, the Bishop of Geneva took his flock to pray on a glacier that was advancing "a musket shot" every day. Even prayers to the Almighty did not change climate and it is extraordinary human arrogance today to think that if we humans twiddle the dials, then we can change climate.

In the Little Ice Age, sea ice surrounding Iceland extended for great distances in every direction, closing many harbours and preventing fishing trips. The population of Iceland halved and the Viking colonies in Greenland died out in the 1400s because they could no longer grow enough food or get through the ice for fishing. In parts of China, warm weather crops that had been grown for centuries were abandoned. In North America, early European settlers experienced exceptionally severe winters. If carbon dioxide is the cause of global warming, then the past global warmings should mirror the rise in carbon dioxide. They don't.

Occasional events such as filling of the atmosphere with aerosols from volcanic eruptions induce short term cooling such as the five years of cool climate after Tambora (1815), Krakatoa (1883) and Tarawera (1886). In fact, 1816 was known as the "year without a summer". This was the time when

Turner painted stormy oceans and skies full of volcanic dust, Mary Shelley wrote *Frankenstein* and Byron wrote *Darkness*.

In the scheme of things, it is not worth getting into a lather about a modelled temperature increase of 0.1°C; this is the temperature change that occurs when you move your head. A complex climate system having been reduced to the impact of one variable (human emissions of carbon dioxide) shows the intellectual vacuosity of the green left environmental activists.

This potted history of warmings and coolings during the last glaciation and the current interglacial show that there has been global warming many times and at a rate far faster than any scary model of the future has predicted. None of these warmings can be claimed as due to human emissions of carbon dioxide. On the basis of the history of coolings and warmings of the Earth, there is every good reason to conclude that current Modern Warming is natural.

Today

The 20th century and early 21st century AD are times of natural post-glacial rebound. Ice sheets, a rare phenomenon in the history of time, still exist. Sea level is relatively low, as are global temperatures and atmospheric carbon dioxide. Between 1920 and 1945, there was a period of warming (0.37°C), then a period of cooling and warming again commenced in 1976 (0.32°C). In 1976 to 1977, global temperatures in the lower atmosphere jumped 0.3°C, sea surface temperature in the equatorial Pacific jumped 0.6°C, sea surface temperature during upwelling increased 1.5 to 3°C but there was reduced upwelling, the heat content of the upper 300 metres of the world's oceans increased, there was increased wave activity in the North Sea and the length of the day changed. The stepwise increase in temperature in 1976 and 1977 shows that there was a major re-ordering of the ocean heat transport coinciding with an orbital change expressed as a change in the length of the day and solar changes. Maybe global warming of the 20th century is just a measure of the variability on a dynamic evolving planet?

To put such measurements into perspective over the history of time, changes in atmospheric temperature in the 20th century can only be considered small and slow. Decades of global coverage by satellites shows a different story to "homogenised" ground temperature measurements and a slight greening of the planet in accord with increased crop productivity. Temperature measurements from balloons agree with the satellite measurements for the period of overlap. Because greenhouse warming is a phenomenon of the atmosphere, significant changes should have been recorded. They have not.

We know from the Roman Warming, Dark Ages, Medieval Warming and the Little Ice Age and the last glaciation that climate changed very rapidly and that climate changes are not driven by the rise and fall of carbon dioxide. Over the last 500 years, the ice cores from Greenland, the position of Northern Hemisphere glaciers and lake sediments show that temperature oscillated about 40 times with changes on average every 27 years. None of these changes could possibly have been driven by changes in atmospheric carbon dioxide or human emissions of carbon dioxide.

Past climate has been driven in the very long term by the position of the continents and 143-million year long galactic cycles,[304] in the not so long term by Milankovic Cycles (100,000,[305] 41,000[306] and 21,000 years[307]), in the medium

---

[304] H-J. Brink, 2014: Singnale der Milchstraße vorbogen in der Sedimentfüllung Zentraleuropäischen Beckensystems? *Z. Dt. Ges. Geowiss.* 166: 9-20.
[305] Orbital eccentricity.
[306] Axial tilt or obliquity.
[307] The combined effect of two precessions.

term by solar (1500-, 217-, 87- and 22-year cycles) and ocean cycles (60-year Atlantic Multidecadal Oscillation, 30-year Pacific Decadal Oscillation) and in the short term by ocean and atmosphere cycles (Lunar tidal nodes, El Niño-Southern Oscillation).

Climate has always changed and always will as the potted history of the Earth above shows. In the past, climate has never changed due to atmospheric carbon dioxide despite monstrous emissions from volcanoes and the oceans. So, what is normal climate? Constant change between cold glaciation and warm interglacials.

Having read the above, just ask me that question again. *"Do you believe in climate change?"*

This is indeed the moron's question.

POPULAR SCARES

Carbon pollution

While we are talking about morons, let's look at a moronic term very often used by politicians and green activists. It is "carbon pollution". There are two natural forms of carbon. One is graphite, the other is diamond. If there was carbon pollution, then we would not be able to see anything because the air would be black due to graphite. The other form of carbon pollution would really be a case of *"Lucy in the sky with diamonds"*. The sky would sparkle, diamonds would be everywhere.

There are tens of thousands of carbon compounds that vary from colourless to black. Carbon black and soot are composed of carbon but this carbon has no crystal structure like graphite. These are common products from incomplete burning and are dangerous. The main source of these materials is from bushfires and home cooking by three billion people in poor countries using twigs, leaves, crop waste, wood and dung.[308] Over 4.3 million of these three billion people die each year from household air pollution from pneumonia, stroke, heart disease and lung cancer. Women and girls are particularly prone to lung cancer because they do the cooking, often in confined spaces. This is certainly carbon pollution and the solution is simple. Provide cheap coal-fired electricity.

All life is based on carbon and carbon occurs naturally. Carbon dioxide is plant food and together with sunlight and water, plants use the carbon dioxide in photosynthesis. Thousands of studies have shown that with increased carbon dioxide in air, crop growth increases, as do yields.[309] Glasshouses pump in warm carbon dioxide exhaust from propane burners to accelerate growth.

How could carbon dioxide be pollution? Carbon dioxide is the food of life. Without carbon dioxide, there would be no plant life on Earth. With no plant life, there would be no animal life. In the geological past, whenever there has been a high atmospheric carbon dioxide content, life on Earth has thrived.

The food we eat contains carbon complexes, our bodies oxidise these carbon compounds to produce carbon dioxide. Humans breathe in air with 0.04% carbon dioxide and breathe out air with 4% carbon dioxide. Once a politician uses the words "carbon pollution" you know you are being conned and that you will pay.

Ignorance and environmental activism have re-labelled carbon dioxide emissions into the atmosphere by humans as "carbon pollution". The solution for activists to "carbon pollution" is simple.

Drop dead.

---

[308] www.who.int/mediacentre/factsheets/fs292/en/
[309] www.co2science.org/about/chairman.php

Sea level

Is sea level rising? Yes and no! The answer is written in history, archaeology and geology.

What seems like a simple question is not simple because nature is complex and fickle. What appears to be a sea level rise may be the sinking of land, the sinking of a measuring station or an actual rise in sea level. Now don't say *"Who knows?"* because there are ways of knowing.

Sensationalist headlines tell us that Pacific Ocean atoll island nations are being inundated because sea level is rising. There is neither evidence nor logic to support this view. The next illogical step is that this alleged sea level rise must be due to climate change. The final illogical step is that someone or something is to blame. Radical activists go one step further and claim that sea level rise must be due to warming, that global warming is a result of burning coal producing carbon dioxide emissions into the atmosphere and hence all coal mining must be stopped.

Such activists are not aware that only 3% of annual emissions of carbon dioxide are human emissions, that it has yet to be shown that human emissions of carbon dioxide drive global warming, that the human emissions of carbon dioxide are massively increasing due to the industrialisation of Asia concurrently with a plateau or decrease of global atmospheric temperature, that a very slight increase in atmospheric carbon dioxide has led to a greening of the planet and increased crop productivity, that the area of Pacific atolls is increasing over the last few decades when alleged global warming and a resultant sea level rise should have inundated these atolls, that coal has brought the West from grinding poverty, that burning coal has given Western activists the life that they now enjoy and that banning coal exports keeps Asian and Indian people in poverty.

During the peak of the last ice age 20,000 years ago, Scandinavia, Russia, Scotland, northern USA and Canada were covered by kilometres of ice. The weight of the ice sheets pushed down landmasses. Ice started to melt 14,700 years ago and the landmasses once covered by ice started to rise as their load had been lifted. This rise continues, there are beaches in Norway some 340 metres above sea level and the 12th century castle at Turku (Finland), once built on an island, can now be reached by foot. The rate of land uplift can be calculated and it is surprisingly fast. As a result of land rise, other landmasses fall as a counterbalance and the Netherlands, northwestern Germany, Denmark and southeastern England are sinking. This sinking has been exacerbated by a post-glacial sea level rise of some 130 metres over the last 14,700 years.

Many substances are both brittle and plastic. If a rock or glass is hit with a hammer, it will break. However, if a rock or glass is loaded for a long period of time, it will bend. A tour of tombstones in a cemetery will show that marble headstones that were originally flat are now bent and, from the date on the tombstone, the rate of bending can be calculated. Ice behaves in the same way. It can be broken with a hammer and it can flow slowly down a glacier under a gentle gravitational gradient. Flow is because ice is constantly recrystallising and the ice crystals become bigger from the head of the glacier to its foot.

Landmasses also rise and fall because of the pushing together and pulling apart of the tectonic plates. Mountains belts such as the Himalayas are rising with Mt Everest increasing in altitude by 2 cm per year. The Rift Valley of East Africa is sinking as are parts of inland Australia.

For more than 1,000 years the Dutch have been building dykes, pumping out water with wind mills and suffering from inundation when storms coincide with king tides. This will continue. The Romans noted that Londinium was sinking and built London on gravel terraces rather than next to the Thames River where malaria was rife. Major engineering works such as the Thames Barrier have been built to deal with river flooding and storm surges resulting from the sinking of London due to geology, the extraction of ground water and traffic.

Ancient cities like Lydia, the birthplace of coinage, are now metres below the water whereas as other ancient port cities like Ephesus, mentioned in the Bible,[310] have risen and are now inland. Subsidence can play some cruel tricks. For example, the tidal measuring station at Port Adelaide is sinking thereby recording a sea level rise. Bangkok, Mexico City, Denver and many other cities are sinking because of compaction after the extraction of groundwater.

Sea level can also rise and fall. This is not only driven by cyclical global cooling and warming. Winds, tides, currents and low-pressure atmospheric systems can push seawater from one side of the ocean to another resulting in sea level change. The loading up of the sea floor with sediment or the adding of water to the oceans can result in the sinking of the plastic ocean floor. The gravitational pull of continents, especially with coastal mountain chains such as the Andes, can pull ocean water to a higher level. Submarine super volcanoes add huge amounts of lava to the sea floor and this results in a sea level rise.

Twice the Straits of Gibraltar was closed and the Mediterranean evaporated. As a result of the northwards drift of Africa, the Straits opened again and water poured into the Mediterranean basin to form the Mediterranean Sea and global sea level fell. Sea level also fell when the Bosphorus was gouged out and the Marmara Sea poured into a basin to form the Black Sea. Major tectonic processes are constantly changing the shape of the ocean floor with the formation of islands, ridges and trenches

---

[310] Acts 19: 1-7.

and these processes produce sea level change. To assume that sea level change is only due to climate is naïve.

We see evidence of land level rise everywhere. Coral atolls in Vanuatu are stranded metres above sea level, dead microatolls along the Great Barrier Reef show that the land level has risen, and rock platforms around Sydney show that the land level had risen before the construction of Fort Denison.

During glacial times, water is locked in continental ice sheets resulting in a sea level drop. At times, there was so much water locked in ice sheets that there was no continental shelf. This suggests that natural sea level changes can be up to 600 metres. During glaciation and the associated sea level drop, there is stress and extinction of life. Coral reefs are exposed and die, shallow water habitats disappear, devegetation occurs, rivers dry up, soils change and the weather becomes colder, drier and windier.

During the peak of the last glaciation 20,000 years ago, there were copses of trees and grasslands in what is now the Amazonian jungle. Sea level was so low that one could walk from Papua New Guinea to Tasmania or France to England. After glaciation, ice sheets partially melted and water was added to the oceans. Sea level rose because of the added water and thermal expansion of water.

The West Antarctic Ice Sheet, once pinned to the land, now has wet feet and is unstable as a result of post-glacial sea level rise. About two-thirds of the ice sheet has collapsed into the oceans and if the last third collapses, sea level will be six metres higher. This is where it was 125,000 years ago. If the current interglacial warming goes to completion and all polar ice melts, sea level will be some 80 metres higher than now.

As a result of sea level changes, there are old beaches under water, drowned valleys (such as Sydney Harbour), beaches above water and kilometres inland and beaches denuded of sediment that later is redeposited. Beaches are constantly changing, again further evidence of an ever-changing planet. The most striking example of a beach is a mineral sands mine in the desert at Pooncarie (NSW), some hundreds of kilometres from the current coast. Rock platforms in eastern and Western Australia and exposed microatolls in the Great Barrier Reef show that the land level has risen.

If humans live at sea level or at the margins of ocean basins, then they are living right at the edge. Why did the great civilisations not evolve in benign areas on the margins of ocean basins or have we already forgotten the 26 December 2004 catastrophic tsunami that killed at least 250,000 people? The south coast of NSW experiences a tsunami every 300 years when continental shelf sediments slide down the continental slope after a minor earth tremor in the Southern Highlands of NSW. Just drop an asteroid into the ocean and see the tsunamic damage. If we live on the coast, we are exposed to known and massive risks.

Australia is, quite naturally, concerned about our neighbours living on atolls in the Pacific Ocean. If there were a sea level rise, we would expect every atoll in every ocean to be inundated. We don't see this. We would expect harbours in the world to record a sea level rise. This is not recorded. Something is seriously wrong with the catastrophist dogma.

In 1830, Charles Lyell published the first of his three volume classic *Principles of Geology: Being an Attempt to Explain the Former Changes in the Earth's Surface by Reference to Causes now in Operation*. Charles Darwin received the first volume in 1830 which he took with him on HMS *Beagle*, the other two were sent to him *en route*. Darwin became fascinated with coral reefs and suggested that they form around the rim of ancient submarine volcanoes, as did Lyell.

Darwin's voyage allowed him to view many volcanic islands,

coral reefs and atolls. In his 1842 book *The Structure and Distribution of Coral Reefs*, Darwin showed that volcanoes were at various heights above the sea floor. If sea level falls or a volcano rises from the sea floor, coral attaches itself to the volcano only to be killed upon later exposure to air. Places such as Vanuatu show coral reefs well above sea level on the sides of volcanoes due to the rise of a local volcano.

If sea level rises, volcanoes became inundated and coral attaches itself to the volcano and grows vertically and horizontally as sea level continues to rise. These produce coral atolls. The same occurs if a volcanic island starts to sink. Sea level rise produces coral atolls, it does not destroy them. Darwin showed this in 1842 and atolls were drilled to test Darwin's theory by Mawson's Antarctic compatriot Professor Sir T.W.Edgeworth David. The drilling was funded by the NSW Government. Darwin's coral atoll theory has now been validated by more than 150 years of independent interdisciplinary science. Why has this been ignored by the climate catastrophists?

Coral atolls can also sink due to compaction of coralline sand, pumping of groundwater or sinking of the volcanic substrate. Again, this is a normal process that induces the rapid growth of coral to reform the atoll. This is what is happening in many Pacific Ocean atoll nations and this subsidence produces an apparent sea level rise. We naughty fossil fuel burners are not causing sea level to rise, some but not all of the Pacific Ocean atoll nations are sinking as part of a normal geological process.

When someone argues that we humans are causing sea level to rise, treat them as a fool. Because they are.

Ocean acidification

There may have been a time in the first 20 to 100 million years of Earth history that there was a warm acid ocean ('Seltzer

Ocean').[311] These were unusual times as the Earth settled down into being a normal rocky planet. The atmosphere was dominated by carbon dioxide, air pressure may have been 100 times higher than now and the oceans would have been at 60 to 110°C.[312,313] It is speculated that the earliest ocean was warm carbonated water that reacted with the crust and in doing so, lost its fizz.

Ocean acidification is a concocted non-issue. To use the word acid with reference to the oceans is fraudulent. The oceans have been alkaline for thousands of millions of years at times when there was far more carbon dioxide in the atmosphere than at present. There is no reason carbon dioxide solubility in seawater would suddenly disobey all the rules of chemistry because we are emitting traces of a trace gas into the atmosphere.

Carbon dioxide dissolves in seawater. About 70% of the planet's surface is covered by water. Solubility of carbon dioxide in water is mainly a function of temperature, pressure and salinity.[314] The colder the water, the more carbon dioxide dissolves.[315] The long-term global average sea surface temperature is 15°C and at this temperature, seawater can dissolve its own volume of carbon dioxide. At 10°C, seawater dissolves 19% more carbon dioxide than at 15°C and at 20°C seawater dissolves 12% less than at 15°C.[316] Not surprisingly, there is a correlation between the sea surface temperature and the atmospheric carbon

---

[311] M. Schoonen and A. Smirnov, 2017: Staging life in an early warm 'Seltzer' ocean. *Elements* 12: 395-400.

[312] N.H. Sleep, K. Zahnle and P.S. Neuhoff, 2001: Initiation of element surface conditions on the earliest Earth. *PNAC* 98: 3666-3672.

313 K. Zahnle, L. Schaefer and B. Fegley, 2010: Earth's earliest atmospheres. *Cold Spring Harbour Perspectives in Biology 2*: doi: 10:1101/cshperpect.a004895

[314] H.S. Harned and R. Davis, 1943: The ionization constant of carbonic acid in water and the solubility of carbon dioxide in water and aqueous salt solutions from 0 to 50°C. *Jour. Amer. Chem. Soc.* 65: 2030-2037.

[315] This is inverse solubility. With many substances (e.g. sugar, salt), the warmer the water the more solid dissolves in water.

[316] L. Endersbee, 2008: Carbon dioxide and the oceans. *Focus* 151: 20-21.

dioxide content. The more saline the water, the more carbon dioxide dissolves. The higher the atmospheric carbon dioxide content, the more carbon dioxide dissolves in seawater. No wonder the oceans degas when the temperature rises.

The oceans continually remove carbon dioxide from the atmosphere, this carbon dioxide is used by organisms to make calcium carbonate shells of aragonite and calcite. When organisms die, these shells accumulate on the sea floor and later become limestone. The solid rock limestone contains 1,000 times more carbon dioxide than the atmosphere. Soils contain carbon-bearing organic compounds and carbonate and, upon erosion, the carbon compounds and carbonates may end up as marine sediments. The wind also pumps carbon dioxide into sea water.[317] In certain circumstances, calcium carbonate precipitates on the seafloor.

In polar areas, the cold surface water absorbs more carbon dioxide than at the tropics. The cool, dense, high salinity polar water sinks and is carried to lower latitudes where it upwells and releases carbon dioxide as the water warms. About 70% of ocean degassing occurs by this process (thermally driven solubility pump) and the other 30% is degassed by life (biological pump).[318] If these biological processes were removed, the level of atmospheric carbon dioxide would increase five-fold.[319] There is a perception that with increased emissions of carbon dioxide into the atmosphere, the atmospheric carbon dioxide content will increase forever. This is not the case as atmospheric carbon dioxide is recycled through the oceans and life. Any slight variation in the floating organisms in the oceans

---

[317] S. D. Smith and E.P. Jones, 1985: Evidence of wind-pumping of air-sea gas exchange based on direct measurements of CO_2 fluxes. *Jour. Geophys. Res.* 90: 869-875.
[318] T. Volk and Z. Liu, 1988: Controls of CO_2 sources and sinks in the Earth scale surface ocean: temperature and nutrients. *Global Biogeochem. Cycles* 2: 73-89.
[319] E. Eriksson, 1963: Possible fluctuations in atmospheric carbon dioxide due to changes in the properties of the sea. *Jour. Geophys. Res.* 68: 3871-3876.

could account for variations far greater than the human input of carbon dioxide into the atmosphere.

At a depth of 10 metres, seawater can dissolve twice its own volume of carbon dioxide. The amount of carbon dioxide that can dissolve in seawater increases with decreasing temperature and increasing pressure. Cold seawater at a high pressure at the bottom of the oceans contains a huge amount of carbon dioxide. When this water rises to the surface, carbon dioxide is released.

The exchange of carbon dioxide between the atmosphere and ocean is well known.[320,321] This gives us an upper limit on how much carbon dioxide concentration in the atmosphere will rise if all available fossil fuel is burned. In order to permanently double the current level of carbon dioxide in the atmosphere and keep the oceans and atmosphere balanced, the atmosphere needs to be supplied with 51 times the present amount of atmospheric carbon dioxide. The total amount of carbon in known fossil fuel resources could only produce 11 times the amount of carbon dioxide in the atmosphere.[322] Unless we change the fundamental laws of chemistry and change the way in which oceans work, humans do not have enough fossil fuel on Earth to permanently double the amount of carbon dioxide in the atmosphere.

If humans burned all the available fossil fuels over the next 300 years, there would be up to fifteen turnovers of carbon dioxide between the oceans and atmosphere and all the

---

[320] R. Revelle and H.E. Suess, 1957: Carbon dioxide exchange between atmosphere and ocean and the question of an increase of atmospheric CO_2 during the past decades. *Tellus* 9: 18-27.

[321] G. Skirrow, G. 1975: The dissolved gases – carbon dioxide. In: *Chem. Ocean. Vol. 2, 2nd Ed.* (Eds J. P. Riley and G. Skirrow), *Academic Press*, 1992.

[322] Z. Jaworowski, T. V. Segelstad and V. Hisdal, 1992: Atmospheric CO_2 and global warming. A critical view, Second Revised Edition. *Norsk Polarinstitutt Meddelelser* 119: 1-76.

additional carbon dioxide would be consumed by ocean life and precipitated as calcium carbonate in sea floor sediments.[323] Chemistry tells us this. Geology does also because it has happened before when the planet had a far higher atmospheric carbon dioxide content.

There is a very small addition of carbon dioxide to the oceans from the burning of fossil fuels. Fossil fuel contains no carbon 14 (derived from cosmic radiation and nuclear bombs) and hence the increase in the carbon 13 and carbon 12 of seawater can be used to calculate the addition of carbon dioxide derived from coal and oil burning.[324] This calculation ignores the contribution of carbon dioxide from other sources such as soil bacteria, volcanoes, floating microorganisms and burning of wood, grass, stubble and dung. Even if it is only the surface of the ocean that contains carbon dioxide of fossil fuel origin and then it is about 3% of the carbon dioxide in the surface water.[325]

The term pH, meaning *pondus Hydronium* or the power of hydrogen, is a numerical measure of the range from extreme acidity to extreme alkalinity. Oceans have a pH of 7.9 to 8.2. This figure is larger than neutral (pH = 7) which means that the oceans are alkaline. The pH scale ranges from 0 to 14, pH 6 is ten times more acid than pH 7 and pH 5 is a hundred times more acid than pH 7. The pH scale is not linear, it is logarithmic hence to acidify seawater from pH 8 to pH 6, an extraordinarily large amount of acid is needed. If acid is present, sediments, rocks and shells become very reactive. These reactions neutralise acid and the oceans return to their normal alkaline state.

---

[323] P.H. Abelson, 1990: Uncertainties about global warming. *Science* 247: 1529.

[324] R. Key, "The dangers of ocean acidification", *Scientific American*, March 2006, 58-65.

[325] Z. Jaworowski, T.V. Segelstad and V. Hisdal, 1992: Atmospheric CO_2 and global warming. A critical view, Second Revised Edition. *Norsk Polarinstitutt Meddelelser* 119: 1-76.

The most alkaline waters in the oceans occur in the centre of ocean circulation patterns whereas less alkaline waters occur at sites of upwelling where deep ocean water is brought to the surface. Deep ocean water brought to the surface at these upwelling sites has a higher carbon dioxide content. As a result photosynthetic microorganisms thrive and become the bottom of the food chain for other abundant marine life. It is no surprise that the great fishing fields of the world occur at sites of upwelling.

If carbon dioxide dissolves in seawater, the oceans should become more acid,[326] so the theory goes. If the oceans become acid (pH <7), then shells of marine organisms would dissolve. This is greatly exaggerated in the popular press as a potential environmental catastrophe.[327] The geological record does not show that shells dissolve otherwise there would be no shelly fossils. Furthermore, the chemistry of fossil shells is used to calculate the pH of oceans and the amount of carbon dioxide in the atmosphere over the last 542 million years.[328] The oceans are saturated with calcium carbonate to a depth of 4.8 km. This means that if any more carbon dioxide were added to the oceans, then calcium carbonate would precipitate.[329,330] Solid

---

[326] $CO_2 + H_2O \rightleftharpoons H_2CO_3$; $H_2CO_3 = H^+ + HCO_3^-$; $HCO_3^- = 2H^+ + CO_3^{2-}$. In the oceans at pH 7.9 to 8.2, CO_2 exists as dissolved gas (1%), HCO_3^- (93%) and CO_3^{2-} (8%). Calcium in seawater binds CO_2 into insoluble carbonates of calcium in shells, coral reefs and mineral precipitates ($Ca^{2+}_{[aq]} + CO_3^{2-}_{[aq]} = CaCO_3$). Furthermore, trapped seawater in sediments precipitates carbonate cement. By these processes CO_2 is removed from the atmosphere and stored in marine sediments as fossils, cement and rock. $CaCO_3$ plankton shells can dissolve back into seawater at a depth of >4.8 km.
[327] S.C. Doney, "The dangers of ocean acidification", *Scientific American*, March 2006, 58-65.
[328] E.T. Rasbury and N.G. Hemming, 2017: Boron isotopes: A "Paleo-pH meter" for tracking ancient atmospheric CO_2. *Elements* 13, 243-248.
[329] W. S. Broeckner, T. Takahashi, H.J. Simpson and T.H. Peng, 1979: The fate of fossil fuel carbon dioxide and the global carbon budget. *Science* 206: 409-418.
[330] Z. Liu, 2013: Review on the role of terrestrial aquatic photosynthesis in the global carbon cycle. *Procedia Earth and Planetary Science* 7, 513-516.

calcium carbonate contains 44% by weight of the gas carbon dioxide. However, buffering of seawater prevents oceans becoming acid.

If there is only a small amount of carbon dioxide dissolved in water, calcium sulphate known as gypsum[331] precipitates instead of calcium carbonate. This has not been found in the ocean but occurs in terrestrial warm fresh water lakes. If the atmospheric carbon dioxide content was extremely high, the calcium magnesium carbonate called dolomite would precipitate from the oceans.[332] This can be shown experimentally, thermodynamically and from geology.

During the Precambrian (older than 542 million years), the atmospheric carbon dioxide content was more than 1% and huge volumes of dolomite were precipitated thereby extracting carbon dioxide from the oceans and air. Solid dolomite contains 48% carbon dioxide gas by weight. This means that the balance of carbon dioxide between the oceans and atmosphere we see today has not changed for thousands of millions of years.[333] This balance was not changed during times of intense sudden release of carbon dioxide from volcanoes.

There is a correlation between increased volcanic production of carbon dioxide and increased sedimentation of calcium carbonate from the oceans.[334] Again, we see that planet Earth has been remarkably stable for billions of years. The geological processes of carbonate precipitation in the oceans that have taken place for billions of years are ignored in the computer climate models of the IPCC.

---

[331] $CaSO_4.2H_2O$
[332] $CaMg(CO_3)_2$
[333] H.D. Holland, *The Chemical Evolution of the Atmosphere and Oceans*, Princeton University Press, 1984.
[334] M.I. Budyko, A.B. Ronov and A.L. Yanshin, *History of the Earth's atmosphere*, Springer, 1987.

Scary Scenarios

A reaction between seawater and minerals on the ocean floor keeps the oceans alkaline[335]. This is called buffering. The floor of the oceans is covered with the volcanic rock basalt. This is a highly reactive rock, especially when it is glassy[336]. Seafloor basalts are fractured allowing the ingress of seawater. Well known reactions between seawater and basalt make the oceans more alkaline. This balances the acid added by hot springs and balances the addition of carbon dioxide to seawater. These reactions can be duplicated in experiments, can be calculated thermodynamically and can be observed in oceanic settings. Again, this is the coherence criterion of science.

Over time, the basalt-seawater reactions have controlled the atmosphere and seawater chemistry[337]. Seawater in contact with ocean floor rocks, especially basalt, also removes carbon dioxide from seawater to form carbonates[338].

There is a fine balance in the oceans where microorganisms consume carbon dioxide as plant food. This makes seawater more alkaline whereas the decomposition of organisms makes the oceans less alkaline. The more carbon dioxide in the atmosphere, the more microorganisms thrive in the oceans.

These mineral and biological processes have taken place for billions of years and modern and ancient seafloor basalts shows that the oceans have been stopping an increase of acidity even

---

[335] For example, the weathering of silicates such as pyroxenes consumes CO_2 and forms carbonates. The same reactions apply for olivines, a far more reactive family of minerals than the pyroxenes. $CO_2 + CaSiO_3 = CaCO_3 + SiO_2$; $CO_2 + FeSiO_3 = FeCO_3 + SiO_2$; $CO_2 + MgSiO_3 = MgCO_3 + SiO_2$

[336] Feldspars are the most abundant minerals in terrestrial and submarine rocks and buffer acidity by reaction to form kaolinite. $2KAlSi_3O_8 + 2H^+ + H_2O = Al_2Si_2O_5(OH)_4 + 2K^+ + 4SiO_2$; $2NaAlSi_3O_8 + 2H^+ + H_2O = Al_2Si_2O_5(OH)_4 + 2Na^+ + 4SiO_2$; $CaAl_2Si_2O_8 + 2H^+ + H_2O = Al_2Si_2O_5(OH)_4 + Ca^{2+}$

[337] R.S. Arvidson, M. Guidry and F.T. Mackenzie, 2005: The control of Phanerozoic atmosphere and seawater composition by basalt-seawater exchange reactions. *Jour. Geochem. Explor.* 88: 412-415.

[338] $Ca^{2+} + H_2O + CO_2 = CaCO_3 + 2H^+$; $H^+ + (OH)^- = H_2O$

when the carbon dioxide in the atmosphere was 25 times the current amount. Despite huge changes in atmospheric carbon dioxide over the last few hundred million years, average global temperature has not changed by more than ±3.5°C, oceans have not become acid and there has been no runaway greenhouse.[339] What a remarkable stable planet we live on where everything is at equilibrium. Fossil shells, algal reefs and coral reefs in ancient rocks show that oceans could not possibly have been acid at times when atmospheric carbon dioxide and temperature were far higher than now. Plants also thrived at those times.[340]

In fact, the higher the atmospheric carbon dioxide in the past, the easier it has been to form shells. If the oceans were acid, shells would dissolve and the oceans would become alkaline. Furthermore, at these times the high temperature and high carbon dioxide content of the atmosphere are unrelated.[341,342,343] Geological history shows us that for the oceans (and land plants) to efficiently fix atmospheric carbon dioxide and store it in the rocks, the atmospheric carbon dioxide content needs to be far higher than at present.

The salinity of the oceans has been almost constant for billions of years[344]. The earliest oceans may have been slightly

---

[339] D.L. Royer, R.A. Berner and J. Park, 2007: Climate sensitivity constrained by CO_2 concentrations over the past 420 million years. *Nature* 446: 530-532.

[340] K.L. Bice, E.T. Huber and R.D. Norris, 2003: Extreme polar warmth during the Cretaceous greenhouse? Paradox of Turonian $\partial^{18}O$ record at Deep Sea Drilling Project Site 511. *Palaeoceanography* 18:1-11.

[341] J. Veizer, Y. Godderis, and L.M. François, 2000: Evidence for decoupling of atmospheric CO_2 and global climate during the Phanerozoic eon. *Nature* 408: 698-701.

[342] C.R. Tabor, C.J. Poulsen, D.J. Lunt, N.A. Rosenbloom, B.J. Otto-Bliesner, P.J. Markwick, E.C. Brady, A. Farnsworth and R. Feng, 2016: The cause of late Cretaceous cooling: A multi-model proxy comparison. *Geology* 44, 963-966.

[343] E. Nardin, Y. Goddéris, Y. Donnadieu, G. Le Hir, R.C. Blakey, E. Pucéat and M. Aretz, 2011: Modelling of early Paleozoic long-term climatic trend. *Geol. Soc. Amer. Bull.* 123, 1181-1192.

[344] W.W. Hay *et al.* 2001: Evolution of sediment fluxes and ocean salinity. In: *Geologic modeling and simulation: sedimentary systems* (eds D.F. Merriam and J.C. Davis), Kluwer, 163-167.

warmer and more saline than the modern oceans.[345] Acid rain leaches salts from rocks on the land. These salts are transported by rivers and accumulate in the oceans and are constantly recycled[346]. In the oceans, seawater chemically reacts with basalts and this adds more salts to the water. The same water-rock chemical reactions that have kept the oceans saline have also kept them alkaline. If the oceans were becoming acid, then they should also become less saline. This we don't see.

In some places, seawater is trapped in basins that have been isolated. For example, the Strait of Gibraltar has been closed twice in its history. During this time, the evaporation rate in the Mediterranean Sea exceeded the rate of input of river water. The sea evaporated and left large salt deposits on the sea floor. These salt layers are still there. The last time the Mediterranean evaporated was between 5.96 to 5.33 million years ago. The Mediterranean was later flooded by water pouring in from the Atlantic Ocean through the re-opened Strait.[347]

Seawater evolves over time. This can be measured using chemical fingerprints[348] in shells of known ages. Although the Earth's atmosphere has been far warmer than at present for most of geological time,[349] changes in shell chemistry may result from an increasing depth of the oceans over the last 500 million years.[350] Seawater evolution can be traced through the

---

[345] L.P. Knauth, 2005: Temperature and salinity history of the Precambrian ocean: implications for the course of microbial evolution. *Palaeogeog. Palaeoclim. Palaeoecology* 219: 53-69.

[346] J.J.W. Rogers, 1996: A history of the continents in the past three billion years. *Jour. Geol.* 104: 91-107.

[347] J. Gargarni and C. Rigollet, 2007: Mediterranean Sea level variations during the Messinian Salinity Crisis. *Geophys. Res. Lett.* 34: L10405.

[348] $\partial^{13}C$, $\partial^{18}O$ and $^{87}Sr/^{86}Sr$.

[349] Apart from the major glaciations.

[350] J.F. Kasting, M. Tazwell Howard, K. Wallman, J. Veizer, G. Shields and J. Jaffrés, 2006: Paleoclimates, ocean depth, and the oxygen isotopic composition of seawater. *Earth Planet. Sci. Lett.* 252: 82-93.

evolution of oxygen and strontium isotopes (driven by plate tectonics and the evolution of continents) and the evolution of carbon and sulphur cycles (driven by biological and chemical cycles).

For the last 500 million years, seawater evolution has been unaffected by atmospheric carbon dioxide, despite the gas being more concentrated and temperature being far higher in the past.[351] There is no reason why these tectonic, biogeochemical and geochemical cycles should change just because humans are now on Earth.

Computer simulations tell a different story and indicate that the oceans will become acid.[352,353] Experiments with seawater are flawed because they are done in laboratories removed from the ocean floor rocks, sedimentation from continents and flow of river waters into the oceans. It is these processes that have kept the oceans alkaline for billions of years. Laboratory experiments have to provide results in a short time to be reported in scientific journals for career advancement. Processes over geological time cannot be that easily replicated.

These limited constrained experiments show that when increasing amounts of carbon dioxide were added to seawater, it became acid and dissolved shells. If a few handfuls of gravel, sediment and clay from the sea floor and some floating

---

[351] J. Veizer, D. Alabc, K. Azmyb, P. Bruckschena, D. Buhla, F. Bruhnad, G.A.F. Cardenae, A. Dieneraf, S. Ebnethag, Y. Godderisbh, T. Jaspera, C. Kortea, F. Pawelleka, O.G. Podlahaai and H. Straussari, 1999: $^{86}Sr/^{87}Sr$, $\partial^{13}C$ and $\partial^{18}O$ evolution of Phanerozoic seawater. *Chem. Geol.* 161: 59-88.

[352] K. Caldeira and M. Wickett, 2003: Anthropogenic carbon and ocean pH. *Nature* 425: 365.

[353] J.C. Orr, V.J. Fabry, O. Aumont, L. Bopp, S.C. Doney, R.A. Freely, A. Gnanadesikan, Gruber, N., Ishida, A., Joos, F., Key, R.M., Lindsay, K., Maier-Reimer, E., Matear, R., Monfray, P., Mouchet, A., Najjar, R.G., Plattner, G-K., Rodgers, K.B., Cabone, C.L., Sarmiento, J.L., Schlitzer, R., Slater, R.D., I.J. Totterdell, M-F, Weirig, Y. Yamanaka and A. Yoo 2005: Anthropogenic ocean acidification over the twenty-first century and its impact on calcifying organisms. *Nature* 437: 681-686.

photosynthetic life had been added to the experiment to simulate real conditions, then the result would be completely different. Computer simulations that ignore observations and natural processes that have taken place over billions of years end up with a result unrelated to reality. Reality is written in rocks, not models based in incomplete information.

Recent research[354,355] suggests that decades of experiments testing the effect of elevated atmospheric carbon dioxide on marine life have failed because of poor experiment design, reporting failures, failure to recognise complexities of ocean chemistry and basic errors in chemistry (especially regarding carbonate solubility). This poor science is driven by the pressure to publish.

In modern active reefs where waters are less alkaline, coral reefs are thriving with more species and greater coral cover.[356] There is bioerosion which occurs on all coral reefs. Many reefs also have mechanical erosion from storms. This is contrary to models and incomplete laboratory experiments and told us what we knew. There is no problem with ocean acidification. There are, of course, the normal uncertainties about pH measurement and the very small number of measurements made in such large bodies of water.

Knowledge of the past puts the ocean acidification scare to bed. It is a non-problem touted by the usual suspects. However, the UN is putting out its hand and shrieking that ocean acidification could cost the economy $US 1 trillion as a result

---

[354] http://www.nature.com/news/crucial-ocean-acidification-models-come-up-short-1.18124

[355] C.E. Cornwall and C.L. Hurd, 2015: Experimental design in ocean acidification research: problems and solutions. *ICES Jour. Mar. Sci.* 118: doi: 10.1093/icesjms/fsv118

[356] H.C. Barkey, A.L. Cohen, Y. Golbuu, V.R. Starczak, T. DeCanto and K.E.F. Shamberger, 2015: Changes in coral reef communities across a natural gradient in seawater pH. *Science Advances* 1, doi: 10.1126/sciadv.1500328

of losses to fishing and tourism. Oyster farm losses were used as an example. However, losses by oyster farmers are normally due to local viruses, bacteria or heavy metal contamination. I'm sure that if Western countries provide large amounts of funding for the UN or IPCC then the problem will be solved and sticky fingers will be waiting in the wings for the next confected crisis.

Although rainwater is slightly acidic (pH 5.6), by the time it runs over the surface and chemically reacts with minerals in soils and rocks, it enters the oceans as alkaline water.[357,358,359] In limestone terranes, we see that rocks dissolve from acid rain. The salts transported down rivers result from rainwater reacting with rocks making river water alkaline and slightly saline.[360] Soils contain far more carbon dioxide than the atmosphere. During weathering, soils release a huge amount of carbon dioxide which ends up in river systems[361] and the total dissolved carbon dioxide in river systems depends upon the season, the position of the water in the river system and whether dissolved carbon has been converted to carbon dioxide.[362]

---

[357] M.A. Velbel, 1993: Temperature dependence of silicate weathering in nature: How strong a negative feedback on long-term accumulation of atmospheric CO_2 and global greenhouse warming? *Geology* 21:1059-1061.

[358] L.R. Kump, S.L. Brantly and M.A. Arthur, 2000: Chemical weathering, atmospheric CO_2 and climate. *Ann. Rev. Earth Planet. Sci.* 28: 611-667.

[359] J. Gaillardet, B. Dupré, P. Louvat and C.J. Allègre, 1999: Global silicate weathering and CO_2 consumption rates deduced from the chemistry of large rivers. *Chem. Geol.* 159: 3-30.

[360] A. Karim and J. Veizer, 2000: Weathering processes in the Indus River Basin: implications from riverine carbon, sulfur, oxygen, and strontium isotopes. *Chem. Geol.* 170: 153-177.

[361] K. Telmer and J. Veizer, 1999: Carbon fluxes, pCO_2 and substrate weathering in a large northern river basin, Canada: carbon isotope perspectives. *Chem. Geol.* 159: 61-86.

[362] J.A.C. Barth and J. Veizer, 1999: Carbon cycle in St. Lawrence aquatic ecosystems at Cornwall (Ontario), Canada: seasonal and spatial variations. *Chem. Geol.* 159: 107-128.

This process, weathering, has been removing carbon dioxide from the atmosphere and soils for billions of years and storing the carbon dioxide in rocks[363,364,365]. All this we have known for a long time. If an addition of carbon dioxide to the atmosphere drives global warming as the green left environmental activists tell us, then removal of carbon dioxide from the atmosphere should trigger glaciation. It doesn't so we must reject the unsubstantiated opinion that atmospheric carbon dioxide drives climate change. The higher the temperature and carbon dioxide content, the quicker the removal of carbon dioxide by calcium carbonate precipitation.[366]

Over time, there has been a balance between carbon dioxide uptake by soils, rocks, water and life and release of carbon dioxide into the atmosphere.[367] This has resulted in the long-term stabilisation of the global surface temperature. Even if this geological stabilisation did not occur, there still would be long term stabilisation by life.[368]

Rainwater runs into and accumulates in freshwater lakes,

---

[363] R.A. Berner, A.C. Lesaga and R.M. Garrells, 1983: The carbonate-silicate geochemical cycle and its effect on atmospheric carbon dioxide over the past 100 million years. *Amer. Jour. Sci.* 283: 641-683.

[364] M.E. Raymo and W.F. Ruddiman, 1992: Tectonic forcing of late Cenozoic climate. *Nature* 359: 117-122.

[365] $CO_2 + H_2O = H_2CO_3$; $H_2CO_3 = H^+ + HCO_3^-$; $2Ca^{2+} + 2HCO_3^- + KAl_2AlSi_3O_{10}(OH)_2 + 4H_2O = 3Al^{3+} + K^+ + 6SiO_2 + 12H_2O$; $2KAlSi_3O_8 + 2H^+ + H_2O = Al_2Si_2O_5(OH)_4 + 2K^+ + 4SiO_2$; $2NaAlSi_3O_8 + 2H^+ + H_2O = Al_2Si_2O_5(OH)_4 + 2Na^+ + 4SiO_2$; $CaAl_2Si_2O_8 + 2H^+ + H_2O = Al_2Si_2O_5(OH)_4 + Ca^{2+}$; $KAl_2AlSi_3O_{10}(OH)_2 + 3Si(OH)_4 + 10H^+ = 3Al^{3+} + K^+ + 6SiO_2 + 12H_2O$; $CO_2 + CaSiO_3 = CaCO_3 + SiO_2$; $CO_2 + FeSiO_3 = FeCO_3 + SiO_2$; $CO_2 + MgSiO_3 = MgCO_3 + SiO_2$

[366] J.C.B. Walker, P.B. Hays and J.F Hasting, 1981: A negative feedback mechanism for the long term stabilization of the Earth's surface temperature. *Jour. Geophys. Res.* 86: 9776-9782.

[367] R.A. Berner, 1980: Global CO_2 degassing and the carbon cycle: comment on 'Cretaceous ocean crust at DSDP sites 417 and 418: carbon uptake from weathering vs loss by magmatic activity." *Geochim. Cosmochim. Acta* 54: 2889.

[368] D.W. Schwartzman and T. Volk, 1989: Biotic enhancement of weathering and the habitability of Earth. *Nature* 311: 45-47.

which are invariably slightly acid because of the lack of major buffering and organic material.[369,370] Lakes commonly contain shells of floating organisms and macrofauna,[371,372] especially those that are alkaline.[373] If lakes become extremely acid, life dies.[374] Freshwater lakes also have an excess of calcium[375] and some lakes and seas (e.g. Black Sea) have oxygen-poor bottom waters. Apart from a few short sharp events,[376] the oceans have remained alkaline for billions of years. The oceans have removed dissolved carbon dioxide by precipitation of calcium carbonate minerals in shells, coral reefs, cement that binds mineral grains together and mineral deposits. Because the oceans have an excess of calcium, if more carbon dioxide is dissolved in the oceans, then more calcium carbonate is precipitated. While the oceans have an excess of calcium, they cannot become acid and will remain in equilibrium. Calcium is continually added to seawater because it is dissolved in river water.

---

[369] R. Dermott, J.R.M. Kelso and A. Douglas, 1986: The benthic fauna of 41 acid sensitive headwater lakes in North Central Ontario. *Water Air Soil Poll.* 28: 283-292.

[370] H.H. Harvey and J.M. McArdle, 2004: Composition of the benthos in relation to pH in the LaCloche lakes. *Water Air Soil Poll.* 30: 529-536.

[371] E. Pip, 1987: Species richness of freshwater gastropod communities in central North America. *Malacol. Soc. Lond.* 53: 163-170.

[372] T. von Rintelen and M. Glaubrecht, 2003: New discoveries in old lakes: Three new species of *Tylomelania* Sarasin & Sarasin, 1897 (Gastropoda: Cerithioidea: Pachychilidae) from the Malili Lake system on Sulawesi, Indonesia. *Malacol. Soc. Lond.* 69: 3-17.

[373] O. Bennike, W. Lemke and J.B. Jensen, 1998: Fauna and flora in submarine early Holocene lake-marl deposits from the southwestern Baltic Sea. *The Holocene* 8: 353-358.

[374] J. P. Nilssen, 1980: Acidification of a small watershed in southern Norway and some characteristics of acidic aquatic environments. *Internat. Revue der Gesamten Hydrobiologie* 65: 177-207.

[375] For example, the weathering of limestone by acid rains produces calcium for accumulation in lakes and oceans: $CO_2 + CaCO_3 + H_2O = 2(HCO_3)^- + 2Ca^{2+}$

[376] T.K. Lowenstein and R.V. Demicco, 2006: Elevated Eocene atmospheric CO_2 and its subsequent decline. *Science* 313: 1928.

Acid enters the ocean from submarine hot springs such that seawater around hot springs can be less alkaline than elsewhere. This is especially the case with hot springs that are close to the coast in populated areas where water clarity, sediment deposition, shelter from waves, runoff of rainwater and human influences (e.g. sewage) change alkalinity. In these cases, short periods of decreased alkalinity lead to a decrease in animals that graze on green algae and hence an increase in green algae.[377]

Extinction

It is touted by environmentalists that we are in the sixth great mass extinction. And we are to blame. Not so, unless at least 60% of species have cunningly disappeared in front of our eyes without us noticing. Some species have become extinct such as the Tasmanian tiger, others are close to extinction such as Mexico's Vaquita porpoises which drown in fish nets and North America's Franklin trees which now survive in parks and gardens. The large terrestrial macrofauna are always good for a sensationalist media story about extinction. In reality, it is the shallow marine life in the oceans which tell us the story of extinction.

For centuries, human beings have been causing other species to become extinct at a rate higher than the normal species turnover rate because of agriculture, infrastructure and habitat loss. There are apocryphal suggestions that Earth is entering

---

[377] J.M. Hall-Spencer, R. Rodolfo-Metalpa, Ransome, E., Fine, M., Turner, S.M., Rowley, S.J., Tedesco, D. and Bula, M.C. 2008: Volcanic carbon dioxide vents show ecosystem effects of ocean acidification. *Nature* 453: doi:10.1038/nature07051

its sixth mass extinction of multicellular life.[378,379,380] This is misleading as the Earth has had five major mass extinctions and many minor mass extinctions. There are the normal scientific problems in ascertaining the number of species that were present at any one time on Earth because not all species become fossilised and new fossils are being found every day. Furthermore, the rate of extinction in mass extinctions is very rapid because most are catastrophic planetary events.

Some 60-70% of all species disappeared in the Ordovician-Silurian extinction events (450-440 million years ago), 70% in the Late Devonian extinction (375-360 million years ago), 96% in the Permian-Triassic extinction event (251 million years ago), 70-75% in the Triassic-Jurassic event (217 million years ago) and 75% in the Cretaceous-Tertiary event (65 million years ago).[381]

Minor mass extinction events occurred at 2,400, 542, 517, 502, 488, 428, 424, 420, 416, 305, 270, 232, 183, 145, 117, 33.9, 15.5, 2, 0.64, 0.07 and 0.0013 million years ago. Extinction is normal and occurs for a great diversity of reasons. No species, including *Homo sapiens*, is on Earth forever. When we shuffle off, we vacate an ecosystem for another species to thrive. Not one previous extinction event was due to global warming. We are certainly seeing a loss of species, this could be by normal species turnover but it is certainly not at the levels of a mass extinction, whether a major or minor mass extinction.

---

[378] S. Pimm and P. Raven, 2000: Biodiversity: Extinction by numbers. *Nature* 403: 843-845.
[379] A. Barnosky, N. Matzke, S. Tomiya, G.O.U. Wogan, Swartz, B., Quental, T.B., Marshall, C., McGuire, J.L., Lindsey, E.L., Maguire, K.C., Mersey, B. and Ferrer, E.A. 2011: Has the Earth's sixth mass extinction already arrived? *Nature* 471: 51-57.
[380] G. Ceballos, P.R. Ehrlich, A.D. Barnosky, A. Garcia, A., R. Pringle and T.M. Palmer, 2015: Accelerated modern human-induced species losses: Entering the sixth mass extinction. *Environ. Sci.* 1: doi 10.1126/sciadv.1400253
[381] David Raup, *Extinction: Bad Genes or Bad Luck?*, W.W Norton, 1992.

One of the main causes of extinctions has been ignored. It is invasive species.[382] Waves of extinctions occurred in the Caribbean in the 1500s, the extinction rate rose again during the first phase of expansion and exploration in the 1700s and then rose again during the age of empires after 1850 that peaked at the beginning of the 20th century. There is now no *Terra Incognita* that hasn't been visited by explorers and hence the worst of the extinctions from introduced predators are behind us.[383]

Predators, parasites, poxes and pests were brought in by explorers and settlers and have had a devastating effect, especially on islands. We really don't know how many species of life exist on Earth. New species are continually being described. Although 1.5 million species have been discovered, described and classified, there may be anything from two million to 50 million species on Earth.[143] This list does not include microorganisms that constitute the greatest biomass on Earth.

Some 77 mammal species (out of a total described species count of over 5,000), 34 amphibians (out of over 6,000 species) and 140 bird species (out of a total of about 10,000 described species) have become extinct in the last 500 years.[384] There may be others but it is the order of magnitude that is important. Not a single rainforest bird or mammal has gone extinct and Europe has lost only one bird species over the last 500 years.

Many other species on the brink of extinction are sitting in the waiting room (e.g. South China tiger, Sumatran elephant, Amur leopard, Atlantic goliath groper, Gulf porpoise, Northern bald ibis, Hawksbill turtle, black rhinoceros, pygmy three-toothed sloth, Chinese pangolin etc). Great efforts are being

---

[382] Matt Ridley, *The Times*, 22 June 2015.
[383] http://wattsupwiththat.com/2013/07/07alexander-the-great-explains-the-drop-in-extinctions/
[384] International Union for Conservation of Nature; www.iucn.org

made by conservationists and zoos to breed threatened species and some species that were thought to be extinct are sometimes rediscovered later (Coelacanth, Bermuda petrel, Chacoan peccary, Lord Howe Island stick insect, Monito del Monte, La Palma giant lizard, takahe, Cuban solenodon, New Caledonian crested gecko, New Holland mouse, giant Palouse earthworm, large-billed reed warbler, Laotian rock rat). There are clearly many more Lazarus species to be found.

Of the total birds and mammals that have become extinct almost all lived on islands with only nine that lived on a continent (excluding the island continent Australia).[385] Island species that became extinct over the last 500 years included the dodo, Steller's sea cow, the Falkland Island wolf, the quagga, the Formosan clouded leopard, the Atlas bear, the Caspian tiger and the Cape lion. On a per unit area basis, the extinction rate on islands was 177 times higher for mammals and 187 times higher for birds than on continents. The continental mammal extinction rate was 0.89 to 7.4 times the background rate whereas on islands it was 82 to 702 times the background rate. Continental bird extinction rate was 0.69 to 5.9 times the background rate and on islands it was 98 to 844 times background rate. Fossil assemblages on islands not only show dwarfism but also accelerated extinction. The increased island extinction rate is mainly due to the introduction of predators (including man), microorganisms and diseases.

The carked continental critters are the Bluebuck antelope, Algerian gazelle, Omilteme cottontail rabbit, Labrador duck, Carolina parakeet, slender-billed grackle, passenger pigeon, Colombian grebe and Atitlan grebe. This extinction rate is not apocalyptic, as we have been led to believe by doomsdayers. To call it a mass extinction is an over-exaggeration.

---

[385] C. Loehle and W. Eschenbach, 2011: Historical continental bird and mammal extinction rates. *Diversity Distrib.* doi: 10.1111/j.1472-4642

The invasion of hunter-gatherer humans with stone-tipped spears across the Bering Strait into North America 12,000 years ago led to the extinction of many grassland macrofauna species. Some threatened species are now being revived. Explorers and settlers brought rats, sheep, goats, cats, dogs, pigs, mosquitoes and avian malaria onto Hawaii, through the Caribbean, the South Atlantic islands, the Indian Ocean islands and the Pacific islands. Young mammals, birds, reptiles and amphibians have been eaten by introduced animals, especially rats and cats. Hawaii has lost about 70 bird species over the last 200 years.

If the green left environmental activists really wanted to make Australia a better place, they would eradicate every introduced feral animal with guns, traps, poisons, genetic engineering, organism specific viruses and pesticides. In Australia, feral cats, foxes and rats kill native birds, mammals, amphibians and reptiles. Feral goats just eat everything in their path. Feral dogs form marauding killer packs that have even killed children, feral horses destroy alpine upland areas in national parks, cane toads wipe out wild life, pigs pollute and destroy waterways, camels kill vegetation in sensitive desert areas, introduced fish replace native fish and introduced plants dominate some native plant habitats.

The time has come to close our ears to green left environmental activist noise. I want to see them killing introduced plants and animals. Where are the green left environmental activist armies in rural and outback Australia conducting genocide of introduced species? Time to leave the comforts of a city life only possible because of coal-fired electricity and actually do something to make the world a better place. Don't hold your breath, it is easier to criticise from a city media studio than to actually do something useful.

In the UK, introduced species such as grey squirrels, mink and signal crayfish have already pushed the endemic red squirrel, water voles and native crayfish populations towards extinction. British native animal populations are decreasing because of ash dieback and introduced zebra mussels, harlequin ladybirds, Chinese mitten crabs, New Zealand flatworms and munutjac.

To argue that economic development should be reduced or stopped because of extinctions is flawed. It is also flawed to spend billions on climate change because climate change might create extinctions.

Introduced organisms create local extinctions. A few decades ago, scientists began noticing a decline in frog and toad populations in Central America. A good example is Costa Rica's golden toad that lived in the cloud forest. It was touted as an obvious example of extinction due to climate change. After all the hysteria died down, the real reason was found. It was an introduced African chytrid fungus. And how did it get to Central America? From the African clawed toad which was used by scientists for laboratory tests.

Even in a developed desert country such as Australia, some 55.8% of the land is used for low impact grazing, conservation and natural environments comprise 36.7%, cropping 3.5%, forestry 1.8% and intensive uses such as cities, roads, railways, houses, office blocks, factories quarries, mines and oil/gas production facilities only occupy 0.4% of the land area[386]. The intense use of land is very efficient. For example, in Victoria the Minerals Council states that both farming and mining each contribute to about 9% of the state's GDP, farming uses 70% of the land whereas mining only uses 0.07% of the land area.

In NSW, mining uses just 0.1% of the land compared to

---
[386] ABARES, 2015.

76% for agriculture, 7.6% for conservation and national parks and 1.8% for homes and urban development. In NSW, the coal industry directly employs 34,000 people hence with the flow on multiplier, about 140,000 people from mechanics to manufacturers and from hospitality to health professionals.[387] Coal also produces more than 70% of the state's electricity needs.

There are major mass extinctions, minor mass extinctions and extinctions due to species turnover. Humans have driven some species to extinction, as do other species. Survivors taking advantage of new opportunities created by humans. Many are spreading into new parts of the world, adapting to new conditions and often evolving into new species.[388] There has been an influx of world travellers, moved by humans, as ornamental garden plants, pets, crops and livestock, or simply by accident, before they escaped into the wild. We are transporting species at an unprecedented rate and some species transported to islands such as rodents have driven island birds to extinction, especially ground-nesting and flightless birds. Escaped yellow-crested cockatoos in Hong Kong thrive while continuing to decline in their Indonesian homeland.

Throughout the history of the planet, species have survived by moving to new locations, especially when there is a cooling event. Most species fit in with limited effects on endemic species. In the UK, nearly 2,000 species have established populations in the past couple of thousand years. There is no doubt that species turnover is taking place and that species are becoming extinct. This is the way the biological world works.

---

[387] Stephen Galilee, New South Wales Minerals Council, 21 July 2015.
[388] Chris D. Thomas 2017: Inheritors of the Earth: How nature is thriving in an age of extinction. https://www.theguardian.com/.../2017/.../chris-d-thomas-conservation-inheritors-of-the-earth-interview

However, extinction leads to species diversity and the current high rate at which new animals and plants are coming into existence is unusually high (e.g. apple fly, Italian sparrow, Oxford ragwort). It looks like the evolution of humans has led to some extinctions and the appearance of new species. The end result is more species, not fewer. Rather than moaning about the disappearance of a species of a slimy toad, maybe we should be celebrating human-induced species diversity and the appearance of new species.

New species and genera of bacteria and species of all three multicellular kingdoms (plants, fungi, animals) have been made in the laboratory.[389] It's easy to generate mutations in bacteria and CRISPR technology now allows more rapid mutation of archaea as well.

There is something self-satisfying for environmentalists about the idea that humans' hubris and short-sightedness are so profound that we humans are destroying a planet. However, this has been called junk science.[390] To make a facile comparison about species turnover (and extinction of specific species) is not the same as a mass extinction.

The marine fossil record shows what mass extinctions are really like. They are real mass extinctions, not the extinction of a species here and there, with more than 75% of species wiped out in a major mass extinction and at least 40% of species wiped out in a minor mass extinction.

At present, there are no geographically widespread, abundant, durably skeletonised marine taxa that have gone extinct. Of the best assessed groups of modern animals (stony corals, amphibians, birds and mammals), somewhere between 0 and 1% of all species have gone extinct in recent human history.[391]

---

[389] www.sciencedaily.com/releases/2017/03/170308092502.htm
[390] Doug Erwin, 2017: *Geological Society of America*, annual meeting.
[391] Peter Brannen, "Earth is not in the midst of a sixth mass extinction: 'As scientists

By contrast, some 96% of species became extinct in the mass extinction event 251 million years ago.

We do not live in the midst of the sixth mass extinction. We live in a time of species turnover and increasing biodiversity and we humans may be driving this process. We humans are also stressing the planet, the best example I can think of is the amount of plastics in the oceans.

Coral reefs

Reefs have been around for a very long time. About 3,500 million years ago, calcareous organo-sedimentary reef-like structures called microbialites appeared. They are composed of calcium and magnesium carbonates with the calcium, magnesium and carbon dioxide extracted from seawater. For the next 2,500 years, photosynthesising blue-green algae (stromatolites) produced shallow marine reefs. They still do (e.g. Lee Stocking Island, Bahamas Banks; Shark Bay, Western Australia). The history of the Earth shows us that stromatolite reefs thrive in warm times, especially when there is a high atmospheric carbon dioxide content.

Coral reef communities started to appear about 600 million years ago and from about 600 to 540 million years ago, sponge-like animals (Archaeocyathids), stromatolites, calcareous cyanobacteria and algae were the main organisms in reef communities. From about 540 to 350 million years ago, reefs were dominated by complex communities of algae-sponge-coral associations. Some of the corals present in these assemblages are now extinct (e.g. rugosa).

From about 350 to 220 million years ago, the reef assemblage became even more complex and was dominated by algae-

---

we have a responsibility to be accurate about such comparisons'," *The Atlantic*, 13 June 2017.

bryzoan-coral communities with minor foraminifera, sponges, stromatoporoids and rudist bivales. For the last 220 million years, reefs comprised dominant scleractinian corals as the main reef builder and modern reefs represent the most complex and developed scleractinian reefs in the history of the planet.

Mass coral mortalities in modern coral reefs have been reported many times in all major reefs since the 1870s.[392,393,394,395,396] In all cases the coral reefs recovered.

Coral bleaching and attacks by predators such as the crown-of-thorns starfish are not new and have happened many times in the past. Just because a reef is bleaching or degrading it does not necessarily mean that climate change is to blame. Coral reefs have been around for 540 million years and are remarkably resilient. In the past, the big killer of coral reefs has been sea level fall, especially during periods of global cooling. In the past, coral reefs have expanded and thrived during periods of global warming.

Water movements, tropical storms, sedimentation, salinity changes, temperature changes, El Niño events, emersion at low tide, changes in light intensity, volcanic ash falls, uplift,

---

[392] S.R. Stoddart, 1969: Ecology and morphology of recent coral reefs. *Biological Reviews* 44, 433-498.

[393] R.E. Johannes, 1975: Pollution and degradation of coral reef communities. *Elsevier Oceanography Series* 12, 13-51.

[394] J.D. Woodley, E.A. Chornesky, Clifford, P. A., J.B.C. Jackson, L.S Kaufman, N. Knopwlton, Lang, J.C., Pearson, M.P., Porter, J.W., Rooney, M.C., Rylaarsdam, K.W., Tunnicliffe, V.J., Wahle, C.M., Wulff, J.L., Curtis, A.S.G., Dallmeyer, M.D., Jupp, B.P., Koehl, M.A.R., Neigl, J. and Sides, E.M. 1981: Hurricane Allen's impact on Jamaican coral reefs. *Science* 214, 749-755.

[395] B.E. Brown, 1987: Coral bleaching: causes and consequences. *Coral Reefs* 16, S129-S138.

[396] M.A. Coffroth, H.R. Lasker and J.J. Oliver, 1990: Coral mortality outside of the Eastern Pacific during 1982-1983: Relationship to El Niño. *Elsevier Oceanography Series* 52, 141-182.

subsidence, earthquakes, phytoplankton blooms, predation, penetration of coral skeletons by organisms, competitive interactions for space and human interactions all kill reefs.[397]

Over the history of time, reefs have mainly died out because of sea level fall or coral-fringed islands and the continental shelf have risen above sea level. Seawater temperature change[398] (especially cooling) and changes in salinity, dissolved oxygen and stability of dissolved chemicals may also kill reefs. Some reefs are killed by inundation by sediment as a result of flooding in the hinterland.

In more modern times, the greatest threats to reefs have been tourists, fishing with cyanide and dynamite, mining of coral for roads, cement and construction, dredging, introduction of competing non-native species, runoff of sewage, nitrates and phosphates, and increased natural and anthropomorphic sedimentation rates.

Populations of super corals that are found in New Caledonian mangrove systems are able to survive in hot, acidic, low oxygen waters.[399] More than 20 species covering 35% of a lagoon were discovered.[400] A new Wildlife Conservation Society study in the marine national parks of Kenya reveals evidence that some corals have adapted to the warmed oceans in the 1998 El Niño event warming ocean waters and that there is now less coral bleaching.[401]

An analysis of more than 600 coral reef islands in the Pacific

---

[397] R. Endean, 1976: Destruction and recovery of coral reef communities. In: *Biology and geology of coral reefs* (ed. O. A Jones), Elsevier.
[398] The Global Coral Monitoring Network in 2000 reported that 16% of the world's coral reefs were "effectively lost" in nine months during the 1997-1998 El Niño. The reefs recovered.
[399] Hannah Osborne, *Newsweek*, 9 June 2017.
[400] Emma Camp UTS; *Journal Scientific Reports.*
[401] T. Goreau, T. McClanahan, R. Hayes and A. Strong, 2000: Conservation of coral reefs after the 1998 global bleaching event. *Conservation Biology* 4, 5-15.

and Indian Oceans shows that some have remained stable (40%) or increased in size (40%). Only 20% have decreased in area yet it is widely promoted in environmental circles that coral islands, atolls and reefs are disappearing with sea level rise. Some islands grew as much as 5.6 hectares in a decade. Tuvulu's main atoll, Funafuti, comprising 33 islands around the rim of a lagoon gained 32 hectares during the last 115 years.[402]

Contrary to popular disaster stories in the media and promoted by local politicians and green left environmentalists, the Pacific Ocean island atoll States are not disappearing due to sea level rise. They are getting larger. This is not new news. We have known for nearly 200 years that coral atolls increase in size with a relative sea level rise.[403] Over the period from 12,000 to 6,000 years ago during the 130-metre post-glacial sea level rise, coral reefs have kept up with sea level rise.[404] The coral sand atoll islands were actually produced by the destruction of reef material during the two-metre sea level fall since the Holocene Optimum. Corals don't have a problem with sea level rise, can adapt to warmer seas and just grow faster. They die when sea level falls.

In South Tarawa, Kiribati's 15 square kilometre island capital, crowded with some 50,000 people, coral blocks are used for seawalls, causeways between islands and creating new land. This has led to greater storm erosion, changes in sedimentation patterns and more common inundation during

---

[402] A.P. Webb and P.S. Kench, 2010: The dynamic response of reef islands to sea-level rise: Evidence from multi-decadal analysis of island change in the Central Pacific. *Global Planet. Change* 72: 234-246.

[403] Charles Darwin, *The structure and distribution of coral reefs. Being the first part of the geology of the voyage of the Beagle under the command of Capt. Fitzroy, R.N.* Smith, Elder and Co, 1842.

[404] R. van Woesik *et al.* 2015: Keep up or drown: adjustment of western Pacific coral reefs to sea-level rise in the 21st century. *Roy. Soc. Open Sci.* doi: 10.1098/rsos.150181

storm surges. The real danger to coral reefs and atolls is sea level fall and human activity such as removal of coral sand for cement, building of roads and airstrips together with ground water extraction, blasting of reefs for shipping lanes and use of reef blocks or coral sand concrete for sea walls.

In Australia, the Great Barrier Reef, the poster child of the green left environmental activists, disappeared during glacial events more than 60 times over the last three million years. It reappeared after every one of these events. The Great Barrier Reef first formed about 50 million years ago and has survived hundreds of coolings and warmings and massive rain events that deposit sediment on the Reef. It has survived hundreds of attacks by predators and events of bleaching.

The sea level fall and lower temperature during glacial events kills higher latitude coral reefs and they continue to thrive at lower latitudes. The geological record shows that coral reefs love it warm, especially when there is more carbon dioxide in the atmosphere. During glaciation events, tropical vegetation is reduced from rainforest to grasslands with copses of trees, somewhat similar to the modern dry tropics inland from the Great Barrier Reef.

The Great Barrier Reef of Australia is 2,300 km in length comprising 3,000 coral reefs, 600 continental islands and 300 coral cays. It has over 600 species of soft and hard corals and a breathtaking range of other species.[405] The reef has migrated eastwards and northwards and even disappeared when sea level dropped during glaciation. Sea level was 130 metres lower at the end of the last glaciation and the Great Barrier Reef did not exist at that time. There was no Amazonian rainforest during the last glaciation.

It is claimed that bleaching of coral in the Great Barrier

---

[405] http://www.gbrmpa.gov.au/about-the-reef/facts-about-the-great-barrier-reef

Reef is due to increased water temperatures driven by human emissions of carbon dioxide driving global warming.[406] We all know that. But is it true? What if the sea level falls, as it does in El Niño events or with rising land levels such as coastal Queensland, expose coral to intense sunlight which bleaches coral.[407,408] The Cairns tide gauge data shows that since 2010, the average low tide dropped by 10-15 cm.[409] This means the land level is rising, sea level is falling or both. The widespread death of microatolls on the Great Barrier and high tide marks shows that over the last 5,500 years, the land level has risen at least two metres.

We are told that the Great Barrier Reef is bleaching and we had better hurry up to see the Reef before it is totally destroyed. Coral reef expert Ove Hoegh-Guldberg told us in 1998 that the reef was under pressure from global warming and that the reef had turned white. In 1999, he warned that global warming would cause bleaching of the reef every two years from 2010. In 2006, he warned that global temperatures meant that:

> between 30 and 40 per cent of coral on Queensland's Great Barrier Reef could die within a month.

He later stated that there had been a minimal amount of bleaching of the reef. He also stated that the reef had made a *"surprising recovery"*. The alleged recovery would only have been surprising for Hoegh-Guldberg if he believed his own ex-

---

[406] T.P. Hughes *et al.*, 2015: Global warming and recurrent mass bleaching of corals. *Nature* 543, 373-377.

[407] P. Larcombe, R.M. Carter, J. Dye, M.K. Gagan and D.P. Johnson, 1995: New evidence for episodic post-glacial sea-level rise, central Great Barrier Reef, Australia. *Marine Geology* 127, 1-44.

[408] E.A. Ampou, O. Johan, C.E. Menkes, F. Niño, F. Birol, S. Oullon and S. André-foulët, 2017: Coral mortality induced by the 2015-2016 El-Niño in Indonesia: The effect of rapid sea level fall. *Biogeosciences* 14, 817-826.

[409] Jim Steele, "Falling sea level: The critical factor in the 2016 Great Barrier Reef bleaching", *Watts up with that*, 5 April 2017.

aggerations. Maybe the reef was not too damaged and maybe recovery is quite normal, as the scientific literature shows. Scaring the community witless by spurious claims about the imminent death of the Great Barrier Reef puts bread on the table for many people.

The Australian Institute of Marine Science showed that in a study of 47 reefs over 1,300 kilometres of the Great Barrier Reef, that the coral cover was stable and there had been no net decline since 1995. The ABC was telling us in 2002 that 10% of the Reef has been lost to bleaching since 1998. In order to persuade voters to support a "Carbon Tax", Australia's Prime Minister told voters that global warming is already killing the Great Barrier Reef. She did not tell us that it has survived warmer times in the past, that coral reefs in far lower latitudes seem to being thriving and that coral reefs come and go. Is the ABC operating as an environmental advocacy organisation in the absence of science? Why did Australia's Prime Minister, with access to a great diversity of scientific advice, not tell the truth?

The story of large long-lived corals is different from what we see on television. A 337-year record shows that there were wetter conditions and higher river flow into the Great Barrier Reef in the late 17th to mid 18th and in the late 19th century. Drier conditions were in the late 18th to early 19th and mid 20th centuries. We are told that by emitting carbon dioxide, we humans are going to make the climate, wetter, drier, colder and warmer. Whatever the weather is, apparently it is our fault. The Great Barrier Reef shows that it was extremely resilient to wet and dry times since well before human industry was emitting carbon dioxide.

In an analysis of the scientific data on the Great Barrier Reef it was shown that corals like it hot, that every now and then abnormally high sea surface temperature create carnage, corals have a biological juggle and have adapted to temperature variability and that there is very little QA/QC when it comes to coral reef research and the resultant extraordinary catastrophic claims.[410] It is healthy for scientists to question the methodology and conclusions of their colleagues. However, when sacred myths such as the health of the Great Barrier Reef are questioned, then host universities attempt to sack the questioner.[411]

For the last 3,500 million years, there have been reefs in shallow marine settings. Some rare isolated corals live in deep cooler waters today. The fact that reefs existed in previous times when it was warmer and the atmosphere had a higher carbon dioxide content, shows that modern coral reefs are in no danger if the atmosphere warms and the carbon dioxide content increases. In fact, it is the exact opposite. The history of time shows us that reefs, be they algal or coral, thrive when it is warmer and when there is a high atmospheric carbon dioxide content. For 3,500 million years, reefs died during cold times when sea level fell.

Can the green left environmental activists provide the evidence as to why reefs die when it is warmer or there is a slight increase of an atmospheric trace gas.

There is 3,500 million years of contrary evidence written in stone.

Polar ice

The current ice age was a long time coming. India collided with Asia 50 million years ago. Local climate was changed and carbon dioxide continued to be drawn down into soils and sediments from the atmosphere as the mountains rose. New

---

[410] P. Ridd, 2017: The extraordinary resilience of the Great Barrier Reef corals, and problems with policy science. In: *Climate Change: The Facts 2017* (ed. J. Marohasy), IPA, 9-23.
[411] JCU Peter Ridd.

soils formed. The Earth's climate has been cooling for the last 50 million years. South America had the good sense to pull away from Antarctica 37 million years ago, a circum-polar current isolated Antarctica from warm water and local ice caps joined to form a single continental ice sheet on Antarctica some 34 million years ago.

As planet Earth cooled, the slight cyclical variations in the Earth's orbit and distance from the Sun started to have a profound influence on climate. However, there were some short sharp periods of global warming unrelated to carbon dioxide or industry (as humans were not around then). Climate changes drove human evolution over the last five million years. In southeastern USA, between 5.2 and 2.6 million years ago, atmospheric carbon dioxide content was more than now as were global temperatures (2 to 3°C higher) and sea level (10 to 25 metres higher). This was probably driven by the regular changes in the Earth's orbit.

With the closure at Panama of the connection between the Atlantic and Pacific Oceans 2.67 million years ago by volcanoes, Earth started to cool because heat could not be transferred between oceans. Coincidentally, there was a supernoval eruption at the same time. The bombardment of the Earth by cosmic rays from this supernova eruption led to the formation of low-level clouds that cooled the Earth's surface. The two coincidental processes accelerated cooling. As a result, the Greenland ice sheet formed.

Climate fluctuated in cycles between warm and cold periods every 41,000 years. This was driven by changes in the Earth's axis. About 1 million years ago, the climate started to fluctuate between cold and warm periods on 100,000-year cycles driven by changes in the orbit from elliptical to circular patterns.

We are currently in the warmer interglacial phase of an ice

age that has been in progress for 34 million years. During the last interglacial, sea level was 4 to 9 metres higher than now and temperatures were 3 to 5°C higher than now. During the current interglacial 6,000 years ago, sea level was two metres higher and temperature 2°C higher than now. We cannot escape the fact that the current orbitally-driven interglacial will end and we will enjoy another 90,000 years of glaciation. Previous glaciations had kilometre-thick ice sheets that covered Canada, northern USA, most of the UK, most of Europe north of the Alps, most of Russia and all of Scandinavia. Elevated areas in both hemispheres such as the Andes, New Zealand and Tasmania were covered by ice. There is no reason why the next inevitable glaciation will be any different.

Upland areas, even in the tropics, had glaciers. In areas with no ice sheets, strong cold dry winds shifted sand and devegetation occurred. Dunes in Australia, North Africa, the Middle East, Asia and North America again moved and great wind-deposited loess deposits covered Australia, Mongolia, China and northern USA. When Earth eventually has another glaciation, there will be mass depopulation, devegetation, dunes, extinctions, sea level fall and the destruction of coral reefs. It's happened before, it will happen again.

Imagine ice sheets covering the same area as in the last glaciation. Most of Europe, Canada and northern USA would be covered by kilometre-thick ice sheets. Other areas (e.g. China, Mongolia, Australia, Africa) would have howling cold winds and shifting desert sands Alpine areas would be covered with snow and ice. We could not grow enough food to feed billions of people.

Ice sheets grow and shrink. At times, they disappear. At other times, ice starts to cover polar areas and high mountains. That's what ice has done over the history of our planet. The

Greenland and Antarctic basins are more than a kilometre deep and deeper in the centres than around the edges so that ice is squeezed uphill like toothpaste out of a tube by the weight of overlying ice. The alarmist media stresses that changing sea ice and continental glaciers indicate rapid global warming. Is this really so?

During our current interglacial, summer sea ice in the Arctic has been far from constant. Sea ice comes and goes without leaving a clear record. For this reason, our knowledge about its variations and extent was limited before we had satellite surveillance or observations from aeroplanes and ships. A huge amount of the Earth's surface water moves alternately between the ice sheets and the oceans. Svend Funder, commenting on his recent *Science* paper stated:

> Our studies show that there have been large fluctuations in the amount of summer sea ice during the last 10,000 years. During the so-called Holocene Climate Optimum, from approximately 8000 to 5000 years ago, when the temperatures were somewhat warmer than today, there was significantly less sea ice in the Arctic Ocean, probably less than 50% of the summer 2007 coverage, which is absolutely lowest on record. Our studies also show that when the ice disappears in one area, it may accumulate in another. We have discovered this by comparing our results with observations from northern Canada. While the amount of sea ice decreased in northern Greenland, it increased in Canada. This is probably due to changes in the prevailing wind systems. This factor has not been sufficiently taken into account when forecasting the imminent disappearance of sea ice in the Arctic Ocean.

In order to reach their unsurprising conclusions, Funder

and the rest of the team organised several expeditions to Peary Land in northern Greenland. Funder said:

> Our key to the mystery of the extent of sea ice during earlier epochs lies in the driftwood we found along the coast. One might think that it had floated across sea, but such a journey takes several years, and driftwood would not be able to stay afloat for that long. The driftwood is from the outset embedded in sea ice, and reaches the north Greenland coast along with it. The amount of driftwood therefore indicates how much multiyear sea ice there was in the ocean back then. And this is precisely the type of ice that is in danger of disappearing today.

The new understanding came from getting away from computer modelling and doing field work in pretty inhospitable areas. Back in the laboratory and again away from computer models, the wood type was determined and dated using carbon 14. This wood came from near the great rivers of present-day North America and Siberia. This shows that wind and current directions have changed. The field study of coastal beach ridges shows that at times there were waves breaking unhindered by ice over at least 500 km of coastline. At other times due to sea ice cover, there were no beaches. This is the present situation.

Even if there is a great reduction in sea ice, all is not lost. Funder stated:

> Our studies show that there are great natural variations in the amount of Arctic sea ice. The bad news is that there is a clear connection between temperature and the amount of sea ice. And there is no doubt that continued global warming will lead to a reduction in the amount of summer sea ice in the Arctic Ocean. The good news is that even with a reduction to less than

50% of the current amount of sea ice the ice will not reach a point of no return: a level where the ice no longer can regenerate itself even if the climate was to return to cooler temperatures. Finally, our studies show that the changes to a large degree are caused by the effect that temperature has on the prevailing wind systems. This has not been sufficiently taken into account when forecasting the imminent disappearance of the ice, as often portrayed in the media.

We had been told that polar bears were heading for extinction because of the retreat of the ice sheets and sea ice. A few poignant pictures of a polar bear on a small iceberg was all that was needed to support the story. The International Union for the Conservation of Nature estimates the current polar bear population at between 22,000 and 31,000 which has been called the "highest estimate in 50 years."[412]

The evolutionary biologist and palaeozoologist Dr Susan Crockford at the University of Victoria wrote:

> Polar bears have survived several episodes of much warmer climate over the last 10,000 years than exists today.

and

> There is no evidence to suggest that the polar bear or its food supply is in danger of disappearing entirely with increased Arctic warming, regardless of the dire fairy-tale scenarios predicted by computer models.[413]

An attempt at climate alarmist niche tourism to Antarctica in January 2014 ended in farce. But it could have been a tragedy with multiple fatalities. It was promoted as an "expedition to

---

[412] Susan Crockford, *The Washington Times*, 9 January 2017.
[413] Climate Depot Special Report. *2016 State of the Climate*, Report, November 2016.

answer questions about how climate change in the frozen continent might already be shifting weather patterns in Australia" by retracing the steps of Sir Douglas Mawson 100 years earlier. The tourists on this largely taxpayer-funded jaunt that cost $1.5 million found no flowers growing in meadows around Mawson's Hut in Antarctica and were not able to return as conquering heroes with the proof of human-induced global warming.

Chris Turney, plus wife and children, mustered paying tourists and a sympathetic free-loading media onto the *Akademik Shokalskiy* to watch with bated breath the heroic planet-saving scientists battling against the elements to measure the thinning ice to obtain their pre-ordained conclusion: Antarctic ice is melting. One short trip to the frozen continent is all that was needed to prove what they already knew. And it is all due to global warming and we sinners are the cause.

Never mind the huge amount of fossil fuels burned to get to Antarctica. The activist "scientists" ignored measurements made far more easily from satellites and history that show that the Antarctic ice sheet is currently expanding. This was activism at its worst and was clearly not science.

However, nature has a sense of humour. The Russian gin palace was trapped in ice, a Chinese ice breaker sent to rescue these heroic adventurers also became trapped in ice, the real Antarctic research on bases was interrupted as an Australian ice breaker supply ship was diverted and the climate tourists were flown by a fossil fuel-driven helicopter to the fossil fuel derived warmth of a ship well away from the ice.

The Americans, Australian and Chinese all ran up huge costs to rescue the passengers from the Ship of Fools. All sorts of excuses were invented to show that the climate science activists on the ship were not ill prepared, incompetent, ignorant or aware of past ground and satellite measurements. When questioned about the failure, they resorted to obfuscation.

The climate activist community was silent, the normal suspects in the media became very creative with excuses (especially those on board) and the journal *Nature* showed that it was a magazine of political activism rather than one of scientific independence. The expedition was to show that this area was warmer than when Sir Douglas Mawson was in the exact same place 100 years ago. The farce showed the exact opposite as ironically Mawson was able to penetrate much closer to land because of the lack of ice.

There was no chance of frostbite, eating huskies or death of companions, as Mawson experienced. It was all a bit of a giggle with a games program organised on the ice because passengers were getting bored. The rescue was conducted by vessels using fossil fuels, not wind or solar power. Dozens of tourist vessels visit the Antarctic without becoming trapped in ice. It appears that the only tourist ship ever to be trapped in ice in summer was one with climate "scientists" trying to show that the ice was disappearing.

If Chris Turney did not live off research grants and was not employed by a university, he would have been sacked for gross incompetence, breaches of safety protocols and misleading and deceptive conduct. He was not and the taxpayer still keeps paying him. The public was not fooled, but activist climate "scientists" showed their true colours and even Turney's university defended him when they should have hidden him in a burrow.

As if this was not exciting enough, on the other side of the world the organisation that hangs banners from everything they can climb was active. Not only was the Antarctic melting due to we sinful humans burning fossil fuels, the Northern Hemisphere ice was also.

Greenpeace activists were arrested by Russians for trying to stop Russians drilling for oil in Russian waters. What did Greenpeace expect? To try to climb onto a drilling rig in Arctic waters is against all safety protocols, endangered the lives of others and was a breach of Russian sovereignty. Does Greenpeace really think that it is so important that it is above the law?

There were government travel warnings that the activists ignored and, after arrest, those imprisoned called upon their governments to help them. Again, this cost the taxpayer a huge amount. These two summer farces were wonderfully entertaining, helped the punter conclude that they will not fund the stupidity of others and showed the punter the shallowness of climate change activists' claims.

The four-year long $17 million BaySys study is to study the "contributions of climate change and regulation on the Hudson Bay system." The 2017 summer study was cancelled because the Canadian research icebreaker was required by the Canadian Coast Guard to rescue fishing boats and supply ships that had been stuck in the "unprecedented ice conditions". A lead scientist from the program claimed that the increased ice was because of climate change.[414] He was right. Greenland and the Arctic are cooling down. Similar trips to investigate global warming in the Artic were also cancelled because of ice (e.g. Caitlin expedition to the North Pole as was the David Hempleman Adams expedition).

The BBC announced to the credulous on 17 August 2017:[415]

> British explorer Pen Hadlow and his crew set sail from Alaska in an attempt to be the first people to sail to the North Pole. With Arctic ice melting at an unprecedented rate, previously inaccessible waters are opening up,

---

[414] James Delingpole, "Ship of fools III – Global warming study cancelled because of 'unprecedented' ice", *Breitbart*, 13 June 2017.
[415] *Watts Up With That*, 31 August 2017.

creating the potential for their planned 5,500 km (3,500 mile) journey for the first time in human history.

The attempt started on 4 August 2017 and was abandoned on 29 August 2017 due to sea ice which had grown at record rates.[416] The area of Arctic sea ice was less in 2012 than now. Apparently that was due to human-induced global warming.[417] Now that the sea ice in 2017 is considerably greater than in 2012, I wonder what excuses will be invented. The Arctic refuses to melt as the models and green activists predicted.

Every time there is a polar expedition to show that humans are warming the planet, nature has a good laugh.

Climate "scientists" and the media state that ice calving off glaciers indicates global warming. Ice always falls off the front of a glacier. If ice did not melt, then the planet would now be covered in ice. Ice drops off the toe of both advancing and retreating glaciers and the melting snout of a glacier is at a point determined by the balance between the forward movement of the ice by gravity and the rate at which it melts.

Ice falling off the front of a glacier means absolutely nothing when the air temperature is less than zero. Ice sheets grow and contract. At times, ice sheets disappear. The story of glacial retreat of is far more complex than a simple television image. Many glaciers that are now in retreat did not exist until the Little Ice Age (which climaxed in the middle to late 17th century). During the Medieval Warming (which peaked at about 900 to 1300 AD), alpine glaciers in the Northern Hemisphere were either smaller or did not exist.

Over much of the Canadian Cordillera, there may have been no glaciers at all during the Holocene Maximum (8,000 to 6,500 years ago), a period when temperatures were considerably higher than now. Records from New Zealand and Norway

---

[416] www.arcticmission.com/follow-arctic-mission
[417] www.theguardian.com/environment/jul/24/greenland-ice-sheet-thaw-nasa

show that glacier retreat commenced in the 18th and 19th centuries. Most of the modern ice retreat is due to post-Little Ice Age warming, changes in humidity and a decrease in ice flow rates.

The idea that a glacier slides downhill on a base lubricated by melt water was a good idea when first presented by de Saussure in 1779. We now know a lot more yet this treasured idea remains. Ice moves by creep, a process of constant recrystallisation of ice crystals. Ice at the snout of a glacier has crystals 1000 times larger than those in snow as a result of growth during recrystallisation. Ice sheets in Antarctica and Greenland are in deep basins and must first flow uphill before flowing down glaciers. The upward flow of ice cannot possibly be due to human-induced global warming producing melting. There are some places in the world today where glaciers are expanding. Ice sheets and glaciers grow and retreat for a great diversity of reasons. This is what ice does. For scientists to argue that ice retreat is due to human activity is simplifying a very complex process. Furthermore, it is too cold in Antarctica and Greenland for ice to melt.

Since the discovery of the Hubbard Glacier (in Alaska) in 1895, it has been advancing 25 metres a year during periods of cooling and warming. The glacier is 122 km long with an ice front is 10 km wide and 27 metres high. What does the ice do at the snout of the glacier? It falls off because it is getting pushed from behind. This has nothing to do with temperature, it shows that ice behaves like a plastic material and that ice sheets are always changing. Studies of the behaviour of tropical glaciers over the last 11,000 years show irregular shrinkage with slower rates in the Little Ice Age and faster rates in the 20th century. Glaciers such as the Bolivian Telata glacier reflects long-term warming during the current 10,500 year-long interglacial and that glacial retreat was in progress thousands of years before industrialisation.

Scary Scenarios

As with all areas of science, there are surprises. It was always thought that ice formed from frozen snow. The science was settled and there was a consensus. Recent work in the East Antarctic shows that the deepest part of the ice sheet contains ice that did not originate as snow. It was melt water that seeped to the base of the ice sheet and then froze. The amount of ice formed by this method is probably greater in volume than all glaciers on Earth outside Antarctica and Greenland.

The computer models predicted that this melt water escaped to the oceans and contributed to sea level rise. Wrong. The volume of water in this ice is larger than Antarctica's sub-glacial lakes. The addition of hundreds of metres of ice at the base of an ice sheet bends the overlying ice and causes uplift of the surface of the glacier. This changes the slope and flow of the ice. The thickest sub-glacial ice was 1,100 metres and this pushed the top of glaciers up 410 metres to reflect the shape of the added basal ice.

As snow falls, it traps air. This air is preserved as the snow becomes an ice sheet. This trapped air remains trapped and uncontaminated in ice and is used to measure past atmospheres. Antarctic ice core (Siple) shows that there were 330 parts per million of carbon dioxide in the air in 1900; Mauna Loa Hawaiian carbon dioxide measurements in 1960 show that the air then had 260 parts per million carbon dioxide. Either the ice core data is wrong, the Hawaiian carbon dioxide measurements are wrong or the atmospheric carbon dioxide content was decreasing during a period of industrialisation.

Measuring stations on Greenland show that temperature has been declining since 2005.[418] The typical July maximum temperature at Summit Station (3,216 metres altitude) is -10°C.

---

[418] T. Kobashi, L. Menviel, A. Jeltsch-Thömmes, B.M. Vinter, J.E. Box, R. Muscheler, T. Nakaegawa, P. L. Pfister, M. Döring, M. Leuenberger, H. Wanner and A. Ohmura 2017: Volcanic influence on centennial to millennial Holocene Greenland temperature change. *Science Reports* 7, doi:10.1038/s41598-017-01451-7

On 4 July 2017 it plunged to a record low of -33°C. Since mid 2016, there was a record accumulation of snow and ice and the rate of snow accumulation is greater than the rate of ice loss from calving. Is this global warming? Some Greenland glaciers are retreating (e.g. Istorvet) and revealing plant remains that show that the glacier was smaller than at present in the Medieval Warming (900-1300 AD).[419] Was this glacier melting in the Medieval Warming due to human emissions of carbon dioxide from vehicles, coal burning and industry? No.

The ice cap summit temperatures have declined by over 2°C from 2005 to 2015 and the RSS and UAH satellite data shows a 0.13°C temperature decrease per decade for Greenland yet 7,000 to 8,000 years ago ice core data shows that temperatures were 3 to 5°C warmer than now. In northwest Greenland, summer temperatures 4,000 to 8,000 years ago were 2.5 to 4°C warmer than now.[420] Some 8,000 years ago, the Greenland ice sheet had retreated some 20 to 60 kilometres behind its present margins[421,422] and reached a minimum area 5,000 to 3,000 years ago.[423]

---

[419] T.V. Lowell, B.L. Hall, M.A. Kelly, O. Bennike, A.R. Lusas, W. Honsaker, C.A. Smith, L. Levy, S. Travis and G.H. Denton, 2012: Late Holocene expansion of the Istorvet ice cap, Liverpool Land, east Greenland. *Quaternary Science Reviews* 63, 128-140.

[420] G.E. Lasher, Y. Axford, J.M. McFarlin, M.E. Kelly, E.C. Oterberg and M. Berkelhammer, 2017: Holocene temperatures and isotopes of precipitation in Northwest Greenland recorded in lacustrine organic materials. *Quaternary Science Reviews* 170, 45-55.

[421] K.H. Larsen, B. Lecavalier, A.A. Bjork, S. Colding, P. Huybrechts, K.E. Jakobsen, K.K. Kjeldsen, K-L. Knudsen, B.V. Odgaard and J. Olsen, 2015: The response of the southern Greenland ice sheet to the Holocene maximum. *Geology*, doi:10:1130/G36476.1

[422] N.E. Young and J.P. Briner, 2015: Holocene evolution of the western Greenland Ice Sheet: Assessing geophysical ice sheet models with geological reconstructions of ice margin change. *Quaternary Science Reviews* 114, doi: 10.7916/D80P0ZDW

[423] J.P. Briner, N.P. McKay, Y. Axford, O. Bennike, R.S. Bradley, S. de Vernal, D. Fisher, Francus, P., Fréchette, B., Gajewski, K., Jennings, A., Kaurman, D.S., Miller, G., Rousten, C. and Wagner, B. 2016: Holocene climate change in Arctic Canada and Greenland. *Quaternary Science Reviews* 147, 340-364.

Elsewhere in the Arctic, the Svarlbad islands were 6°C warmer 9,000 to 10,000 years ago.[424] Cooling led to temperatures 2 to 3°C colder than at present some 1,200 years ago. At Illulisat and Jakobshavn in Greenland, temperatures were 2 to 3°C warmer than at present 6,000 to 5,000 years ago.[425] A study of sediments in Greenland shows that the ice cap was similar to or smaller than at present throughout the last 8,000 years.[426] On Milne Land in east Greenland, past temperatures were 3 to 6°C warmer than at present[427] and fossil assemblages in the fjords of Scoresby Sound show that Greenland was much warmer than today some 6,000 years ago.[428,429] On Baffin Island in Canada, temperatures were about 5°C warmer than at present 6,000 years ago.[430,431]

The rate of melting of Greenland glaciers has been examined

---

[424] J. Mangurud and J.I. Svendsen, 2017: *The Holocene* Thermal Maximum around Svarlbad, Arctic North Atlantic; molluscs show early and exceptional warmth. The Holocene, doi 10.1177/0959683617715701
[425] A.E. Carlson, K. Winsor, D.J. Ullman, E.J. Brook, D.H. Rood, Y. Axford, A.N. LeGrande, F.S. Anslow and G. Sinclair, 2014: Earliest Holocene south Greenland ice sheet retreat within its late Holocene extent. *Geophysical Research Letters*, doi:10.1002/2014GL060800
[426] A.R. Lusas. B.L. Hall, V.T. Lowell, M.A. Kelly, O. Bennike, L.B. Levy and W.M. Honsaker, 2017: Holocene climate and environmental history of East Greenland inferred from lake sediments. *Journal of Paleolimnology* 57, 321-341.
[427] A.D. Schweinsberg, J.B. Briner, G.H. Miller, O. Bennike and E.K. Thomas, 2016: Local Glaciation in West Greenland linked to north Atlantic Ocean circulation during the Holocene. *Geology,* doi 10.1130/G38114.1gy
[428] D.E. Sugden and B.S. John, 1965: The raised marine feature of Kjove Land, East Greenland. *Geographical Journal* 131, 235-247.
[429] O. Bennike and B. Wagner, 2013: Holocene range of Mytilus edulis in central East Greenland. *Polar Record* 49, 291-296.
[430] E.R. Thomas, E.W. Wolff, R. Mulvaney and T. Popp, 2007: The 8.2 ka event from Greenland ice cores. *Quaternary Science Reviews*, doi: 10.1016/j.quascirev.2006.07.017.
[431] Y. Axford, J.P. Brinner, C.A. Cooke, D.R. Francis, N. Michelutti, G.H. Miller, J.P. Smol, E.K. Thomas, C.R. Wilson and A.P. Wolfe, 2009: Recent changes in a remote Arctic lake are unique within the past 200,000 years. *PNAC* 106, 18443-18446.

for the period 1870 to 2005.[432] The average total retreat was some 1,334 metres and, of this, 1,062 metres of retreat had taken place before 1946 and the retreat after 1946 was only some 272 metres. Glacier melting slowed down in 1946 and only 20% of the retreat was in the years between 1946 and 2005. Something is seriously wrong with the global warming story. Post-World War II industrialisation from 1945 accelerated emissions of carbon dioxide. This should have warmed the planet and led to a more rapid rate of glacier melting. This is the theory. The measurements shows the opposite.

The current warming at the poles is very minor compared to recent past warmings. Are these past warming events due to human emissions of carbon dioxide? No. What makes the current warming so special? Nothing. Wherever we look in the world, we see evidence of coolings and warmings. Over the past 500 years, Greenland temperatures have fluctuated between warming and cooling about 40 times with changes every 25 to 30 years. At least three warming events were 20 to 24 times the magnitude of warming over the past century and four were six to nine times the magnitude of warming over the past century.[433] Why is it that the latest warming event is due to human emissions of carbon dioxide yet all previous warmings could not have been from humans emitting carbon dioxide?

Climate "scientists" do not mention previous warmings and, if they do, they don't attempt to explain why the modern warming is different from previous warming events. This sin of omission is not the behaviour of scientists searching for an

---

[432] J.M. Fernández-Fernández, N. Andrés, T. Saemundsson, S. Brynjólfsson and S. Palacios, 2017: High sensitivity of North Iceland (Tröllaskagi) debris-free glaciers to climate change from the 'Little Ice Age' to the present. *The Holocene*, doi: 10.1177/0959683616683262

[433] D. Easterbrook, *Evidence-Based Climate Science*. 2nd Edition, Elsevier, 2016.

explanation of a phenomenon. It is the action of fraudulent political activists.

Some glaciers are expanding, others are retreating. This is normal and is no cause for alarm and certainly can't be used to give scary predictions. Ice sheets and sea ice show a long history of expansion and contraction, again no cause for alarm.[434] The amplitude of seasonal variation of sea ice in both the Arctic and Antarctic dwarfs the changes in sea ice that are meant to show catastrophic global warming. Furthermore, the extent of global sea ice shows remarkably little change over the past 35 years[435] and even the IPCC shows that there is little or no warming in Antarctica.

President Obama and Secretary of State John Kerry visited Glacier Bay in Alaska in late August 2015. They pointed to the receding glaciers as evidence that humans are the cause of dangerous catastrophic climate change. A bit of history rather than hype would have been more honest. The glacier started to retreat in 1750 AD. When Captain George Vancouver visited in 1794, the ice filled most of the Bay and had only retreated a few kilometres. When the founder of the Sierra Club, John Muir, visited Glacier Bay in 1879, he found that the ice had retreated 50 km from its 1750 position. In 1900, it was almost free of ice. This shows that there is climate change and that it is unrelated to human emissions of carbon dioxide since the Industrial Revolution. Maybe it is related to the 300 years of warming we have enjoyed since the Maunder Minimum.

Sea ice has often and fraudulently been used for climate scares. In 2007, the media and green left activist "scientists" were telling us that the Arctic would be free of ice by 2013.[436]

---

[434] R.L Braithwaite, 2002: Glacier mass balance: The first 50 years of international monitoring. *Prog. Phys. Geog.* 26: 75-95.
[435] http://arctic.atmos.uiuc.edu/cryosphere/
[436] http://news.bbc.co.uk/2/hi/7139797.stm

The year 2013 has been and gone so this prediction can be evaluated. Projections were wrong because of garbage-in garbage-out computer modelling[437] and some modellers made flawed assumptions.[438]

> Some models have not been taking proper account of the physical processes that go on. The ice is thinning faster than it is shrinking and some modelers have been assuming the ice was a rather thick slab. Wieslaw's model is more efficient because it works with data and it takes account of processes that happen internally in the ice.

The year 2013 is been and gone so this prediction can be evaluated. The Arctic still has sea ice and, far from disappearing, its extent has increased. The problem is that in recent years there has been more ice in the world at any time since satellite records began in 1979 and its thickness is increasing. The Arctic summer ice, which was retreating recently, has now started to advance. These are measurements, not models. And by the way, in case you are worried, polar bear numbers are increasing and temperatures in Greenland have shown no increase for decades. Concurrently, there has also been a rapid growth of sea ice in Antarctica to 20 million square kilometres.

Arctic sea ice area should be related to the regular 60 to 70-year Atlantic Multidecadal Oscillation (AMO)[439] which peaked with its warm cycle in 2012 hence post-2012 we could be facing a more significant decline in global atmospheric temperature as predicted by the total solar intensity decrease.

The US submarine *Nautilus* was able to surface at the North Pole through thinned ice in 1956 and during the previous AMO

---

[437] Fall meeting 2007, American Geophysical Union; paper by Wieslaw Maslowski.
[438] http://news.bbc.co.uk/2/hi/7139797.stm
[439] N.G. Andronova and M.E. Schlesinger, 2000: Causes of global temperature changes during the 19th and 20th centuries. *Geophys. Res. Lett.* doi:10,1029/2000GL006109

thinning, Amundsen was able to navigate the Northwest Passage by sail in 1906 during AMO thinning. Trillions of dollars are being spent in case there is global warming driven by human emissions of carbon dioxide. Maybe we need to worry about the next cycle of cooling.

The Pacific Decadal Oscillation (PDO) is a switch between two circulation patterns that occurs every 30 years. It was originally discovered in 1997 in the context of salmon production. The warm phase tends to warm the land masses of the Northern Hemisphere.[440] The AMO and PDO data sets are not similar and cannot be added or averaged. In the 1930s, the AMO and PDO warm phases were coincidental. This was a period of time of record temperatures and dustbowls in the USA. Both cold phases came together in the 1960s and 1970s, the time of scares that the planet was entering another ice age.

For a decade from the mid 1990s, the warm phases again coincided and this was the last time the Earth was warming. The PDO has now turned cold, the AMO has peaked in its warm phase and temperatures are neither increasing nor decreasing. From the mid 2020s to the mid 2030s, both the AMO and PDO will be in their cold phases and planet Earth could experience a similar cooling to that of 1964 to 1979. Solar activity is declining as well and some solar physicists are predicting a cold Dalton Minimum or even a colder Maunder Minimum level. Time to buy a warm coat, thermals and Ugg boots.

Polar ice is another poster child for the climate catastrophists. When the Dark Ages ended (~900 AD) and the planet started to warm naturally in the Medieval Warming (900 to 1300 AD), the Vikings were the first to feel it. They enlarged their fishing grounds and invaded countries to the south as illegal boat people. Although climate catastrophists ignore the Medieval

---

[440] http://www.droyspencer.com/global-warming-background-articles/the-pacific-decadal-oscillation/

Warming, their narrative is that if the planet were warming due to human emissions of carbon dioxide, then the poles would be the first to feel it. However, the poles are cooling.

A good Gaia-fearing decent green left environmental activist would face heroic hardships and go to the poles to get first hand evidence. That would be proof. The deluded demonic denialists would be finally silenced. Such trips have attempted to view the reduced polar ice and were to regularly report back horror stories. Such trips were so well planned that ideology was all that was needed. To view the Cryostat satellite data[441] and polar weather station temperature records would not be necessary because this was the sort of data that denialists used.

We are told that the area of Arctic sea ice had been decreasing for some years. We were all doomed. The European Space Agency[442] reported that the Arctic sea ice had increased by a staggering 33% in 2013 and 2014. Weather stations in western Greenland (Godthab Nuuk) and eastern Greenland (Angmagssalik) have measured temperature since 1900. These stations are on opposite sides of Greenland hence give a good coverage of the island. As expected, the unadjusted temperature data show some years are colder or warmer than others but over the 117-year record of measurement, there has not been a significant warming of Greenland. This is public record GISS data freely available from NASA and NOAA.[443]

For some 80% of time, the Earth has had neither ice sheets nor sea ice. We are now in one of the many interglacials during a long ice age. We know that the current interglacial started 14,700 years ago. We don't know when it will end but we do know it will end badly.

---

[441] http://www.esa.int
[442] http://www.esa.int/Our_Activities/Observing_the_Earth/Cryosat/Arctic/sea-ice-up-from-record-low
[443] 1900-2010: http://data.giss.nasa.gov/gistemp/station_data_v2/; 2011-2015: http://data.giss.nasa.gov/gistemp/station_data/

Those playing with computer climate models need to get outside, collect new data and take into account far more factors than they currently feed into their computer models.

The Antarctic is gaining ice, contrary to widespread propaganda that human-induced global warming is melting the continental ice cap. The land mass has been gaining ice[444] at the rate of 82 billion tonnes per year since 2003. The sea surface temperature around Antarctica has been declining for 40 years,[445] in contrast to the model-based predictions of the IPCC.

Each year, there are countless news stories about heat waves in the Arctic and record ice melt. Data shows a different story. The April ice extent has been stable since 2004,[446,447] Arctic air temperatures are currently below average[448] and the Greenland ice sheet continues to grow at record rates.[449]

Work published in May 2017 by Takuro Kobashi[450] showed that almost everything presented about the melting of Greenland ice is exaggerated or wrong. Mean annual temperature at the summit of Greenland is decreasing, in accord with the northern North Atlantic cooling. In 2017, Greenland recorded the coldest July ever (-33°C) at the same time as ice cover is increasing. If we are enjoying dangerous global warming, how can that be? The Swiss newspaper *Baseler Zeitung* didn't mince its words:[451]

---

[444] https://www.friendsofscience.org/index.php?id=2190
[445] https://www.friendsofscience.org/index.php?id=221
[446] http://ocean.dmi.dk/arctic/icecover.uk.php
[447] http://ocean.dmi.dk/arctic/icecover_30y.uk.php
[448] http://ocean.dmi.dk/arctic/meant8on.uk.php
[449] http://www.dmi.dk/en/groenland/maalingeer/greenland-ice-sheet-surface-mass-budget/
[450] T. Kobashi, L. Menviel, A. Jeltsch-Thömmes, B.M. Vinter, J.E. Box, R. Muscheler, Nakaegawa, T., Pfister, P.L., Döring, M., Leuenberger, M., Wanner, H. and Ohmura, A. 2017: Volcanic influence on centennial to millennial Holocene Greenland temperature change. *Science Reports* 7, doi:10.1038/s41598-017-01451-7
[451] Eugen Sorg, *Baseler Zeitung*, 11 July 2017.

Singing from the same hymn sheet like a church choir, the news media recently reported about the Rhone Glacier shrinking by 10 cm per day. No mention, however, about Greenland. Why? The evident ice-growing and cooling phenomenon should not exist. It cast the central prophesy of a continuous and ultimately lethal global warming, for which we are ourselves to blame, into question.

and

... most journalists and media leaders are active or passive members of the green-socialist Climate Church and the new religion of the post-Christian western world.

Most interglacials last about 10,000 to 12,000 years. Earth is currently at the end of an orbitally- and solar-driven interglacial that should end on a Thursday. From Friday on, it will be cold and we don't know for how long. Unless our climate activists, government and media change the behaviour of the Sun and the Earth's orbit, we will enter the next inevitable glaciation.

For a long time it has been known that there were at least 20 active volcanoes beneath the West Antarctic Ice Sheet,[452,453] volcanoes beneath ice in Iceland, the Yukon, and British Columbia are also well documented.[454] Recent data shows that the heat flow beneath the Antarctic ice sheet has probably been underestimated.[455] Some areas underneath the ice sheet have local volcanic heat sources and others have a widespread area

---

[452] W. E. LeMasurier, 1976: Intraglacial volcanoes in Marie Byrd Land. *Antarctic Jour. U.S. 11*: 269-270.

[453] H.F.J. Corr and D.G. Vaughan, 2008: A recent volcanic eruption beneath the West Antarctic ice sheet. *Nature Geoscience* 1: 122-125.

[454] Ian Plimer, *Heaven and Earth. Global Warming: The Missing Science*, Connor Court, 2009, 279.

[455] A.T. Fisher *et al.* 2015: High geothermal heat flux measured below the West Antarctic Ice Sheet. *Sci. Adv.* 1: e1500093, 9pp.

of high heat flow from deep in the Earth. This is not unusual and has happened many times before.[456] Along an 8,000-km long rift system in Antarctica, there are 61 active and semi-active volcanoes, at least three glaciers (Pope, Smith, Kohler) lie on hot rocks[457] and huge amounts of carbon dioxide are released.[458]

More recent work[459] has shown that there are now 138 volcanoes beneath the West Antarctic Ice Sheet, 91 of which have just been found. More and more volcanoes are being found. Clearly when scientists state that "the science is settled" on any issue, they have closed their minds to new discoveries and thrive on dogma. This is the current position of climate "science". Volcanoes and hot rocks in the Antarctic Rift have clearly had a great and unknown influence on ice sheet growth and decay and this has never been considered in climate models of the Antarctic ice sheet.

The last big eruption in Antarctica was in Roman times and Mt Erebus is continually restless. Addition of heat from below could cause massive melting and detachment of a large block of ice. It's happened before and it will happen again. The instability of the ice sheet could well be related to variable flows of geothermal heat beneath the ice sheet rather than sea or air temperature changes.

A big chunk of ice broke off the Larsen C ice shelf in the

---

[456] E.J. Steig *et al.* 2013: Recent climate and ice-sheet changes in West Antarctica compared with the last 2,000 years. *Nature Geoscience* 6: 372-375.

[457] climatechangedispatch.com/west-antarctic-glacial-melting-from-deep-earth-geological-heat-flow-not-global-warming/'

[458] H. Lee, J.D. Muirhead, T.P. Fischer, C.J. Ebinger, S.A. Kattenhorn, Z.D. Sharp and G. Kianji, 2016: Massive and prolonged carbon emissions associated with continental rifting. *Nature Geoscience* 9, 145-149.

[459] M. van Wky de Vries, R.G. Bingham and A.S. Hein, 2017: A new volcanic province: an inventory of sib-glacial volcanoes in West Antarctica. *Geol. Soc. Lond. Special Pubs.* 461: doi:org/10.1144/SP461.7

southern summer of 2017. It was a quarter the size of Wales, twice the size of Luxembourg or the size of Delaware. And, wait for it, it weighs 1.3 trillion tons. That's scary. Or is it?

In 1927, an iceberg four times the size of the 2017 iceberg broke off and in 1956 and iceberg six times the size of the 2017 iceberg broke off. Moving glaciers grinding across the surface pick up rocks, some of which are ground into powder (rock flour). When a glacier calves and an iceberg breaks off, the ice plus rocks and rock flour go for an ocean voyage. The iceberg starts to melt, layers of rocks and rock flour are dropped onto the sea floor in what are called Heinrich events. Sea floor sediments show that we have had Heinrich events with each interglacial and large icebergs in interglacials are the norm. However, the current iceberg is touted as unprecedented and a result of human emissions of carbon dioxide. The media and climate activists have milked it for all its worth and used as many superlatives as they could exploit.

In summary, Antarctic ice melting is based on irrational fear, not data. The overall ice mass and extent of ice has increased over the last 35 years.[460] NASA found that Antarctica was not losing ice, was not contributing to sea level rise and was actually reducing sea level rise. Antarctica was gaining more ice than it lost.[461] Satellite measurements showed that summer sea ice in the Arctic was 22% greater than it was in the low point in 2012[462] and the Arctic sea ice minimum is now at a 10-year pause with "no significant change in the past decade".[463]

---

[460] climatechangedispatch.com/west-antarctic-glacial-melting-from-deep-earth- geological-heat-flow-not-global-warming/'

[461] NASA Study *Mass gains of Antarctic ice sheet greater than losses*, 30 October 2015.

[462] Danish Meteorological Institute, 15 September 2016.

[463] David Whitehouse, the Global Warming Policy Foundation, 16 September 2016.

We are told that we'll all be doomed and that the ice caps are melting. The exact opposite is the case. If planet Earth is cooling, which the evidence suggests, then we have a bigger problem. Especially if energy to warm us is unreliable, inefficient and expensive.

If, on the other hand, planet Earth is warming, then why is warming such a problem and how can a tax make warming go away anyway?

Extreme weather

The climate activists claim that increased carbon dioxide in the atmosphere will drive global warming and cause abnormally extreme weather events and increased frequency of them. However, reality is different from predicted climate catastrophes.

In 2017, Roger Pielke testified to Congress that there was no evidence that hurricanes, floods, droughts and tornadoes are increasing.[464] He stated: "It is misleading and just plain incorrect, to claim that disasters associated with hurricanes, tornadoes, floods or droughts have increased on global timescales either in the United States or globally."[465] Every day somewhere on Earth has wild weather. Years ago we only read about this much later in newspapers. Now with modern communications systems and a highly competitive 24/7 news cycle, we learn of such wild weather as it is happening. In the past, cyclones, tornadoes, bushfires, rainstorms, snow storms, ice storms, floods and droughts were due to the weather. As communication systems improved and media and green groups needed to feed off

---
[464] Prof. Roger Pielke Jr., *Congressional Hearing: House Science Committee on Science, Space and Technology*, 29 March 2017.
[465] Prof Roger Pielke Jr., *Testimony to U.S. Senate Environment and Public Works Committee*, 18 July 2013.

disasters, this wild weather (now called extreme weather) was then deemed to be due to man-made global warming driven by man's sinful emissions of carbon dioxide.

Screaming babies, crying mothers, grieving fathers, concerned politicians and alarmists apportioning blame to others makes good television. Never let the truth get in the way of a good story. The truth just does not sell. There is a large body of evidence to show that we live in rather boring times, that the frequency of many of these events has actually decreased and that many past wild weather events were more severe than anything we view today from the comfort of our lounge rooms. With increased wealth and population, more people are building at waterside locations and more expensive residences elsewhere and hence the cost of property damage becomes higher.

From 1910 to 1940, many parts of the Earth were warmer than now. Temperature has been increasing for the last 300 years hence it is no surprise that the highest temperatures are towards the end of a warming period before we enter another solar-induced cooler period. To suggest that the future will be hotter, drier and more vulnerable to floods ignores the past empirical evidence. The numerous recent failed predictions have all been wrong and there is no reason to believe that a repeated prediction will be correct. We ignore the past at our peril.

In Australia, the major natural hazards have been droughts, floods, bushfires, storms and cyclones. Hundreds of people have been killed in cyclones, floods and bushfires over the last 200 years. We have had one fatal earthquake and can expect more. There have been deaths from landslides. The last volcanic eruption was 500 years ago and every few hundred years the south eastern coast Australia is hit by a tsunami. Like all other countries, we have more premature deaths from cold weather than from warm weather.

The US National Weather Service has kept statistics on weather-related deaths since 1940. The annual number of deaths from tornadoes, floods and hurricanes varies. For example, the number of people in the US killed by extreme weather events in 1972 was 703. In 1988, it was 72. There is a clear long-term trend: the number of weather-related deaths has dropped dramatically despite the fact that the US population has more than doubled since 1940.

Studies on floods show that intense floods are less common in recent years and that flooding cannot be related to global warming. Major flooding in the US and Europe is not increasing.[466] The authors state:

> The results of this study, for North America and Europe, provide a firmer foundation and support the conclusion of the IPCC that compelling evidence for increased flooding at a global scale is lacking.

Increased flood costs may well be because we have become wealthier and more expensive buildings are now built on flood plains. The proportion of runoff (and flooding) increases when areas are covered by housing and roads.

We keep getting told by climate activists that global warming is responsible for more intense rainfall because more water vapour can dissolve in warm air. When rainfall records are studied, the measurements show the opposite with rainfall decreasing over the last five decades.[467] The same records show that during hurricanes in Texas, the greatest rainfall was in Galveston (1871), Woodward Ranch (1935), Thrall (1921) and Alvin (1979).

The media got into a great lather about Hurricane Harvey

---

[466] G.A. Hodgkins *et al.* 2017: Climate-driven variability in the occurrence of major floods across North America and Europe. *Journal of Hydrology* 552, 704-717.
[467] http://www.nws.noaa.gov/ohd/hdsc/record_precip/record_precip_us.html

that hit Texas in the August 2017. About 50 people died, there was massive inundation by extremely heavy rain because the tropical storm was moving very slowly and then stalled. Harvey hit after a 4,323-day period of no major hurricanes.[468] Houston has a history of flood disasters dating back to the mid 1800s. The flood water level in August 2017 Category 4 hurricane was five metres lower than a Category 5 hurricane in 1935. One unbalanced Fairfax journalist[469] tried to claim that this was retribution for Texas being the world capital of the oil and gas industry and that there was a connection between carbon dioxide and extreme weather despite the EPA warning that climate "scientists" were trying to politicise the hurricane.[470]

Sociologist Kenneth Storey from the University of Tampa went one step further and suggested that Hurricane Harvey is retribution for Texans who voted Republican.[471] He was not aware that in Houston, more people voted Democrat than Republican in the 2016 elections.

No climate activist journalist seemed to be aware that the greatest natural disaster in US history, the 1900 hurricane, hit Galvaston (Texas) and resulted in at least 6,000 to 12,000 deaths. This was before Texas was the oil and gas capital of the world.

Tropical storm Harvey was concurrent with far more devastating massive rains and floods in Nepal, India and Bangladesh that killed hundreds of people. There was no mention of this in the unbalanced media as being due to oil and gas production in the subcontinent.

---

[468] CNC News, 26 August 2017.
[469] Peter Hannam, "Houston, you have a problem, and some of it is of your own making", *Sydney Morning Herald*, 29 August 2017.
[470] Reuters, EPA Says Climate Scientists Trying To Politicize Texas Storm, 29 August 2017.
[471] *New York Daily News*, 29 August 2017.

An increase in the number and intensity of tornadoes predicted from global warming scenarios has proven false. Big tornadoes have become less frequent since the 1950s. The years 2012, 2013, 2014, 2015 and 2016 all saw at or near record low tornado counts in the US.[472] Actual NOAA tornado data for 2016 showed that it was one of the quietest years since records began in 1954. It was below average for the fifth consecutive year.

NOAA data shows that 2016 was the eleventh consecutive year without a major (Category 3 or above) hurricane strike. This is the longest period without a major hurricane since record-keeping started in 1851.[473] On a global scale, there has been no trend in the global accumulated cyclone energy over the past 30 years, a time when there was alleged global warming and increased cyclonic activity.[474] In 2017, the nail was driven into the coffin by NOAA by reporting:

> It is premature to conclude (that man-made global warming has) already had a detectable effect on hurricanes.[475]

The 2016-2017 cyclone season in the Southern Hemisphere was the quietest on record.[476]

Drought

The geological history of the planet shows that huge planet-wide mega droughts occurred when there was an ice age. Little

---

[472] NOAA Tornado data: *2016 "one of the quietest years since records began in 1954" – Below average for 5th year in a row*, 12 November 2016.
[473] CNSNews.com *NOAA data*, October 2016.
[474] Dr Philip Klotzbach, 20 September 2016, www.tropical.colostate.edu
[475] NOAA/Geophysical Fluid Dynamics Laboratory. *Global warming and hurricanes – An overview of current research results. Has global warming affected hurricane or tropical cyclone activity*. Last revised March 2017.
[476] Dr Ryan Maue, 4 April 2017, https://wattsupwiththat.com/2017/global-temperatures-plunge-in-april-the-pause-returns/

water vapour can be held in cold air and hence there was little rain. These great periods of drought during the six major ice ages are characterised by salt deposits and red sandstones that were once dune sands.

NASA's climate data from 23 model simulations between 1995 and 2005 underestimated rainfall compared with measurements. What a surprise. They even underestimated rainfall in some pretty dry parts of the Earth such as the Sahara.

The IPCC in its 2012 and 2013 reports show that there is no evidence of increased extreme rainfall events due to human activity. The 250 years of rainfall records in Britain shows little change as does the 100-year record in the US. Furthermore, deaths from extreme weather are currently at an all time low despite increases in the atmospheric carbon dioxide and population.[477]

In their 2013 report, the IPCC predicted that California would have droughts nearly every year from 2030 onwards and with increasing severity until 2100. It appears that because 22 climate models were used to make such a prediction[478] out to 2100, then the predictions must be true. It appears that the reduction of carbon uptake during the 2000 to 2004 drought in California is the villain.

It's charming to think that what happens in California stays in California.

The normal suspects had screaming headlines such as "California's terrifying climate forecast: It could face droughts nearly every year."[479] However, new studies showed that rainfall

---

[477] Global Warming Policy Foundation 2014.
[478] C.R. Schwalm, C.A. Williams, K. Schaefer, D. Baldocci, T.A. Black, A.H. Goldstein, B.E. Law, W.C. Oechel, T.P.U. Kyaw and R.L. Scott, 2013: Reduction in carbon uptake during the turn of the century drought in western North America. *Nature Geoscience* 5, 551-556.
[479] Darryl Fears, *The Washington Post*, 2 March 2015.

Scary Scenarios

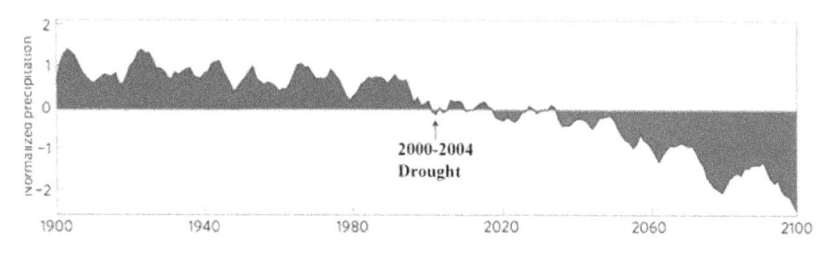

Figure 25: Western North American precipitation, 1900-2100 as used by the IPCC (2013)

is expected to increase in California.[480] There were no screaming headlines for this news. It now appears that there will be no drought in California due to human-induced global warming and that it's now all due to warmer surface temperatures in the tropical eastern Pacific Ocean, 4,000 km west of the US, that will encourage larger numbers of storms in California.[481]

In the US, Prof Roger Pielke noted:

> Droughts have, for the most part, become shorter, less frequent, and cover a smaller portion of the U.S. over the last century.[482]

In 2017, drought conditions in the U.S. dropped even more as they were limited to only 1.6% of the continental U.S. and what was touted by climate activists as California's permanent drought came to an end.[483]

A 2015 study found that megadroughts in the last 2,000 years were worse and lasted longer than current droughts.[484]

In Australia, global warming alarmists panicked Labor governments into creating subsidised wind and solar power and

---
[480] University of California, Riverside, 6 July 2017.
[481] Robert Allen, *Nature Geoscience Communications*, July 2017.
[482] Roger Pielke, Jr, Drought analysis, 24 September 2012.
[483] Paul Dorian, *Vencore Weather*. 10 April 2017.
[484] The Earth Institute, Columbia University, 6 November 2015.

building massive desalination plants during a drought. Four desalination plants built by former State Labor governments in Australia that have since been mothballed and still will cost taxpayers nearly $1 billion *per annum* in maintenance.[485] There are no plans for their use. They were built at times when green left environmental activists were screeching about how Australian cities would run out of water.

Australia has droughts. It is a sunburned country. It also has flooding rains. In 1911, Dorothea Mackellar was correct. Drought has been and will continue to be a major feature of Australia. Green left environmental activists, the media and city-based politicians are unaware of drought because it does not affect them directly. Maybe they should travel in outback South Australia and see the abandoned homesteads and villages where dreams were broken in the 1880-1886, 1888 and 1895-1903 droughts.

There have been ten major droughts in Australia over the last 150 years.[486] Some droughts lasted a decade, some are in clusters and all had a severe effect on rural Australia. Primary production decreased, banks foreclosed businesses, unemployment rose, rural towns died, farmers committed suicide and government assistance was too little and too late. City life blissfully continued unaware of the tragedy on their doorstep and the only impact was that food prices of selected products increased slightly. Many droughts are broken by flooding rains.

The Australian Bureau of Statistics[487] and Bureau of Meteorology[488] document the major Australian droughts (1864-1866,

---

[485] *The Weekend Australian*, 12-13 September 2015.
[486] M.J. Couglan, 1986: *Drought in Australia. Natural Disasters in Australia.* Australian Academy of Technological Sciences and Engineering.
[487] http://www.abs.gov.au/AUSSTATS/abs@.nsf/lookup/1301.0Features%20Article151988
[488] http://www.bom.gov.au/climate/drought/

1868, 1880-1886, 1888, 1895-1903, 1911-1916, 1918-1920, 1939-1945, 1958-1968, 1982-1983, 1999-2009, 2012-2015); less severe droughts (1922-1923, 1926-1929, 1933-1938, 1946-1949, 1951-1952, 1970-1973, 1976) and droughts where most Australian live (i.e. SE Australia; 1888, 1902, 1914-1915, 1940-1941, 1944-1945, 1967-1968, 1972-1973, 1982-1983, 1991-1995, 2002-2003, 2006-2007, 2013-2015).

The three biggest droughts were 1895-1903, 1939-1945 and 2002-2007.[489] However, rather than take advice from government departments that deal with historical information, the last drought led to political panic aided and abetted by green left city-based environmental activists and cheered along by the hysterical ABC and Fairfax media networks. Australia's most unsuccessful climate forecaster (Tim Flannery) was able to state without critical questioning from the media:

> So even the rain that falls isn't actually going to fill our dams and our river systems, and that's a real worry for people in the bush ... I think there is a fair chance Perth will be the 21st century's first ghost metropolis ... Perth is facing the possibility of a catastrophic failure of the city's water supply ... I'm personally more worried about Sydney than Perth. Where does Sydney go for more water?

This was a great scare campaign and neither the media nor Flannery bothered to tell people that drought is normal, that there have been many long hard droughts in Australia before and that previous droughts occurred before Australia was industrialised and emitting plant food (carbon dioxide) from transport and industry. No one in the mainstream electronic or print media challenged Flannery's scary claims or qualifications. Commentators who questioned Flannery's catastrophist

---

[489] http://www.australia.gov.au/about-australia/australian-story/natural-disasters

claims were ignored. Flannery has degrees in English Literature and a PhD on tree kangaroos hence is the perfect media go-to person for expert opinion on drought. In 2007, he opined:

> In Adelaide, Sydney and Brisbane, water supplies are so low they need desalinated water urgently, possibly in as little as 18 months.

The Queensland Labor Premier spent more than $1.2 billion of taxpayers' money on a desalination plant on the basis that the lower than usual rainfall would continue. It didn't. The expensive desalination plant has been mothballed, dams are full and Brisbane has since been flooded twice. Not to be outdone, the Victorian Labor government spent $2 billion on a desalination plant and, at the time of its opening, catchment dams were more than 80% full. The plant was mothballed. Sydney also spent horrendous amounts of money on a desalination plant, since mothballed. It has produced no water since 2012.

In 2008 Flannery stated:

> The water problem is so severe for Adelaide that it may run out of water by early 2009.

It didn't. At times, Adelaide water is character-building and, when mixed with sand and aggregate, can make excellent concrete. Adelaide also got into the act and spent $2.2 billion on a 100 Gigalitre desalination plant that has since been mothballed. For the poorest mainland state in Australia with the highest unemployment to spend billions on a white elephant shows how green ideology has infiltrated politics and, no matter what the cost, the public pays.

The south eastern States of Australia recently spent more than $10 billion building desalination plants. All were mothballed. This was a shocking waste of public funds resulting directly from a green left environmental activist scare campaign. Politicians, media, academics and green left environmental activists were not held to account, they all fell for the global warming scare and have since moved on to try to scare us with other aspects of their predicted global warming disaster. Why should we ever take notice again of green left environmental activists? No person or company in the productive part of the economy has ever made such horrendous mistakes based on propaganda and without a full due diligence process.

In Perth, it was different. There had been a slight decline in rainfall (especially in winter) over the last 40 years in south western Western Australia.[490] Such cycles are common all over Australia. Furthermore, the slightly lower runoff increased salinity of waters captured in dams (e.g. Wellington Dam) and much dam water became unsuitable for drinking. Perth was a rapidly expanding metropolis, much domestic and industrial water was from a shallow aquifer and new water supplies were needed.

A desaliniation plant was built near Kwinana (Perth Seawater Desalination Plant) and supplied Perth with 17% of its water needs. The second and bigger plant was built further south at Binninup (Southern Seawater Desalination Plant).[491] Some 50% of Perth's water is now from desalination plants and the energy for these plants derives from natural gas. Water Corporation WA also purchases the entire output from the 10 MW solar plant at Greenough and from the 55 MW wind plant at Mumbida and claims that the Binninup plant is carbon neutral (except when the wind does not blow and the Sun doesn't shine).

Perth residents did not suffer from climate change. They had a population explosion. Residents also did their bit and changed from well-watered European green lawns to native gardens that

---

[490] www.cawcr.gov.au
[491] www.watercorporation.com.au

needed less water. The additional costs have been absorbed into the Water Corporation WA costs and sandgropers are just happy to have water.

The normal suspects were up in arms about building a desalination plant at Kwinana. They claimed that the sea grass meadows off Kwinana were fragile, and that there would be an increase in salinity in the Indian Ocean. The sea grass was going to die, this would create an ecological collapse and that everything off the coast of Western Australia would die. The sea grass did not die. A simple mass balance calculation would show that there would be no measurable increase in salinity in the Indian Ocean off Kwinana as dilution is effectively infinite. As usual, science and engineering solved a problem. As usual, the green left environmental activists were hopelessly wrong and just moved onto the next perceived disaster story.

We are told that global warming will lead to more drought. Australia is the land of drought and there are good records of droughts. Surely the regular droughts in Australia are due to increases in temperature? No. They are due to decreases in local rainfall. Australian rainfall is highly variable but has not decreased over the last 100 years, the longest measured drought in Australia lasted 69 years and, because there is no water to evaporate and operate as an air cooler, temperatures rise. Warmer temperatures do not create a drought but droughts create warmer temperatures.

The Murray-Darling catchment in SE Australia now contains three times as much water as was held naturally because of water management by dams and irrigation. This has drought-proofed the food bowl of Australia. Recent changes in management of the Murray-Darling catchment resulting from green pressure have destroyed much agriculture and sent many farmers broke. Well done greens. You've done your bit for the world and destroyed a few more productive enterprises.

7
HOW TO STOP THE RORTS

CLIMATE GREENS

The green movement is like a swimming pool. All the noise comes from the shallow end.

Environmentalism

We are all environmentalists and do not want to pollute the air, water, soils and life. We want the next generations to have better lives than ours but this can only be done if the economy is strong. We all want a clean planet and a high standard of living. We want the freedoms that have taken generations of struggle to achieve. However, we are not all greens. Some of us very strongly object to an organised, noisy, unelected, negative minority dominating a positive poorly-organised majority that just wants to get on with a peaceful life.

However, the environmentalism that we support is not the environmentalism of activists. One arm of the environmental movement tries to change policy in a democratic society by political thuggery, fraud and deliberate breaking of the law.

On 29 May 2017, Greenpeace activists illegally trespassed at the Port of Newcastle and painted "CommBank's coal kills" on a coal stockpile. Sixteen people were arrested. This is not the first time that Greenpeaces's actions have contravened the law and it calls into question their eligibility for the Deductible Gift Recipient status. Two employees of Greenpeace broke into the CSIRO experimental farm at Ginninderra (ACT) and destroyed a crop of genetically modified plants. Greenpeace celebrated the illegal activities and illegal behavior of their employees.

Donations to Greenpeace are tax deductible and such gifted funds should not be used to deliberately and consistently break the law. There should be no special tax benefits for organisations that plan to break the law.[492] An inquiry into the Register of Environmental Organisations in 2016 showed that there was unsafe protest activity and protests were designed to interfere with commercial operations. The NSW Police complained about the waste of police and emergency services time and Ports Australia submitted that the protest activity involved serious safety risks to employees and members of the community. The committee heard that environmental Deductible Gift Recipient organisations were soliciting tax deductible donations to pay off fines for their criminal activities.

It has become a trend that green activists with no scientific training have a very strong view about human-induced global warming. This is a political view. Any scientists who, on the basis of science, have doubts about human-induced global warming are threatened with violence[493,494,495] and compared to Holocaust deniers.[496] This is not a process of civilised debate in a democratic country. The environmental movement has allowed itself to be taken over by thugs.

On all scientific matters, there are opposing theories and

---

[492] Brendon Zhu, 2017: Greenpeace: criminal rabble or charity? https://spectator.com.au/2017/06/greenpeace-criminal-or-charity/
[493] www.climatedepot.com/2016/11/03/watch-schwarzenegger-again-threatens-climate-skeptics-i-wouild-like-to-strap-their-mouth-to-the-exhaust-pipe-of-a-truck-turn-on-the-engine
[494] www.breitbart.com/big-hollywood/2017/03/17/delingpole-climate-change-deniers-should-be-executed-gently-says-eric-idle/
[495] www.americanthinker.com/articles/2012/12/professor_calls_for_death_penalty_for_climate_change_deniers.html
[496] www.dailymail.co.uk/sciencetech/article-2566659?Are-global-warming-Nazi-People-label-sceptics-deniers-kill-MORE-people-Holocaust-claims-scientist.html

opinions. This is the case with the alleged human-induced global warming.[497] Since when have matters of science been solving by threatening to kill those who have an opposing view. Why haven't the various green political parties around the world distanced themselves from these fellow travellers. Why haven't people in the community with environmental sympathies stood up and objected to such totalitarian political thuggery? It is because of such stances and the silence of green political parties that many of us have concluded that the environmental movement has nothing to do with the environment and is the political arm of totalitarian thugs. As George Orwell stated:

> The further society drifts from Truth, the more it will hate those that speak it ...

The climate change scare has been a huge business opportunity. Green left environmental activists who seem to detest employment-producing businesses have handed a massive business opportunity on a platter to the climate change industry. With the long-established contradictory climate change policies between the two political parties, any business that takes a risk and invests in wind or solar power is taking a huge gamble. If the policy changes for the correct fiscal reasons, just wait for the screams and emotional language.

The environmental activist Jonathan Moylan is an anti-coal activist and led a group called Front Line Action in Coal. He published a fraudulent media release to 295 journalists and 98 media organisations that announced that the ANZ bank was divesting $1.26 billion from the Whitehaven Maules Creek coal project in NSW. In a matter of hours, $314 million was wiped off the value of Whitehaven Coal on the stock market. When journalists rang, Moylan claimed he was the genuine

---

[497] C.D. Idso, R.M. Carter and S.F. Singer, 2015: *The NIPCC report on the scientific consensus.* http://climatechangeconsidered.org

ANZ employee Toby Kent. He was convicted in the NSW Supreme Court, received a one year eight months suspended prison sentence on condition of two years good behaviour and a $1,000 bond. Moylan claimed he was contrite and remorseful yet his action involved a high degree of planning and premeditation.

Many investors, particularly retirees, lost a lot of money through no fault of their own. If anyone in the corporate world manipulated markets as Moylan did, they would be in gaol for a long stretch because the maximum penalty is 10 years, a fine of $765,000 or both.[498] The courts are very soft on green left activists who deliberately break the law, trespass, vandalise or engage in premeditated fraud. The Australian Securities and Investment Commission (ASIC) stated at the time that they were "satisfied with the sentence".

Coal is essential for the prosperity of Australia and provides jobs and cheap energy. Furthermore, the high levels of health, education and welfare in Australia have been financed by trade, principally in the resources sector. Public opinion is equivocal and has difficulty in opposing the ideological campaign to stop fossil fuels and replace them with renewables. The collective of the Greens and Greenpeace; various green left "think tanks"; the predominantly green left private- and public-funded media; climate change activists in schools, universities, government research bodies, unions and political parties; wealthy individuals and the Christian churches have used constant noise, social media and money to push for renewable energy to replace fossil fuels.

On 18 September 2013, Greenpeace activists tried to scale Russia's Priraziomnaya drilling platform in the Arctic Ocean. This act of piracy endangered others in the sovereign

---
[498] Section 1041E of the Corporations Act.

waters of the Russian Economic Zone again demonstrated that Greenpeace is an anarchist organisation with no concern for sovereignty. The Russian authorities had earlier banned Greenpeace from entering the Russian Economic Zone.

Greenpeace ignored the ban, obviously because their ideology is superior to Russian sovereignty. The Greenpeace activists and crew of *Arctic Sunrise* were lucky not to be incarcerated for life or treated energetically in true Russian style. The Russians were uncharacteristically patient and gentle.

For major new industrial developments in Western countries, there is a due process procedure that has been established by democratically elected governments. Greenpeace and any other group have the opportunity to argue their case against such developments. However, this relies on the law, logic and validated information. If the process is not to the liking of Greenpeace, there may be protracted legal challenges. However, this again relies on the law, logic and validated information so the green left environmental activists try to keep away from the courts.

If a decision is not to their liking, then anarchy becomes the order of the day with demonstrations, trespass, property damage and thuggery. Greenpeace has morphed from an organisation that hung banners from tall buildings to an international business organisation dependent upon donations from the public, unions, foreign foundations and tax donations in order to conduct anarchy. They have discovered that climate change activism is a massive global business.

The ecologist and co-founder of Greenpeace Dr Patrick Moore has left the organisation because they no longer are an environmental group, rather a far left anarchist collective who resort to violence. And what does Greenpeace do? They are now trying to erase him from history.

Just remind me again, where has this happened before?

Changing the world's politics

Some of us strongly object to this noisy minority who want to strip us of our freedoms in order to impose their unproven and impractical ideology. Some of us object strongly to a noisy minority wanting to direct an economy wherein their contribution is, at best, modest. I am yet to see greens that have skin in the economic game. Some of us do not want pressure groups funded from outside Australia putting Australians out of work.

In 2015, Christine Figuerres, executive secretary of the UN's Framework Convention on Climate Change showed us that green politics has nothing to do with science or the environment. It is about gaining control through a world anti-capitalist government and redistributing wealth, for and by noisy people who neither earn nor deserve it:

> This is the first time in the history of mankind that we set ourselves the task of intentionally, within a defined period of time, to change the economic development model that has been reigning for at least 150 years, since the Industrial Revolution.

The environmental movement started with noble sentiments and morphed into the green movement without noble motives. This was well enunciated at Bertrand's Russell's 11 December 1950 Nobel Prize[499] acceptance speech on "What desires are politically important":

> And among those occasions on which people fall below self-interest are most of the occasions on which they are convinced that they are acting from idealistic motives. Much that passes as idealism is disguised hatred or disguised love of power. When you see large

---
[499] Frenz, Horst 1969: *Nobel Lectures, Literature 1901-1967*. Elsevier.

masses of men swayed by what appear to be noble motives, it is as well to look below the surface and ask yourself what it is that makes these motives effective. It is partly because it is so easy to be taken in by a façade of nobility ...

I argue that generating electricity by so-called renewable energy sources such as wind, solar or biomass cannot provide enough energy for a modern society and creating renewable energy harms the environment, does not reduce carbon dioxide emissions and pushes your electricity bills through the roof. It is not renewable energy, it is ruinable energy. Power engineers can improve our electricity generation, supply and costs, not green ideology. Your high electricity costs are because greens ultimately want power over everything you do, say and think.

The greens have kindly taken it upon themselves to save us from global warming and consequently want emissions of carbon dioxide reduced or stopped. This is a proxy for stopping industries that give us light, heat, cooked food, refrigeration, air conditioning, transport, holidays and consumer items. This is the action of authoritarian socialists commonly known as watermelons (green on the outside, red inside). This is not an environmental view, it is a political view contrary to that of most of the community who are aspirational and just want to get on with their busy lives with the least amount of bother.

The concept of green as applied to politics started as a laudable movement to prevent the destruction of areas of natural beauty. This happened in Western countries and is happening in the developing world. It has morphed into an authoritarian, anti-progress, anti-democratic, anti-human and anti-environmental monster. Now the greens destroy the environment with our scenic ancient majestic landscapes speared with bird- and bat-chomping wind turbines that are a danger to health.

Nirvana

More idealistic naïve greens want to create a mythical idyllic lost world by destroying thousands of years of thought, innovation, science, engineering and culture. This mythical *Nirvana* never existed. The past was pretty bleak with disease, starvation, early death, few freedoms and poverty. During periods of global cooling, people died like flies. In periods of global warming, societies thrived. A romantic view of that mythical long lost wonderful past is ignorant of history.

My geological field work in the very remote parts of Africa, South America, Papua New Guinea, Turkey, Russia and Central America has exposed me to this mythical *Nirvana* where there is subsistence living with no electricity, potable water, communications, employment, schools, welfare, health care and transport. People live off a few animals, small gardens and bush tucker. These are wild places, revenge is a large part of the culture and there is no rule of law. The greens would call this sustainable living and ignore the grim realities of it.

The environment is totally destroyed in these parts of the world by illegal clear felling of timber and illegal gold mining. What were once jungles were reduced to areas where clear felling has initiated huge landslides. What were once pristine clear rivers are now toxic mud flows filled with mud and mercury. Illegal mining in mountainous areas has resulted in massive erosion. If an illegal miner is inundated by a fall of rock or an inrush of mud underground, in a creek or on a precipitous slope, there they stay. Illegal miners have no safety equipment, there is no mines rescue system, there are no safe mining conditions and in one place I worked there are still more than 300 bodies buried underground after a mining-induced landslide from illegal mining.[500]

---

[500] Nambija mine disaster, Zamora Canton, Ecuador, 9 May 1993.

I have seen the sustainable living lauded by the greens. It's not pretty and all I can do is to try to provide employment in areas where previously there was not one job. These folk want work and need electricity, water, health care, schools and real employment. These sustainable settlements have permanent deep mud, skinny mangy dogs and kids everywhere. In Western countries, we used to live like this and it was only a short time ago. It was cheap energy that changed our world. Now in the West, we live in an unprecedented time of peace and prosperity when food and energy are abundant.

We hear a lot about environmental sustainability. I don't really know what this means as it has not been defined. I've seen sustainability in the developing world and it's not good for the environment. However, I am all for sustainability. Fiscal sustainability. This can be easily defined as keeping out of the red. Although there may be an argument for government funds to be used as venture capital seed money for new technologies, there is a time when such monies should dry up. There were no government seed funds for 19th century inventions. Either they worked or you went broke.

The greatest period of innovation in the history of the world was driven by capitalism and government was on the sidelines not getting in the way with over regulation and micromanagement. The withdrawal of subsidies for wind and solar electricity generating schemes is a good example of how it should be done. There has been enough time to demonstrate that such alternative energy is sustainable and efficient and yet there is a perception that such schemes should have permanent subsidies at the expense of the consumer in case something happens well after we are dead.

If human-induced climate change is real and needs to be tackled, the best way is to use free enterprise and competition to drive down energy costs and to develop new technologies.

Tried and proven inefficient unreliable old technologies such as wind and solar for electricity generation have had their day. In the 19th century, inefficient technologies that did not live up to expectations as advertised fell by the wayside.

Henry Ford's motor cars didn't become best sellers because they were subsidised, because horses were heavily taxed or because of government decree. They were cheap, efficient, reliable, faster, did the job as advertised, were better than horses and didn't fill cities full of hazardous poo and smells thereby creating an environmental problem. Wind- and solar-generated electricity are not efficient, not reliable, not environmentally beneficial, don't do the job as advertised, bad for human health and fiscally unsustainable.

Over time, Henry Ford's mass-produced cars became better whereas neither wind- nor solar-generated electricity has become more efficient. It is time to stop taxpayers' money propping up inefficient unreliable ideological-driven electricity generation. It's called common sense and fiscal responsibility.

Making sure that each generation does not amass monstrous debt to burden the next generation is fiscal sustainability.

The bleak green world

Much of our wonderful pleasant modern world has been a consequence of human adaptability, ingenuity, risk taking and enterprise and the acceptance of wholesale unquestioned green utopianism would be disastrous for our species and our planet. We are witnessing the tip of the iceberg of green utopianism which touts "alternative" sources of electricity from wind and solar energy regardless of the effects on the environment, communities and the bottom line. "Alternative" energy has become an alternative to energy. Our quality of life depends on humans being able to think, adapt and deal with the threats to our world. The real threat to our world comes from fanatical political green ideology and not from environmental degradation.

We see ignorance hiding behind slogans. In the world of mindless marketing, we hear that if it's green, it must be good. In this book, I argue that this is not the case, despite the vacuous rent-a-celebrities attaching their names to causes they don't understand. The greens say they want to save everything and also the world. And so do I. From greens, green-initiated unemployment and green policies that create high electricity costs. The green political parties claim that they are parties for people, progress and protest. Actions show the exact opposite.

Greens prove over and over again that they underpin their ideology by a lack of knowledge, hypocrisy and dreamy impracticable solutions. Greens are committed to problems, not solutions. They object to any solution and angrily reject any evidence showing that a problem may not be as bad as they purport to fear but, in reality, for which they hope.

Greens erroneously claim that emission of traces of a trace gas by humans is the driver of climate change and accordingly want to change society by totalitarian means. This is underpinned by near-universal scientific fraud as we've seen from the IPCC, scientists, scientific journals and government organisations. After more than a century of collecting weather data, the original contemporary record has been changed so much that it is nigh on useless. Greens have no skin in the game yet want to control society without having faced the electorate.

The normal greens' answer to criticism is *ad hominem* attacks. These attacks are in the absence of cogent alternative arguments based on validated evidence to support their causes.

Rather than killing off industry with indulgent energy schemes and by pretending that Australia is actually globally relevant, there should be some serious work on education, health and welfare reform before Australia is demographically capable of doing anything meaningful on a global scale. Maybe greens could actually go to sea and spend their donations on cleaning plastic refuse from the oceans.

Jobs

There are a huge number of jobs at stake if the theory of global warming from human emissions of carbon dioxide is wrong. Think of all those bureaucratic jobs, research grants and scientific careers that would just evaporate. Think of all the banks, trading houses, investment groups and wind and solar energy companies that would miss out on skinning us alive.

The most prominent and radical green activist groups are partially funded by taxpayers. The same groups try to stop the economic development that employs people and provides funds for government by creating a taxation base. While State governments should be building dams and coal-fired power stations to keep the population fed and watered, green groups are using moneys given to them by State governments to object to the building of basic infrastructure.

The climate industry, academies, professional societies, NGOs, insurance companies, banks and alternative energy companies are not engaged in independent transparent research to add to knowledge or to satisfy intellectual curiosity. Almost every university has a climate institute obtaining funds from a diversity of government departments, NGOs and foreign foundations seeking to destroy our economy.

Every single one of my few noisy scientific critics is either a direct recipient of huge amounts of money or is in an institution that receives masses of cash because they support the government ideology. Just look at the amount of taxpayer money provided via Federal government funds (Australian Research Council, universities, CSIRO) and by government departments, all of which waste money on climate change research, jaunts, conferences and publicity. Furthermore, these climate "scientists" are not in the productive sector of the economy and yet their actions have a direct impact on the average productive worker who pays them.

Not one great scientific discovery has ever been made by consensus or a committee. They have all been made by researchers going out on a limb and challenging the popular contemporary knowledge. The grant-awarding political policies guarantee that there will be no great discoveries in Australia about climate. There is only grant money support for the current political dogma and no genuine interest in the environment, energy efficiency or your electricity bill. This does not appear to be a wise way to create policy or prepare for the future.

Almost every government department has a climate change group. Why? How do these people add value? Even local government has climate change officers. Why? Shouldn't they concentrate on parks, gardens, garbage and pot holes. You pay for this as well as your horrendously high electricity charges. And still, it has not been shown that human emissions of carbon dioxide drive global warming. Independent thought, knowledge and advice have been sacrificed to the god of self-interest for money.

If Australia had cheap electricity, a nuclear industry and more heavy industry requiring cheap electricity, there would be long-term productive jobs created that have a multiplier of about four. Rather than have greens kill off jobs, cheap power would create real and productive jobs.

PARIS

Lip flapping

Climate conferences are held at your cost in pleasant or exotic parts of the world. I am sure that it feels good for you to know that your taxes funded bureaucrats to go to Rio, Kyoto, Copenhagen, Cancún, Durban, Marrakesh and Paris. These talk fests were really about passing control of your energy to unelected opaque bodies, raising your electricity costs and threatening your employment.

Tens of thousands of participants jetted in to these conferences, stayed in expensive hotels, preached doom and gloom and unsuccessfully tried to organise economic suicide pacts. It was so comforting to have folk sacrifice themselves to go to an expensive resort like Cancún and try to work out ways to take money from us. According to the UK Taxpayers' Association, Copenhagen cost as much as the GDP of a small African nation. The Australian taxpayer funded over 100 bureaucrats to attend the Copenhagen group-think. And what was decided? Nothing. What did we learn? Nothing. And who paid? You. It is an absolute disgrace that Australia wasted so much money on jaunts to work out ways to bankrupt Australia. Bureaucrats have advised politicians of the necessity of such jaunts and are living for the moment. For themselves.

This is not a new phenomenon. In *Extraordinary Popular Delusions and the Madness of Crowds* (by Charles MacKay, 1841), we read:

> Men, it has been well said, think in herds; it will be seen that they go mad in herds, while they only recover their senses slowly, and one by one.

I suspect that we are in one of those periods when people are recovering their senses one by one after a period of herd madness. Who needs to fund the pyramid scheme of extortion created by the UN Paris Agreement and the G20?

Why is Australia participating in nation-destroying rather than building a nation based on cheap reliable energy?

Twiddling the dials

The COP21 Paris Agreement on Climate Change (Paris Agreement) aimed to limit global warming to below 2°C above pre-industrial levels and to pursue efforts to limit the temperature increase to 1.5°C above pre-industrial levels. Just how a talkfest will change the way planet Earth works is beyond me.

The Paris Agreement relies on the UN's IPCC as its authority. However, can the IPCC be trusted as an authority? Climategate, leaders in the climate industry and many government institutions show that the data underpinning the IPCC's reports are fraudulent. The IPCC claims that human emissions will raise the Earth's temperature to dangerous levels unless constrained. The IPCC warming scenarios rely on computer model projections that have been shown to be wrong. The IPCC claims that since 1980, there would have been a warming of 0.7° to 1.6°C yet satellite measurements since 1980 show a warming of 0.3°C.

In 2016, the Paris Agreement committed the global community to stop global temperature rise. The assumption underpinning the Agreement is that human emissions of carbon dioxide increase the planet's warming and that all other drivers of climate play no part.

How does the climate elite think that by sitting around together sipping Bollinger in Paris and negotiating an Agreement gives them the power to smugly twiddle the dials and keep global temperature rise to less than 2°C? The arrogance and ignorance are breathtaking. And we pay.

Costs

I know costs is a pornographic word for greens but it is pretty important for running a business or a nation.

The Paris Agreement results in the world committing to spend $US 125 billion on wind and solar subsidies this year alone. Where does the money come from? The Paris Agreement was an attempt for poor countries to squeeze money out of more developed countries. With money for jam, it's no wonder 196 countries signed. Wind will provide 0.5% and solar 0.1% of global energy needs. Maybe this money could be better spent providing potable water and coal-fired reticulated electricity for the Third World. Maybe a few tens of thousands of medical centres and schools here and there in the Third World would be more useful for the planet.

Despite four decades of detailed observation of climate, the lack of warming over the last two decades, the lack of correlation between human emissions of carbon dioxide and climate and the crippling costs of unreliable renewable energy, the bureaucrats at Paris are now telling us that renewable energy is cheap. They are totally out of contact with reality and those that pay them. Why does renewable energy need to be subsidised and why have electricity costs risen so much at the same time more and more renewable energy is installed?

Statistician Bjorn Lomborg has shown that the Paris carbon dioxide emission promises would cost $US 100 trillion over the next 25 years and, using the current climate science models, this would only make a 0.05°C temperature difference to the Earth's 2100 temperature. In Australia, surely we should be more worried about paying more than $1 billion a month servicing debt and about unemployment, poverty and our electricity bills rather than a modelled speculative 0.05°C temperature change.

Others claim that the $US 100 trillion would only make a 0.02°C temperature change. Whichever, a temperature change of 0.02 to 0.05°C is insignificant and is what we experience by moving a metre. Far greater changes in temperature are experienced in everyday life and this does not seem to be catastrophic. At $US 100 trillion, this is a steal. The carpetbaggers are stealing from us. The Paris Agreement maybe would hold fast if all parties actually adhered to their voluntary pledges over the next 83 years. Fat chance of that.

Just to make the whole matter even more meaningless, the Paris Agreement is a non-binding accord dependent upon voluntary commitments unlike the Kyoto Protocol that set targets with legal force.

Paris questions

Did any taxpayer-funded delegate on the Paris jaunt have nagging doubts or ask a few key questions such as:

"Why do we need to subsidise renewable energy if it is so cheap?"

or

"Why are we doing this when we have enough evidence from Germany, Denmark and Australia to show that renewable energy creates poverty?"

or

"Why has there been no temperature increase for two decades yet human emissions of carbon dioxide have increased greatly?"

or

"Just run that past me again, can you please show me that human emissions of carbon dioxide drive global warming?"

or

"Why should I give away my nation's sovereignty, money and economic future to a faceless unelected UN group?"

All the major developing nations, led by China and India, paid lip service to the conference's intentions, showing how they would be investing in "renewables" so long as they were generously subsidised by the developed nations out of a Green Climate Fund worth $US 100 billion a year. Imagine the number of sticky fingers this money will pass through.

They then explained how, to keep their economies growing, they planned to build huge numbers of new fossil fuel power stations, which would lead to a massive increase in their carbon dioxide emissions. Go figure.

Paris Agreement escapees

China, the world's largest carbon dioxide emitter, is planning to double its yearly emissions, by an extra 10.9 billion tonnes. India, the third largest emitter, will treble its emissions, adding 4.9 billion tonnes. All the other major "developing" nations, plus Japan and Russia, are equally planning to build more coal-fired power stations. Thirteen of the countries which signed the G20 communique intend to contribute to what the Intended Nationally Determined Contribution (INDC) shows will be a 46% decrease in global emissions within 13 years. Do we really know what China does? Thousands of emitters of sulphur and nitrogen oxides, volatile chemicals and particulates in northern China submit fake emissions data and resist checks by authorities.[501]

The only G20 countries left committed to carbon dioxide

---

[501] *South China Morning Post*, 31 March 2017.

reductions by 1.7 billion tons are now those in the EU, Canada and Australia, responsible for just 11.3% of global emissions between them. Most of the remaining 88.7% is emitted by countries which plan to increase them. Is it surprising that President Trump wanted no part in such a grotesque display of international hypocrisy?

Without the participation of the big players such as China and India, global action on climate change is dead. All the big players at the 2017 G20 meeting in Hamburg will be increasing carbon dioxide emissions. The final communique took "note" of Mr Trump's decision to immediately cease efforts to enact former President Obama's pledge to curb US greenhouse gas emissions 26 to 28% below the 2005 levels by 2025. *The New York Times* and many other media networks used lead stories to criticise President Trump.[502] However, readers were not told what other G20 nations agreed to do as part of the Paris Agreement and how many large emitters such as China, India and Indonesia have continued business as usual.

China, through its government-owned companies, is planning to build over the next decade twice as many coal-fired electricity generators than the US has today. The INDC allows China to increase carbon dioxide emissions unabated until 2030 and then to level its emissions. By that time everyone in China will have cars and electricity. By that time its emissions will be triple those of the US. By that time no country would have been able to fulfil its Paris obligations and will be in a great huff and puff about another concocted issue. This happened with the Kyoto Protocol. Signatories failed to reduce emissions and non-signatories such as the USA performed far better.

India's INDC admits that it intends to triple its coal-fired generating capacity before 2030 and to expand other forms

---

[502] "World leaders move forward on climate change, without U.S.", *New York Times*, 9 July 2017.

of energy generation. Russia's INDC claims that it will reduce emissions 25 to 20% below the 1990 level before 2030. After the collapse of the Soviet Union in 1991, inefficient dirty industries were shut down and by 2000 emissions were 40% of the 1990 level. In effect, Russia's emissions reduction is an increase from today's level of emissions. Germany is supposed to reduce emissions by 40% below the 1990 levels by 2030. Germany has won the triple. Its renewables are about 30% of total electricity generated, its emissions have risen over the last few years and it is unable to reduce emissions to pre-1990 levels. Germany is now reducing subsidies for renewable energy.[503]

Most of Indonesia's emissions (63%) come from burning down forests with fossil fuel burning contributing to 19% of emissions. Their INDC does not bind them to 1990 or 2030, they promised to burn down less rainforest and will have business as usual and burn more fossil fuels.[504]

The G20 was an international farcical attempt to bully the US into crippling its economy while all other major emitters could do as they please. President Trump has shown he is impervious to bullying and *The New York Times* is so blinded by their hatred of Trump that they missed the real story.

Mrs Merkel criticised President Trump for failing to endorse the Paris Agreement. However, the Deutsche Presse-Agentur, a press agency, reports that more than 330,000 German households have been cut off from the electricity grid because of an inability to pay the massively prices. They are due to increasing climate subsidies for renewables. So is it any wonder Mr Trump has doubts? As a result of Mrs Merkel's climate policies, energy poverty is engulfing Germany. Maybe Mrs Merkel should sort out her own mess and not lecture others.

---

[503] https://www.thegwpf.com/german-wind-energy-market-threatening-to-implode/
[504] Francis Menton, "Looks like global action on climate change is dead", *Manhattan Contrarian*, 10 July 2017.

Trump's decision gives every other country a chance to think again, especially as previous agreements did not stop a rise in carbon dioxide emissions. The Paris Agreement will be the world's most expensive global agreement. Ever. Cutting emissions without affordable effective replacement means more expensive replacements for fossil fuel means more expensive power and less economic growth. The price tag of all of the Agreement's promises would reach $US 1-2 trillion every year from 2030. Without US involvement, the rest of the world must cough up between $US 800 billion and $US 1.6 trillion annually. The Agreement also hinges on the delivery of $US 100 billion "climate aid" to developing nations from 2030. These huge costs have threatened the Agreement since signing.

The Agreement's actual carbon cutting promises, which are not legally binding, only go to 2030. Green energy will be nowhere near ready to take over from fossil fuel by then. Green energy is so inefficient that its deployment is almost entirely dependent upon subsidies. Spain was paying almost 1% of its GDP on subsidies until it hit the wall.[505] This was more than it spent on education. When it reduced subsidies, new wind energy production entirely collapsed. If the Paris Agreement stays in place, after spending $US 3 trillion in direct subsidies, the IEA expects that wind and solar will provide 1.9% and 1% respectively of global energy. Why bother?

In the US after the Global Financial Crisis, people are not concerned about the relentless doom and gloom propaganda about climate change, carbon dioxide emissions and over-exaggerated threats about sea level rise. They are concerned about the environmental impacts on their jobs and a loss of their lifestyle. People remember SARS that was going to kill all of us. They remember swine flu, bird flu and ebola that were

---

[505] https://www.economist.com/.../21582018-sustainable-energy-meets-unsustainable-cost...

going to kill off mankind. They remember Y2K which was a fizzer. There was no great catastrophe and people are now tired of being told what the next deathly threat will be and how they have ruined the planet.

China installed 53 GW of coal-fired generating capacity in 2015. This is more than the whole of Australia's generating capacity. India installed 18 GW of new coal-fired generators after it submitted its Paris climate change target and is building another 50 GW. India has closed some high ash, high sulphur coal mines and inland government coal-fired power stations and is importing more and more coal for its privately-owned coastal power stations. Whether Australia builds up its coal-fired generating capacity, increases renewable generation or becomes "carbon neutral" makes no difference whatsoever globally. India, China and others are ignoring the concepts of the Paris Agreement. Australia should join them, as has President Trump.

This massive increase in coal-fired power stations is because people are improving their standard of living and hundreds of millions of people are moving into cities. Some 1.2 billion people on this planet do not have access to reticulated electricity and 2.7 billion people burn oils, dung, kindling and crop waste to cook meals and heat their homes. Meanwhile, the toxic fumes from low temperature burning of dung, twigs, leaves and wood in homes kills people, especially women and children.

The Paris Agreement did not seem to be concerned about killing people and energy poverty.

RENEWABLE ENERGY TARGETS

Surely no one can object to clean, green, renewable energy? I don't. It's called nuclear. Wind, solar, solar thermal and biomass are unreliable, consumer-subsidised, destructive to the environment and, if the intention were to reduce carbon dioxide emissions, then these energy-generating systems fail. They are neither clean nor green.

We just can't afford to power an industrial economy with renewables. Annual renewable subsidies are now $150 billon *per annum* for production of 3% of the world's electricity. One day there will no money left in the pockets of countries, just debt resulting in a loss of sovereignity.

Advocates of renewables need to be mindful of the fact that 85% of global energy is from fossil fuels and, together with nuclear and hydro, 97% of global energy is from these conventional sources.[506] Why spend billions on the other 3% of energy, called renewable but in fact is uneconomic ruinable energy. Coal, gas and nuclear deliver electricity 24/7 at over 90% of their capacity for at least 50 years. Typical wind and solar plants operate at about 25% of their capacity and, if the wind does not blow or it is dark, they provide no energy. Renewable energy spruikers talk about capacity and not the actual generation of energy which is still intermittent and unreliable. Just because a wind project can provide power to 30,000 homes doesn't mean it does. We are not told that such figures are misleading because the wind does not blow all day and all night. Electricity is not generated by interconnectors or batteries.

Why not reduce our emissions by a small amount? This is a political solution that would keep greens happy and provide only a little pain for the consumer.

A little pain, 5% emissions reduction

If Australia reduces "carbon emissions" by 5% by 2020 or has a tax on carbon dioxide emissions, will it have an effect on planet

---
[506] *BP Statistical Review of World Energy*, June 2016.

Earth? This can be calculated. Humans emit 3% of annual total carbon dioxide emissions, the rest are natural and mainly result from ocean degassing. Australia emits 1.5% of the total global human emissions. If the atmospheric measurements of carbon dioxide are projected from 2011 (389 parts per million) to 2020 (412 parts per million), then a 5% emissions reduction by Australia will lower the atmospheric figure to 411.987 parts per million. This would lower modelled global air temperature about 0.0007°C by 2050. Why should we bother?

Even if Australia stopped emitting carbon dioxide today, the decrease in global temperature would be 0.0145°C by 2050. Temperature is constantly changing by far greater amounts than this. In fact, if a Tasmanian moved from Hobart to Darwin, the average temperature rise the Tasmanian would experience would be 18°C. If a Finn moved from Helsinki to Singapore, the average temperature rise the Finn would enjoy would be 22°C. Variation between summer and winter temperatures in the outback is more than 30°C and in high latitude areas more than 50°C.

Let's just ignore the annoying little fact that there has been no global warming since 1998 yet carbon dioxide emissions have been slightly increasing. If every person on Earth was to try to stop the planet warming by the modelled predicted 2°C temperature rise, it would cost every man, woman and child on the planet $US 60,000. This is 60% of global GDP and 22 times the maximum estimates for doing nothing about possible global warming. Are you prepared to pay $US 60,000 for less than a quarter of a degree temperature change? I don't think so. Especially as there has been no warming for two decades.

Now you can see why the climate alarmists, governments, industry and banks don't want their policies questioned because it could stop the flow of money from your pocket into theirs. The same people who brought us the Global Financial Crisis are salivating at the prospect of an even bigger and better scam. It is the young people, many of whom have green sympathies, who will pay. Government action on human-induced global warming has nothing to do with climate or the environment. It is a method to take money out of your pockets, to keep the big banks happy and to take away some of our basic freedoms. It is a game of power (and I don't mean electricity).

If Australia had a 5% decrease of carbon dioxide emissions, the increase in Chinese emissions alone is expected to be over 100 times as large as Australia's reduction hence, whatever Australia does, it will have no effect on the planet. Australia, like the USA, has coal-fired power stations because we have hundreds of years of coal reserves.

We have retrofitted our older coal-fired power stations. They are now cleaner but maybe not as clean as a new HELE generator. It is crazy to think that what is sinful in Australia is good in China. The politics of the Green Party clearly are not pragmatic as they do not involve global solutions to a perceived global problem. But then again, history tells us that ideology and economics have never been good bedfellows.

Scientifically illiterate politicians face huge pressures to do something to save the planet. Instead of educating the public about the realities of climate past and present and seeking a diversity of opinions, politicians capitulated under pressure. We should be mindful that in 1910, Argentina was *per capita* the wealthiest nation on Earth. To live in a lucky country is just not good enough. Argentina went out backwards through bad policy. Australia is a lucky country but bad decisions and poor policy could very quickly push Australia into poverty, as happened with Argentina.

Fund managers have invested in wind energy to make money, not to save the environment. Their due diligence would have shown that wind industrial complexes are a costly unreliable subsidised high-risk method of ruining the environment and that a Renewable Energy Target was unobtainable. Rather than plead to the government for even more money, fund managers should be sacked. It is not the role of government to bail out high risk investors who follow fads, fashions and frauds and who spend more money on advertisements in newspapers than they do on due diligence.

Link between wind and solar power and your electricity bill

There is a direct link between the installation of wind and solar generators and the costs of energy.

We had been told too many times that renewable energy was cheap and that it would lower our electricity bills. Not so. The cost of electricity has risen so much that intermittent expensive renewable energy is hurting you.

No matter how many weasel words politicians and renewable energy suppliers use, renewable energy worldwide is the main reason for high electricity prices. It is no different in Australia. The escalation of renewable energy wind and solar generating plants has resulted in a huge increase in Australian electricity costs, especially South Australia.

The Finkel Report

The Federal government commissioned a report by the chief scientist, Alan Finkel. The aim was to give the government some breathing time and to take the heat off while electricity prices continued to sky rocket. The report is done and dusted and governments in Australia still don't know what to do. And your electricity bill keeps increasing.

It's claimed that the Chief Scientist Finkel's report to COAG aims to address the "trilemma" of achieving lower prices, greater security and a 28 per cent reduction in emissions by 2030. Wrong. The report is about meeting the emissions reduction aspiration (which it converts into a commitment) at the lowest cost without major interruptions to supply. It's not about affordable, reliable power. It's about maintaining the current mantra of "climate change". The Finkel report suggested a 42% Renewable Energy Target reboxed and rebadged as a Clean Energy Target (CET). South Australia cannot survive on a similar target and if the rest of Australia even attempted to get anywhere near this target, there would be mass social disruption. The end result will be that your electricity prices will continue to rise because the heart of the problem had not been addressed.

The incentives provided to produce electricity with lower carbon dioxide emissions have already led to a disaster because of an over reliance on wind and solar energy. It will get worse because of the closure of coal fired power stations in South Australia and Victoria. It will get worse with the additional costs of carbon capture and storage. And it will get even worse as there are back door attempts such as the Finkel report to block coal and even gas with what is essentially a tax on hydrocarbon-sourced energy. And all this because it is claimed that human emissions of carbon dioxide drive global warming. The most sensible solution is to release carbon dioxide into the atmosphere where it is used as plant fertiliser.

In a 2014 review of the Renewable Energy Target (RET), Jacobs (formerly SKM) and ACIL Allen predicted where Australia's energy market would be in 2017. Despite the bleeding obvious, they claimed that the electricity price would fall to around $50 per MWh. It didn't. It tripled. The average wholesale cost of electricity has doubled in the last 12 months and it will continue to rise because of the must-take, RET and other consumer-damaging policies and it will rise even more with the closure of coal-fired power stations such as Hazelwood. NSW might gleefully look at South Australia's own goal. When Liddell (2022) and Vales Point (2028) close, 3,200 MW will disappear from NSW and they will have critical power shortages. A year ago, the Australian Energy Market Operator (AEMO) was predicting a fall in electricity prices. This prediction was based on maintaining base-load coal-fired electricity, increasing renewable energy output and abundant cheap gas.

But the price of gas went through the roof. Three years ago, the same Australian Energy Market Operator predicted that scrapping the carbon tax, falling electricity demand and increased capacity would cause retail electricity prices to fall. They didn't. They doubled because the AEMO did not anticipate the rise in the gas price and the bans on coal seam and conventional onshore gas exploration. At the time contracts were written for the export of gas, the international price was far higher than the domestic price. No one objected. Now that the domestic price is far higher than the international price, agitators are wanting gas suppliers to break contracts and supply cheaper gas locally. Such an action would internationally telegraph that Australia is untrustworthy and that a contract is not worth the paper it's printed on. Although tearing up contracts might solve a short-term gas problem, it presents a long-term risk problem which translates as less future investment and fewer jobs.

The Finkel report did not address the core of the problem:

(a) Australia cannot survive on sea breezes and sunbeams.

(b) Subsidies and forced purchasing of renewable energy has destabilised the market and the grid resulting in the closure of cheap, efficient, reliable coal-generating plants.

(c) Unless Australia has large scale coal, nuclear, gas or hydro, electricity will become even more unreliable and expensive.

The Finkel report has stimulated a special kind of insanity. Proposed Federal renewable energy subsidies paid by the taxpayer will be \$60 billion by 2030.[507] This is the cost of going green. The end result will be more unreliable and increasingly expensive electricity and electricity retailers will be forced to buy 33 TWh of renewable electricity. The \$60 billion would be enough to build half a dozen nuclear reactors which could provide Australia with an excess of cheap energy for 50 years. The South Australian royal commission into nuclear power estimated that a large-scale reactor would cost \$9.3 billion. Australia is the only G20 country without nuclear power and the perfect place for future reactors is South Australia. However, the Finkel review totally ignored nuclear power as an option and pushed harder for more and more renewable energy. Victoria is looking at 25% renewables by 2025, South Australia 50% and the ACT 100%.

Australia is going for broke on renewables while other countries have a mixture of energy sources and are building new coal-fired power stations. At present, planned or under construction coal-fired power stations are China 583, India 217, Indonesia 145, Turkey 71, Vietnam 84, Japan 43 and Australia 0. Most of these power stations will use Australian coal.

The best place for the Finkel report is in some dusty cupboard in Canberra with all the other costly reports and Royal Commissions that were undertaken for political purposes.

---

[507] Adam Creighton, "The cost of going green: Taxpayers hit with \$60 billion power bill", *The Australian*, 1 September 2017.

Subsidies and must-take

The combination of excessive regulation and subsidies is creating the perfect storm over our power grid. This has already led to skyrocketing electricity prices and blackouts. Our way of life is threatened. The war on fossil fuels is denying the poorest people and nations of the world a chance of better paying jobs, cleaner energy and more abundant food.

The government-mandated requirement that renewable energy generation has priority must-take status whenever offered to the electricity grid, thereby exempting it from free energy market competition. This has created extensive energy market distortions which are unacknowledged and concealed from the public by renewable energy activists, owners and their biased mainstream media supporters.

These government-mandated market price distortions cannot overcome the predictable load variation, cannot provide the spinning reserves required for system regulating and load changes and do they provide for synchronous generators needed to stabilise electric grid system frequency. If there is a blackout, renewable energy has no black start capabilities. Renewables cannot provide grid reliability and stability that manage both load and frequency variations. Because of the must-take provisions of renewable energy, the number of operating hours available for dispatchable power plants is significantly reduced which drives up the unit costs of production.

Open market pricing of non-dispatchable power sources would be deeply discounted, not subsidised, if any significant value were to be placed on the reliability of the network. Currently, prioritising non-dispatchable power sources leads to situations like South Australia. The defenders of renewable energy, wind and solar, will claim it does work anyway, and dismiss all evidence to the contrary.

The government has forced the electricity market to take subsidised wind and solar power when it is generated. However, large coal-fired power stations take about 24 hours to turn on and so they cannot just shut down and re-open depending upon the weather and sunlight. Some gas and hydroelectricity units can stop and start quickly. This means that coal-fired power stations must operate 24/7 but can only sell some of the electricity they generate.

Melbourne, Adelaide and Sydney normally have 10 to 15 hot days each year when the wind does not blow. If dependent upon wind, then they are almost certain to suffer blackouts. Each and every summer. There is now not enough power for peak summer demand. Can ageing generators and power distribution networks operate without breakdown? No investor will maintain or build a new coal-fired power station that produces electricity 24/7 but, as soon as there is just one electron from a subsidised wind or solar generator, the coal fired power stations cannot sell their product.

The playing field is not level and this was created deliberately by governments. The grim picture facing Australia is that electricity costs will transform Australia. You, the average consumer are the front line, you are feeling it first with your high electricity costs and Australia will suffer. Mind you, no politician will ever say that they were wrong. When there are deaths of children in hospitals because of power failures, then there will be action. If there is increased mortality of old people, nothing will happen because old people don't count.

Politicians and public servants don't really have any idea of what to do. This is typified by Victoria where the government announced it was erecting some 5,500 MW of renewable power generation (mostly wind) to offset the closure of Hazelwood. Green groups gave them great praise and the overall community felt good. The transmission network is structured around transmitting electricity from the Latrobe Valley to Melbourne whereas wind generators are in remote areas with no transmission infrastructure. Without Hazelwood, Victoria will be struggling on a hot summer day when the wind is not blowing.

South Australia and now Victoria have blundered into generating large amounts of wind power without reliable backup and with a network that is not equipped for it. The backup and transmission economics alter the economics of wind and solar. The South Australian blackouts were made worse by failure of poorly maintained equipment. Victoria and NSW have not learned and are going down the same track as South Australia and Queensland is planning major investments in wind.

Gas is required for back up for renewables for when the wind does not blow and for night time when solar generation is not possible. Victoria has hundreds of years of gas supply waiting to be tapped onshore but Victoria has kicked a wonderful own goal by banning onshore gas exploration yet allowing offshore gas exploration. Most offshore gas from Victoria goes some 2,000 km by pipeline to the Gladstone LNG export facility in Queensland and by-passes Victoria.

South Australia has closed coal-power stations and is dependent upon wind. They have realised that for much of the time South Australia will not have enough electricity, especially as they can no longer draw power through interconnectors from the Hazelwood plant in Victoria. South Australia has decided to install diesel backup. The cost of running a diesel generator, compared to an efficient coal-fired power plant on a dollar per MW hour basis, is eye watering. A modern diesel plant will at its near optimal 60-70% loading generate 3 KWh per litre. If diesel is at $1.30 per litre and 333 litres is needed for 1 MWh, the cost is $433/MWh. Coal fired power generated day-in-day out provides electricity at $50/MWh to deliver.

South Australia has also realised that wind will not keep the lights on. South Australia's power problem is those who wield it. A fully functioning electricity system with 70 to 90% of energy from wind cannot exist because wind generation rarely gets much above 30%. South Australia has taken a step into the darkness and will install batteries to hold electricity generated from wind that is not used. However, wind capacity needs to be quadrupled. What do we need to pay for four times excess capacity in wind generation, tens of millions of Teslas worth of batteries and massive new transmission capacity? Oh dear, so many unanswered questions for which its government either does not know the answers or won't provide them if it does. And every dopey policy slugs the wounded taxpayer even more.

The tooth fairy subsidies mean that some people with solar panels on their roof think that they are getting cheap electricity and they are helping the environment when in fact someone else is paying part of their bill. That someone else is normally someone who could not afford the capital cost of roof solar panels hence the poor are subsidising the rich.

Wind power generation is massively subsidised in Australia. Wind generators harvest subsidies and not the wind. Without subsidies out of your pocket, there would be no wind turbines. The annual report of a small rural cooperative[508] states that 62 % of their annual revenue came from Renewable Energy Certificate sales priced at $64.54 per MW and these are forcibly purchased by energy retailers. The remainder (38%) from the actual sale of wind-generated electricity (at $42.73/MW), probably to those same retailers. The net profit of this co-op was a tidy $213,961 (or 18% of gross revenue), for generating

---

[508] www.hepburnwind.com.au/wp-content/uploads/2014/06/FY16_Hepburn-Wind-Annual-Report-.pdf

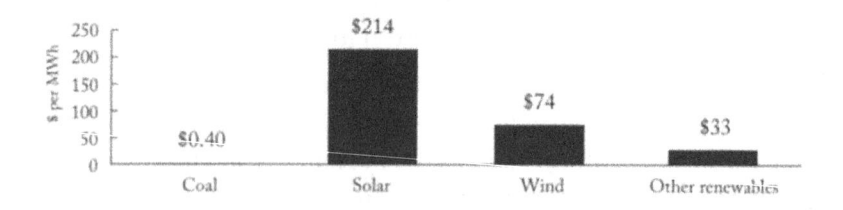

Figure 26: Subsidies per unit of electricity 2015-2016 (from BAEconomics, electricity subsidies in Australia, March 2017)

9872 MWh of non-dispatchable electricity over the year at a capacity factor of 27.5% of nameplate. No wonder there are so many snouts in this trough!

There is a very strong correlation between governments putting their nose into the energy market and electricity costs. Since the mid 1950s in Australia, each State was an isolated electricity entity. They generated power for themselves, tried to attract business by having cheap power and there were no interconnectors between the States. For example, South Australia established an industrial base under the leadership of Sir Thomas Playford and had a car and whitegoods manufacturing base, attracted ten pound Poms as workers and built new towns such as Elizabeth to accommodate them. The State generated its own reliable and cheap electricity and this attracted new businesses to South Australia. The end result of competition, a lack of subsidies and increased efficiency was that electricity prices fell.

The end result of competition between States, a lack of subsidies and increased efficiency was that electricity prices fell. Most coal mines, generators and grids were State owned. Once States sold their mines, poles and wires, generators and marketing agencies and a National Electricity Market was established, it all went pear-shaped. After nearly 50 years of falling electricity prices, the cost of electricity rose. This was exacerbated by wind and solar power generation, mandated must-take, subsidies, fragmentation between coal mines, generators and grids. Once a government interferes in a market, there is a disaster looming. It's a sad story. Since the advent of renewable energy, South Australia has now the highest household electricity prices in the world[509] and relies on electricity from Victoria to keep the lights on 24/7. Businesses are closing or moving interstate or abroad.

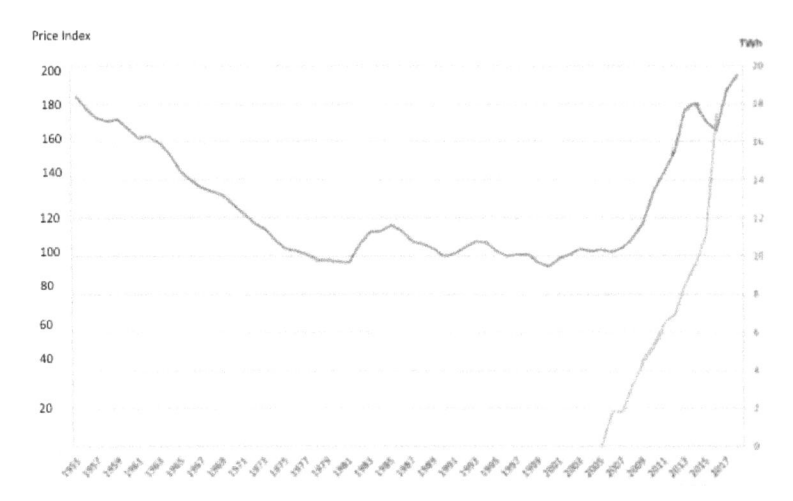

Figure 27: Indexed real consumer electricity prices 1955-2018 (1990=100)[510] *(heavy line) and increasing contribution of wind and solar energy (light line).*

Wind power subsidies occur in most Western countries. Investors come to wind and solar power generators like bees to honey because these industries are subsidised, the subsidy contractual arrangements are long term and governments have been compliant and commercially backward.

---
[509] Michael Owen and Meredith Booth, "State leads the world – for power prices", *The Australian*, 28 June 2017.
[510] Frank Brady, 1996: *Electricity in Australia*, Australian National Committee of CIGRE, CPI.

The world's renewable energy subsidies are increasing and the IEA estimates they are now $US 150 billion. In Australia, they were $US 3 billion for 2015-2016.[511] Up until very recently, Germany and Denmark had the most expensive electricity in the world. South Australia now leads the pack.

Germany's Renewable Energy Levy started in 2000 at €0.02c/kWh, the then environment minister, a Green MP, said that the cost to consumers would be no more than a "scoop of ice cream". The Renewable Energy Levy has now risen to €0.068/kWh and is still rising, despite renewables producing only 30% of the energy in Germany. Funds from the Levy are used to construct new renewables energy plants, only making matters worse.

The high proportion of renewables in the system has forced Germans to pay over €1 billion to stabilise major powerlines and the total cost for transforming to renewable energy would reach €520 billion by 2025 and will keep rising.[512] When the wind does not blow or the Sun does not shine, Germany imports nuclear-, coal-, gas- and hydro-sourced power from its eight neighbours. The whole exercise is stupidly pointless.

If you are starting to get the impression that you are being defrauded, you are right. When will a government put out a remotely honest effort to calculate the real cost of the mostly wind-and-solar-generation system that they are busy trying to force onto people.

What happens if I am totally wrong about the influence of carbon dioxide on warming and the fraud of the climate industry?

How much are you prepared to pay for electricity? Less? Double your current bill?

---

[511] Minerals Council.
[512] Dusseldorf's Institute for Competition Economics, 2016.

There is a limit to what people are prepared to pay for vanity signalling and moral posturing and, unless you are a wealthy green, the limit has already been exceeded. There's a price for no power at all or unreliable power. Freezing or boiling in the dark.

Fossil fuel subsidies

According to the International Energy Agency (IEA), world fossil fuel subsidies dropped from costing $US 500 billion a year in 2014 to $US 325 billion a year in 2015.

The US, Europe and Australia are not the culprits. Russia, Iran, Saudi Arabia and other Gulf States used taxpayer money to artificially reduce the cost of production or the cost at the pump for consumers. My experience in oil-producing countries such as Saudi Arabia and Ecuador is that diesel can be bought at the pump for 8c/litre and 25c/litre respectively.

Nuclear power subsidies

Subsidies are not only constrained to wind and solar power generation.

In New York State, the New York Public Service Commission voted to impose a new Clean Energy Standard (CES) for the entire State. It requires that 50% of New York's energy must come from "carbon-neutral" sources by the year 2030.[513] The plan openly subsidises financially struggling nuclear power plants in upstate New York through Zero Emission Credits (ZECs). Other power utilities would be compelled to purchase ZECs from a government bureaucracy, which the government first obtains from the company operating the struggling nuclear power plants. This amounts to a wealth distribution

---

[513] Timothy Lee, "Cuomo's energy boondoggle triggers bipartisan rejection", *The Daily Caller*.

from financially healthy power plants to financially struggling plants to satisfy carbon-free green energy regulations. This is little different from the wealth transfer from efficient coal fired generators in Australia to inefficient wind and solar generators.

The costs of the ZECs, not surprisingly, will be paid by New York consumers, households and businesses. The scheme guarantees $US1 billion to the struggling plants in the first two years alone with an estimated cost of $US8 billion over the life of the CES until 2030. Moody's warned investors that the costs for the duration of the program are $US127.48/MWh of production. Despite the subsidies for the nuclear generators it estimates a 45% price increase. The subsidies will benefit a single company (Exelon) which controls the struggling nuclear plants that qualify for the subsidy (Ginna, Nine Mile Point).

At best this is crony capitalism, at worst it is a brown paper bag. This is essentially a tax to keep ageing nuclear plants operating, little different from the tax that keeps unreliable inefficient wind and solar plants alive in Australia and elsewhere.

Subsidies, whether for wind, solar or nuclear, allow crony capitalism and all sorts of sweetheart deals at your expense.

Cheating (again)

In the UK, the 2010 environmental and social costs components amounted to 4% of the average electricity bill. Today it is nearly 15%. Subsidies for renewables are now £4.5 billion and will double by 2020. Even if the UK reduces emissions, the amount of carbon dioxide entering the atmosphere is not reduced. The UK is now importing the high-energy produced goods that they once produced. British jobs are lost and global emissions of carbon dioxide are not reduced.

In Europe, 65% of the renewable energy is from biofuels. Most of this involves burning trees from USA to feed power stations to make electricity. Under the EU rule, burning wood is actually considered to be carbon neutral. The EU rules allow us to destroy forests in the name of saving the environment.

There have been some wonderful renewable schemes established by the wizards in government. Northern Ireland's Renewable Heat Initiative gave a vote-buying £160 for every £100 spent on wood chips.[514] And this was going to save the planet? The invitation to make a government-sanctioned 60% return was so attractive that wood burners operated 24/7 in vacant factories, sheds, piggeries and garages. No wonder many of us think that governments should be kept well clear of markets, commercial deals and business negotiations.

In June 2017, the Bloomberg New Energy Foundation estimated that the global push for renewables would cost $US 7.4 trillion between 2016 and 2040 and an additional $US 5.3 trillion would be needed to hold the planet to its 2°C warming trajectory as promoted by the Paris Agreement.[515] This is madness.

Many Western countries already have dangerously high debt and the 2°C target is based on the erroneous assumption that human emissions of carbon dioxide drive global warming and that future climate can be modelled by just considering carbon dioxide and not the numerous other drivers of climate.

UK's Department of Business, Energy and Industry Strategy was written by consultancy firm Frontier Economics. The report has long been known to be in progress and supposedly was to address the "total system costs" of variable renewable

---

[514] www.irishnews.com/news/politicalnews/2016/10/26/stormont-renewable-energy-scheme-one-of-biggest-scandals-since-devolution--757665/

[515] Brett Hogan, 2017: The destruction by government of Australia's electricity market. Students for Liberty Annual Conference, 1 July 2017.

energy generators. The report was delayed for a year. The Global Warming Policy Foundation stated:

> The study is not only very late, but contains no quantitative assessments of additional system costs.

Wasn't that the whole point of the exercise? In the released material, there are some peer review comments from which one can infer that quantitative estimates of those additional costs were in the drafts but have been deleted from the final. The GWPF's comment, entitled *"Is the UK concealing very high renewables system cost estimates?"* stated:

> After an unexplained delay of a year since completion of the UK's Department of Business, Energy and Industry Strategy (BEIS) has published (24.03.17) a report a report by Frontier Economics on the total system costs of uncontrollably variable renewable generators, a topic of crucial importance in understanding the cost-effectiveness of current climate policies. The study is not only very late, but entirely qualitative, and contains no quantitative estimates of additional system costs per megawatt hour (i.e. pounds/Meh), figures which would normally be considered the principal output of such work. However, an examination of the peer reviews, which are published with the study, reveals an entire table of numerical cost estimates, some of which were described by the external reviewer as "very high", were in fact present in the version sent out for comment in mid 2015, but have been subsequently removed. This does not smell right and BEIS should release the original draft.

If you are starting to get the impression that you are being defrauded, you are right.

When will a government somewhere put out a remotely

honest effort to calculate the real cost of the wind- and solar-generation system that they are busy trying to force onto people.

Accountability

We are in a new dark age when scare campaigns promoted by dark greens with missionary zeal impute that it is sinful to have air conditioning, heating, cars, houses and a high standard of living. Of course, sin has draconian consequences and doomsday garbage-in-garbage-out modellers try to frighten us with projected temperature increases of 2 to 4°C and a projected sea level rise of 50 centimetres in 100 years. Dark greens want materialist Western countries to feel guilt about unproven global warming.

In order to avoid misinformation, I suggest that the lobbyists, unionists, climate "scientists" and green activists be required to fulfil the same provisions of the law as company directors. Sections 180(2), 1317S and 1318 of the Corporations Law and the former Section 52 of the Trades Practices Act come to mind. These would require the greens to have done appropriate validated due diligence, to act in good faith and for proper purpose and to tell the truth. Directors of energy companies are required to fulfil the provisions of the Corporations Act, why can't the anti-coal and anti-nuclear lobbyists have the same ethical standards? It is only when the scientific misinformation ceases that Australia can have a knowledge-based economy and public debates based on validated information.

Lobby groups such as the ABC[516] claim that "facts are facts". Really! Are temperature or carbon dioxide measurements factual? Is the "homogenised" temperature the real fact whereas

---
[516] *The Barrier Daily Truth*, 17 August 2017.

the raw data is not the fact. Do human emissions of carbon dioxide drive global warming? Many critical facts can just be omitted. This is fraud. It happens all the time with the climate industry breathlessly announcing a new finding but just happening to ignore the past.

For example, an article by Larry Hamlin summarises a paper that:

> exposes and debunks the contrived claims of a recent renewable energy study which falsely alleged that low cost and reliable 100% renewable energy electric grids are possible.

The new paper evaluated the study and found that it:

> used invalid modeling tools, contained modeling errors, and made implausible and inadequately supported assumptions.

Both papers were published by the Proceeding of the National Academy of Science. The new paper points out failure of the prior study to *"deal with critical electric system reliability requirements"*.

Missing from the model are the ability to model frequency regulation, transmission requirements and operating reserves. The study *"ignores transmission capacity expansion, power flows, and the logistics of transmission constraints."* The reliability needs of the grid must be provided by dispatchable electrical generation from fossil fuels and hydro power plants. Countries that have high installed solar and wind capacity have more than double the power cost of other countries that do not, indicating that renewable energy is not cost effective. The paper's abstract stated:

> Policy makers should treat with caution any visions of a rapid, reliable, and low-cost transition to entire

energy systems that relies almost exclusively on wind, solar, and hydroelectric power.[517]

We live in times when media organisations, lobby groups, green activists, bureaucrats, politicians and climate "scientists" can just announce anything in public and take no responsibility for the factual basis of the statement. Such lack of accountability means that poor policy cannot be critically evaluated. This has happened with electricity generation in Australia.

WHAT CAN I DO?

Politicians are terrified of the next trend on Twitter, they are enslaved by group think and are followers, not leaders. The only good politician is a frightened politician and every few years their job comes up for review by the electorate.

The solution is to remove the cause of the problem. The causes are that there is no science to show that human emissions drive global warming, that the measurements of temperature are fraudulently changed and that governments have created subsidies and mandates to allow rorting and the destruction of cheap reliable electricity in Australia.

Don't sullenly pay your electricity bill without doing something about it. Put daily pressure on those elected to represent you. Politicians monitor Letters to the Editor, newspaper articles, electronic media news and letters from constituents to get a mood of the electorate. It is only selected blogs that politicians follow.

---

[517] C.T.M. Clack, S.A. Qvist, Apt, J., Brazilian. M., A.R. Brandy, K. Caldeira, S.J. Davis, V. Diakov, M.A. Handschy, P.D.H. Hines, P. Jaramillo, D.M. Kammen, J.C.S., Long, M.G. Morgan, A. Reed, V. Sivaram, J. Sweeney, G.R. Tynan, D.G. Victor, J.P. Weyant, and J.F. Whitacre, 2017: Evaluation of a proposal for reliable low-cost grid power with 100% wind, water and solar. *Proc. Nat. Academy Sci,* www.pnas.org/cgl/doi.10.1073/pnas.1610381114

Urge the following actions:

1. *Cancel*:

- RET, LRET, REC, CET and any associated funding that subsidises generation of electricity.[518]
- Must-take mandate for wind and solar energy.
- Funding for climate research institutes, climate conferences, CSIRO, carbon capture schemes.
- Paris Agreement.[519]
- Charitable status of feral activist groups such as Greenpeace.

2. *Continue*:

- Electricity generation by black coal, brown coal, bagasse, gas and hydroelectricity.
- Continue wind and solar power generation but with neither subsidies nor must-take mandates.

3. *Change*:

- Corporations Law such that green activists, environmentalists, unionists and lobbyists have the same conditions of probity as company directors.

4. *Purchase*:

- Nuclear submarines and ships that can protect Australia's borders and, when infrastructure of a town is destroyed by a cyclone, use the modular

---

[518] Those with snouts in the trough would take Court action. However, it is noted that the UK Court of Appeal upheld the government's right to cancel the climate change levy for renewables [Global Warming Policy Foundation 6th November 2016]. If the Courts would not allow the cancellation of subsidies, then legislate for a 100% tax on the profits from electricity generation by wind and solar. If the REC cannot be contractually cancelled, then change the REC price to $0.01 per MW of electricity generated.

[519] Do nothing and join all the other nations that will not abide by the Paris Agreement.

nuclear power to provide power until systems can be repaired.
- Modular nuclear power plants for remote towns and mines.

5. *Establish*:
- A business environment with a 50-year level playing field whereby modern coal-fired power stations, nuclear power stations and HELE power stations could be built if economically viable.
- Mandate that electricity providers have 95% availability 24/7.
- Mandate that up to 20% of Australia's superannuation funds be invested in new non-subsidised coal, gas and nuclear power plants.

It will be a long and difficult path of change because *"It's easier to fool the people than to convince them they've been fooled."*[520]

Once your electricity bill is too high, you will know that you have been conned. A very noisy minority of green activists have controlled the agenda and all kinds of political smokescreens cover the front pages. It is now time for the majority of the population not to be silent and to force major structural changes.

The choice is yours. You can either keep struggling to pay higher and higher electricity costs or force a political change.

---

[520] Mark Twain.